Secrets of State

SECRETS OF STATE

The State Department and the Struggle Over U.S. Foreign Policy

BARRY RUBIN

OXFORD UNIVERSITY PRESS
New York Oxford

Oxford University Press

Oxford New York Toronto
Delhi Bombay Calcutta Madras Karachi
Petaling Jaya Singapore Hong Kong Tokyo
Nairobi Dar es Salaam Cape Town
Melbourne Auckland

and associated companies in
Beirut Berlin Ibadan Nicosia

First published in 1985 by Oxford University Press, Inc.,
200 Madison Avenue, New York, New York 10016

First issued as an Oxford University Press paperback,
with a new Preface, 1987

Library of Congress Cataloging in Publication Data
Rubin, Barry M.
Secrets of state.
Includes index.
1. United States—Foreign relations—1945-
2. United States—Foreign relations administration.
3. United States. Dept. of State. I. Title.
JX1417.R83 1984 327.73 84-20703
ISBN 0-19-503397-3
ISBN 0-19-505010-X (pbk.)

2 4 6 8 10 9 7 5 3 1

Printed in the United States of America

To Jennifer Noyon

Alas! Neither in politics nor in domestic life has it yet been ascertained whether empires and happiness are wrecked by too much confidence or too much severity.

HONORÉ DE BALZAC
Honorine

The United States is a strange country full of secrets and motives, and no rule applies to it.

Akhbar al-Khalij (Bahrein)
April 1, 1981

So we'll live . . . and laugh
At gilded butterflies, and hear poor rogues
Talk of court news; and we'll talk with them too,
Who loses and who wins, who's in, who's out . . .
And we'll wear out,
In a wall'd prison, packs and sects of great ones
That ebb and flow by the moon.

WILLIAM SHAKESPEARE
King Lear

Preface to the Paperback Edition

The crisis of the U.S. foreign policymaking system predicted by this book exploded in November 1986. It was a scandal involving secret arms sales to Iran, money laundered through Swiss bank accounts to Nicaraguan guerrillas, a dubious slush fund for covert operations, and apparent violations of several U.S. laws.

The controversy encompassed several issues: management of the policy process; the proper way of freeing Americans held as hostages; the administration's tough public stance against negotiating with terrorists or selling arms to Iran, despite its secret willingness to do both; an embattled Central American policy; and the executive branch's failure to consult or heed Congress.

The policy process deserves special attention. After all, the crisis would never have happened if the decision-making process had not been so flawed, and unless improvements are made, similar problems will recur. The culpability of President Ronald Reagan, White House Chief of Staff Donald Regan, and other top officials may be argued, but the president's responsibility is beyond dispute. The results, and any lack of knowledge on Reagan's part, were due to his own system of governance. In part, Reagan's administration followed some historic trends:

• Growing U.S. global responsibilities and interests required diverse instruments and multi-level involvement that put high-stress, high-stakes demands on the policy process.

• Presidents' mistrust of the State Department bureaucracy often undermined relations with their own secretaries of state as well. Liberal administrations, like those of Kennedy and Carter, saw State as too conservative; conservative administrations, like those of Nixon and Reagan, thought State too liberal. None of these administrations considered the permanent staff sufficiently in tune with their own plans, priorities, and the urgency of events.

• Pluralism is implicit in the American style of decision-making. A critical media unrestricted by secrecy laws, a broad variety of interest groups, and a large community of academics and analysts judge every issue and action. The U.S. Congress has more power than do other countries' parliaments. Executive branch political appointees in the most seemingly homogeneous administration wage heated debates and struggles based on institutional interests, ideological shadings, and their personal relations or abilities.

Presidents who mistrust State—as too careerist, stodgy, slow, unimaginative, and too eager to appease foreigners—seek other agencies to do the job. Each succeeding administration expanded the number of challengers to State's primacy. Eisenhower elevated the CIA; Kennedy, his White House staff; Johnson, the Defense Department; and Nixon, the national security adviser.

Reality and State's shortcomings demanded a variety of channels and bodies. The fault was not in refusing to make State supreme, but in failing to preserve clear lines of authority. Instead, a growing number of voices competed to lead, pulled in opposite directions, subverted decisiveness, and blocked any coherent line of strategy.

The great exception was the Nixon administration in which Henry Kissinger held so much power. Kissinger's role illustrated not so much his personal mystique or anything implicit in his office of national security adviser so much as it did the president's ability to create a powerful lieutenant and a centralized system for foreign policy.

The concentration of power has advantages and disadvantages. It allows an administration to command the bureaucracy and, freed from the normal grinding constraints of compromise, to follow a coherent approach. This very effectiveness in pursuing its goals, in the era of Vietnam and Watergate, also intensified the reaction of those who opposed its specific policies.

The Carter and Reagan administrations both rejected centralism. Carter allowed free debate and freewheeling competition between Secretary of State Cyrus Vance and National Security Adviser Zbigniew Brzezinski. Since Carter hesitated to settle disputes and make decisions, U.S. policy drifted aimlessly in coping with revolutions in Nicaragua and Iran.

The Reagan administration exacerbated the trend toward internal indiscipline and conflict. The new right-wing counter-establishment brought to high posts an even larger number of inexperienced people who felt an even greater rejection of the career bureaucracy and past patterns of policy. The president was temperamentally incapable of providing consistent, well-informed leadership, preferring general, abstract vision over strategy. Neither the secretary of state nor the national security adviser was designated or allowed to fill the gap. Secretary of State Alexander Haig understood the problem, but his attempt to aggrandize power without presidential sanction led to his fall.

With no one in charge, power shifted from issue to issue and from week to week. People changed jobs with dismaying speed. Lines of authority became more entangled. Officials able to implement orders but incapable of analyzing problems pursued utopian goals—military superiority over the Soviets, overthrow of the Nicaraguan regime—rather than trying to gain the best possible solution in the face of constraints.

On most issues, the imperfect discipline of reality forced the Reagan administration to moderate its behavior. Budgetary limits, domestic needs, and the military's own inefficiency circumscribed its arms buildup; public opposition and political considerations prevented any invasion of Nicaragua. Yet as the administration benefitted politically from its moderation in practice, pressure from the right and its own ideology made it chafe at these limits. Its leaders were all the more eager to take extreme action on issues where Reagan's strong feelings let them escape the hated straitjacket of establishment conformity.

Aside from Reagan himself, the main responsibility for the situation rests not with the NSC staff but with his personal advisers. As the White House's obsession with the president's image made it wary about unpopular public actions, support for covert operations became a substitute for the direct assertion of U.S. strength the administration had favored and promised.

The White House advisers knew less about foreign policy and were more prone to the blind arrogance of an isolated in-group than any other sector of government. After all, the State and Defense departments—even the NSC staff—must respond to conditions in foreign countries and in Congress. The White House staff's sole consideration is the president's will.

Secretary of State George Shultz could not win a debate with the president, Vice-President George Bush, and Regan—all of whom favored the Iran arms deal and covert aid to the Nicaraguan Contras. Reagan ordered the NSC to carry out questionable actions. By using the NSC staff rather than the CIA (to avoid Congress's oversight), Reagan lost the CIA's insulating role of protecting him from blame.

These choices overrode the fail-safe systems of the policy process. Anyone who opposed the decisions—including those at State who had a better understanding of Iran, Central America, and the relevant strategic issues—was not consulted and thus had no chance to urge another course. The decision-makers only half understood their briefings and were instead spurred to act too quickly and incautiously in the hopes of gaining both a dramatic diplomatic success and public acclaim for freeing the hostages.

In short, the Reagan White House was left with the Nixon–Kissinger regime's problems, but not its strengths. Unable to formulate or effectively implement a diplomatic strategy, its apparent hypocrisy and contradictions over Iran, Central America, and terrorism and its lack of respect for law were as much a result of a policy structure incapable of consistency as of any bad faith.

The Reagan administration's experimentation had failed as surely as that of the Carter administration because it never attempted to create a clear line of command with a clever and qualified person in charge. The United States does need a vicar of foreign policy, but whether he serves as secretary of state or national security adviser is secondary.

Non-government institutions and the public also bear some responsibility. It has become increasingly hard for any president to mobilize support for an active foreign policy. No administration can gain the benefits of international leadership without the cost or compromise it entails. There can be no security without expense, no purist morality ignoring the world's roughness or the brutality of democracy's enemies, no quick release of hostages without concession, and no deterrence of Soviet expansionism without active international engagement. To pretend otherwise encourages governments to seek the "quick fix" of adventurist irresponsibility.

The Reagan administration's debacle provides no startling revelation on how to improve the policymaking system. The answer is simple to describe, if harder to implement: appoint competent people to key positions and create clear lines of authority. Congress is not an arrant interferer in the policy process but a rightful participant whose position has been made necessary by past White House misdeeds. The State Department has lost leadership partly through its own failings. The Reagan administration was not the victim of a rogue NSC staff but of a policy system that had lost the ability to either control its institutions or work out a consensus among them. The results were damaging compartmentalization and bitter resentment that created unprecedented public dissociation and recrimination among the government agencies. Consequently, the Reagan administration was unable to formulate and promote a diplomatic initiative on any issue—be it arms control, Central America, or South Africa.

Washington, D.C.
December 1986

B.R.

Preface

A number of previous studies have narrated the history of U.S. foreign policy, explored the principles of decision making, or provided prescriptions for reform. This book is only partly interested in those subjects. Its purpose is to give both the policy community "insider" and the interested observer a picture of how the policymakers and policymaking system have actually functioned.

Perhaps the greatest difficulty in understanding U.S. foreign policy arises when the process itself is left out of consideration. Anyone who has dealt directly with international affairs knows that these human elements and bureaucratic considerations cannot be ignored. I have tried to follow a middle course between two extremes: the dry diplomatic history that presents decisions as clear-cut and inevitable by omitting the clash and blend of motives, personalities, abilities, and even accidents that occur in the policy process, and the journalistic account focusing on gossip and personalities to the exclusion of fundamental issues and options.

In some ways, this is a frustrating endeavor. There are so many factors and people involved that it is easy to oversimplify the thoughts and activities of some of those active on any given issue. Calling for a clear and consistent U.S. foreign policy is one of the hoariest clichés in American politics. It is no accident that this goal is virtually never attained. A myriad of interests and countries, rapidly changing situations, necessarily conflicting relationships, and the need to bargain or bluff with other states require a certain lack of public clarity and a measure of apparent inconsistency. But these are part of the tactical plane. At the same time, an administration should be capable of developing a strong line of strategy on any given issue, a goal to attain, and a way of getting there.

A fickle world demands flexibility, making it unprofitable to apply doggedly any one strategy for a four-year term or for all issues. Nevertheless, it is reasonable to expect an administration to set a well-defined goal on specific questions and pursue it for twelve, eighteen, or twenty-four months. Such constancy requires coordination, leadership, and a clear chain of command capable of charting and sustaining a course of action. An administration lacking a power center or effective leadership will produce a series of conflicting actions and initiatives every few months or even simultaneously. Problems of unity and responsiveness to events plagued U.S. foreign policy in the Carter and Reagan administrations; in contrast, the Kissinger era faced difficulties arising from isolation at the top.

At the same time, it is wrong to believe that the United States has no enemies but only faults. The wills and actions of others, purposely or not, often create or sustain problems and crises. As columnist Meg Greenfield suggested, American thinking is "to an amazing degree, premised on the assumption that we alone are real. . . . We invariably emerge (in our own meditations) as the only agents and doers in the world. All else is reaction." There is a strong tendency to see America as possessing unbridled free will and strength. To "carrot" liberals, America could wipe out poverty and repression by acts of cooperation and generosity; to "stick" conservatives, it could easily defeat enemies by showing determination and power. But a much smaller degree of U.S. government control differentiates foreign from domestic policy, a distinction often lost on policymakers and observers alike.

The scientific law that observation of a phenomenon also affects it is certainly true of contemporary U.S. foreign policy, which has been increasingly monitored, reported, and critiqued by commentators, think tanks and analysts, the media, interest groups, Congress, and the public. These factors, irritating to policymakers who tend to see outsiders' involvement as damaging interference, often form a useful counterweight to leaders who are not always so expert or correct. Many events in the last twenty years undermined public trust in the politicians and policymakers, understandably leading to the conclusion that foreign policy is too important to be left to the annointed professionals.

Yet criticism can only be effective if it is based on knowledge of the limitations and pressures on the policy process itself and the nature of the world with which it must cope. The influence of outsiders on officials is also easily exaggerated. Congress has been able to modify marginally the executive branch's actions only on the few occasions it mobilizes its power; corporations or lobbies gain benefits on specific issues of special interest to them, often because there are no countervailing pressures or strategic priorities.

A typical survey taken in 1976 showed that 52 percent of Americans believed the United States should maintain a dominant world position "at all costs, even going to the brink of war," while the same people, in the next breath, opposed overseas interventions and put "keeping our military and defense force strong" in only eleventh place among their political priorities. They want results without cost, ideal solutions defying the reality of constraints and mutually exclusive choices. Policymakers must move from an abstract infinity of possibilities through a limited set of realistic options and resources into a realm of specific decisions; critics, on the other hand, have the benefit of being able to maintain contradictory ideas as to what should be done and may reverse their views without being held accountable.

Observing U.S. foreign policymaking as a writer, analyst, and marginal participant as well as researching and interviewing for this book further convinced me that the greatest secrets of state are the techniques and failures of the policymaking process. Since holding high office does not necessarily endow one with skill or knowledge of international affairs, the quality of individual officials plays a vital role in the success of U.S. diplomacy and, for that matter, in the fate of the world.

Chapters 1 through 5 are informed by work with archival materials, government documents, presidential libraries, and memoirs. Chapters 6 through 11 use extensive interviews as well as written accounts by participants. To encourage respondents to speak freely, I decided not to attribute individual interviews and resulting quotations. All statements used or examples quoted from books and articles have been carefully considered for accuracy and typicality. To preserve the clarity of the text, a number of comments, anecdotes, and background points have been put into the notes. Readers less familiar with the structure of the U.S. government and the nature of policymaking might prefer to read first Chapters six, ten, and eleven.

This book also benefits from the generous help of scholars and former or current government officials. Well over 150 people provided interviews and helpful comments on parts of the manuscript. Since most of them would prefer not to be identified, I can only thank them collectively for their time and assistance. Several researchers have also contributed and I would especially like to thank Bruce Plotkin, Eric Fredell, Susan Meisel, and Rachel Ekeroth. David Thomas provided valuable assistance on Chapters 1 and 2.

B.R.

Washington, D.C.
January 1985

Contents

Secrets of State

Abbreviations

ACDA	Arms Control and Disarmament Agency.
AF	Bureau of African Affairs, U.S. State Department.
AID	Agency for International Development.
ARA	Bureau of Inter-American Affairs, U.S. State Department.
CIA	Central Intelligence Agency.
EA	Bureau of East Asia and Pacific Affairs, U.S. State Department.
EUR	Bureau of European Affairs, U.S. State Department.
INR	Bureau of Intelligence and Research, U.S. State Department.
NEA	Bureau of Near East and South Asian Affairs, U.S. State Department.

1

Foundations of State

Although the founders of the United States understood well the workings of eighteenth- and nineteenth-century European power politics, they did not at all admire them. The secret and conspiratorial diplomacy of oppressive monarchs fighting over bits of border territory, making and unmaking alliances, and indifferent toward the costs of such maneuvers for their subjects, seemed unjust to the American revolutionaries. Diplomacy was, in Thomas Jefferson's words, "the pest of the peace of the world, as the workshop in which nearly all the wars of Europe are manufactured."[1]

For the United States, these dynastic games were particularly dangerous. As an infant nation, jealous of a fragile independence and regarding itself as a difficult and courageous experiment, the United States was a mouse among elephants. Yet this situation did not lead to self-quarantine, as has been the case with so many later revolutions, for although the United States had no overseas territorial objectives, it did seek peaceful commerce. And so it forged a policy of activist isolation, playing off the European states against each other for its own maximum advantage, convincing each of the importance of American friendship and making all compete for it. As in so many other areas, the founders of the Republic were also pioneers in the field of nonalignment.

This policy not only made a virtue of necessity but also a necessity of virtue. America's uniqueness was the justification for this neutrality—a new type of nation required a new type of foreign policy, different from the amoral European game of nations. Religion and morality required "good faith and justice toward all nations," said President George Washington in his Farewell Address of September 17, 1796, and it was necessary to "cultivate peace and harmony with all."[2]

Such sentiments were not based on abstract principles alone but also on the need to avoid making foreign powers into either enemies or overbearing protectors. America's message would be expressed not by interventionism but by example, as Washington explained, and its independence preserved by flexibility.

These principles were to dominate American thinking toward foreign policy, and those who practiced it as a profession, for the next century and a half, and in some ways they continue to exert influence down to the present day. The United States accepted diplomacy as a necessary evil to keep foreign predators at arm's length. This craft involved its participants in a dirty game, and the curse could only be removed by attempts to use it for implementing and promulgating democracy and American political morality.

George Washington's views were nonetheless pragmatic. He warned against permanent antipathy or attachment toward any one nation in order to avoid becoming a client for any great power; friendship toward all would avoid the "frequent collisions, obstinate, envenomed, and bloody contests" that were produced by competing alliance systems. This was a formula for being both courted and left alone—diplomats were supposed to work hard to minimize foreign involvements.

European interests, causes, and controversies were foreign to this country, Washington concluded; America was "detached and distant. . . . Why forego the advantages of so peculiar a situation? Why quit our own to stand upon foreign ground?" Why interweave "our destiny with . . . European ambitions, rivalship, interest, humor, or caprice?"[3]

Along with a mistrust of foreign entanglements, the skepticism of European theories of power politics, and the shrewd blend of self-interest and idealism conveyed by Washington, the Founding Fathers developed other lasting American ideas toward foreign policy. Important among them was belief in a unique American mission involving a global struggle between democracy and despotism, in which every victory of the latter posed a danger to the survival of America's own system.

"The Royalists everywhere detest and despise us as Republicans," wrote John Quincy Adams shortly after the triumph of European reaction in the 1815 Congress of Vienna. America's political principles "make the throne of every European monarch rock under him as with the throes of an earthquake." America's growth and prosperity would naturally arouse jealousy and antagonism abroad because of its role as a democratic light unto the nations.

"Our institutions form an important epoch in the history of the civilized world," wrote President James Monroe in 1823. "On their preservation and in their utmost purity everything will depend." Americans must "be thankful to the kindness of Providence," said Senator Henry Clay that same year, "for having removed us far from the power and influence of a confederacy of kings, united to fasten forever the chains of the people."[4]

Monroe's Secretary of State, John Quincy Adams, who succeeded him as president, spoke in similar terms. In a July 4, 1821, speech in Washington before an audience that included the European diplomatic corps, Adams characterized the British system as having been "founded in conquest" and "cemented in servitude." The Declaration of Independence, which he read on the occasion from an original manuscript, was the cornerstone of a new type of government, "destined to cover the surface of the globe. It demolished at a stroke, the lawfullness of all governments founded upon conquest. It swept away all the rubbish of accumulated centuries of servitude."[5]

Despite this boisterous confidence, Americans saw their experimental republic as fragile and perishable. Like its Grecian and Roman forebears, it might easily be subverted by luxury, foreign wars, or the concentration of power empires produced, as well as by attempts at subversion by foreign powers. Adams thought it an "inevitable tendency of a direct interference in foreign wars, even wars for freedom, to change the very foundations of our own government from liberty to power."[6]

Consequently, Adams concluded, America "goes not abroad, in search of monsters to destroy. She is the well-wisher to the freedom of all. She is the champion and vindicator only of her own."[7] Interventions could be justified only in defensive terms. "It is only when our rights are invaded or seriously menaced that we resent injuries or make preparation for our defense," said President Monroe in his speech proclaiming the Monroe Doctrine, which placed the Western Hemisphere off limits to European colonialism. Attempts to spread that imperial system would be "dangerous to our peace and safety."[8]

This concept of action in response to a foreign, antidemocratic threat would always form the ultimate basis for U.S. activism abroad. President Woodrow Wilson called for a world made safe for democracy, not a democratized world. In the months before Pearl Harbor, President Roosevelt and the State Department tried to convince the American people of an Axis threat sufficient to warrant a decisive response. At the outset of the Cold War, a return to isolationism on the part of the American people was countered by presenting Soviet expansionism as a direct threat to our own freedom. "The only way we can sell the public our new policy," wrote

Joseph Jones, one of President Truman's advisers, "is by emphasizing the necessity of holding the line: communism vs. democracy should be the major theme."[9]

Dealing with a people so suspicious of diplomacy made life constantly difficult for the State Department. If American diplomacy itself was a practice designed to quarantine foreign contagions, those on the front lines were deemed particularly prone to corruption. From earlier days it was feared that diplomats might succumb to European foppishness and manners. During the McCarthy era, State Department officials responsible for studying and dealing with Communism were thought especially subject to subversion. The well-dressed, prep-school-educated Secretary of State Dean Acheson drew both right-wing political hatred and populist sartorial criticism. Diplomats, like scientists, became objects of suspicion because of their arcane interests as well as their cosmopolitan behavior and manners.

Nevertheless, the United States was always prepared to defend its national rights and interests when they seemed under challenge. As Alexander Hamilton wrote in 1787, "No government could give us tranquility and happiness at home which did not possess sufficient stability and strength to make us respectable abroad."[10] The State Department, however, had to cope with threats more quietly than did the armed forces and without the prospect of clear-cut victory expected in military operations.

The early American leaders' reservations about diplomacy also did not prevent them from securing many international successes. They won independence from Britain, military aid from France, and recognition from other European powers, freed the Mississippi Valley from foreign control, maneuvered the Spaniards out of Florida, bought the French out of Louisiana, and drove the Mexicans out of the Southwest. They purchased Oregon from Britain and Alaska from the Russians. The Spanish were forced out of Cuba, Puerto Rico, and the Philippines. The Monroe Doctrine initially aided (though it later seemed to oppress) Latin American freedom and the Confederacy was isolated from European help. It is easy to forget that most Americans today live in territory that became part of the United States through diplomacy.

Only slowly and gradually did a State Department emerge to play a role in these events. Congress created a Department of Foreign Affairs in January 1781, with offices and a four-man staff headquartered in a small, three-story Philadelphia house.

The U.S. Constitution, ratified eight years later, gave the federal government sufficient authority to deal vigorously with foreign affairs.[11] The initially confused structure had to be replaced by some permanent system.[12] The notable success of American diplomacy in the Revolutionary

War era had been due chiefly to the great personal skill of the first American diplomats, including John Adams, Benjamin Franklin, John Jay, Thomas Jefferson, and James Madison, not to the machinery they were obliged to use, which was weak and inadequate for its purpose.[13]

During the War of Independence, the Continental Congress had directed the conduct of foreign affairs through the Committee of Secret Correspondence and its successor, the Committee of Foreign Affairs. Neither performed with notable efficiency.[14] The Committee of Foreign Affairs had no formal powers and there was no sentiment in the Congress for giving it any. John Jay, the American representative in Madrid in 1780, registered this typical complaint about the work of the committee: "[T]il now I have received but one letter from them, and that not worth a farthing."[15] Later, the Congress of the Confederation established a Department of Foreign Affairs, whose secretaries, Robert Livingston and John Jay, displayed conspicuous ability. While power was at first withheld from the department, the very necessities of government gave it more authority as time went by until it came to resemble a real foreign office.

The new Department of State was created by a September 1789 act of Congress that stipulated that the principal officer should thereafter be called the secretary of state. Thomas Jefferson was commissioned as the first secretary that same year. Although their appointment required confirmation by the Senate, the secretaries of state and war were soon made responsible to the president alone and subject to his direction. When the first question of dismissal from office arose, the Senate decided that the president could remove officials without its consent. This precedent made the cabinet departments and the diplomatic service responsible to the chief executive.[16]

The Senate also acknowledged that the president was completely responsible for the actual conduct of foreign relations and that the Senate itself ought not to participate directly in these affairs, a conception of presidential responsibility and senatorial prerogative that now appears somewhat quaint.[17] The secretary of state, as custodian of the seal of the United States and as the first cabinet member designated by Congress, occupied a position of higher standing than that of any other department head and in theory enjoyed a first-among-equals position in the cabinet.[18]

Under President Washington, little of the business of the new Department of State was transacted without the president's sanction, and Thomas Jefferson, the first secretary of state, consulted with the president on all the department's important business. In this respect, President Washington established a pattern that would be followed by nearly every successor. There was no doubt, however, that the secretary of state was to be the sole intermediary for communications with American diplomatic agents abroad and with foreign government representatives in the United States.[19] President Washington helped establish this convention when he declined direct correspondence with the French minister in the period just prior to Jefferson's appointment as secretary.[20]

Even before Jefferson arrived in New York to take up duties, those against the department's establishment resisted a bill before Congress to provide pay for diplomatic officers. Opponents of a diplomatic department were motivated by a spirit of prosaic frugality and republican egalitarianism, of antipathy to courts and courtiers.[21] Similarly, a later editorial in a New York newspaper declared that the proposed appropriation of $40,000 would cause ambassadors "to be maintained at splendid courts," and then asked rhetorically: "Has America ever realized any substantial advantage from foreign ministers?" One senator opined that the money would be thrown away since, as he put it, "I know not a single thing that we have for a minister to do at a single court in Europe."[22]

Thomas Jefferson himself held no exalted opinion of either foreign courts or of diplomats, but he did grasp the importance of maintaining an adequate foreign establishment on a liberal allowance, and appreciated intuitively the art and importance of diplomacy.[23] In 1790, Jefferson drew up reforms for the diplomatic bureaus.[24] Jefferson also appointed 16 consular officers and formally instructed them as to their responsibilities, which in a farsighted manner were framed to include the collection of political, military, and commercial intelligence.[25]

The goals of the State Department, as envisioned by President Washington and the first secretary of state, were modest and restrained, as befitted the capacities of the new institution.[26] State's main job, however, was to collect information from diplomats, who watched political developments, and from consuls, who helped American citizens abroad and gathered commercial data as well as intelligence on any war preparations. William Palfrey of Massachusetts, a former Revolutionary War officer, was the first American consul and the first to lose his life in the line of duty—in a shipwreck on the way to his post in France.

At first, European court protocol caused some problems for the zealous revolutionaries. Jefferson refused to follow traditional rules in seating foreign diplomats at dinners. The low pay of American representatives abroad was supposed to signal that they were to live like their compatriots at home. Nevertheless, the need to move in aristocratic settings without standing out as too threadbare meant that American diplomats needed personal wealth to perform their jobs. Indeed, a number of consuls were merchants already living abroad.

Later, President Franklin Pierce's secretary of state, William Marcy, told envoys to wear "the simple dress of American citizens," but amidst court finery, their dark suits were sometimes confused with mourners' garb. As minister to Great Britain, future president James Buchanan added flair by wearing a sword, and during the Civil War, his successor in London, Charles Francis Adams, moved even farther away from the "undertaker look." "I am thankful," Queen Victoria is said to have remarked, "we shall have no more American funerals."[27]

The department's organization at home was also kept on spartan lines. As secretary of state in 1819, John Quincy Adams personally wrote out three copies of a treaty by hand. President Madison considered, Adams drily remarked, "that the public duties of the department were more than sufficient for one man." So great was the overwork, Adams complained, that important letters frequently disappeared; at one time he had on his desk 11 unanswered dispatches from the minister to Great Britain.[28]

Adams began an indexing procedure for papers, but as late as 1870 there was still no comprehensive system; officials often had to rely on staff memories. Since messages to American ministers abroad, dispatches from those missions, and notes to foreign diplomats in Washington were each bound in their own separate volumes, it was difficult to assemble material on any one subject. Shortly after the death of Secretary of State John Hay in 1905, one assistant had to delve among the papers in Hay's cellar to find an important note to a Russian representative of which no official record had been made. Only in that year was an effective filing method started.

From 1815 to 1898 foreign policy was largely in eclipse. National energy was devoted to industrialization, the Civil War, and the westward migrations. Secretary of State Edward Livingston sadly commented in 1833 that Americans thought of their diplomats as privileged characters "selected to enjoy the pleasures of foreign travel at the expense of the people; their places as sinecures; and their residence abroad as a continued scene of luxurious enjoyment."[29] A quarter-century later, Rep. Benjamin Stanton of Ohio said he knew of "no area of the public service that is more emphatically useless than the diplomatic service—none in the world."[30]

It is not surprising that such a maligned department was allowed to grow so slowly. By the time of the Civil War it had 30 officers and 27 support personnel at home and 281 abroad. The Washington and foreign sections were completely separate, and officials in the capital handled only correspondence and inquiries—there was no policymaking. An assistant secretary of state was not authorized by Congress until 1833, in the first of many reorganizations. Seven units were established, the most important of which were the Diplomatic Bureau and the Consular Bureau.

The former had one clerk responsible for England, France, Russia, and the Netherlands, a second for the rest of Europe and all of Asia and Africa, and a third for the Western Hemisphere.[31]

As early as 1796, passports were issued for ships bound to the Mediterranean. The detachable top portion of the document was forwarded to Algiers. If an American ship was stopped the captain produced the matching lower section, thereby proving that the vessel was under American protection. The department gained the exclusive right to issue passports in 1856, though they were generally used only in wartime and not required until 1914. Overseas missions increased from 15 in 1830 to 33 in 1860. In 1835 the first four Marines were assigned to protect the consulate in Lima, Peru. Not until 1948 was the comprehensive Marine Security Guard program established.*

Even without any threat of attack, life was not all parties in striped pants, especially at consulates located in the tropics. From Recife, Brazil, in 1858 consul Walter Stapp reported that the post's reputation was so bad that his predecessor resigned before arriving; four other consuls died there.

The American author Nathaniel Hawthorne served as the U.S. consul in Liverpool from 1853 to 1857. In his diary he noted the timeless complaints of the consul, "All penniless Americans or pretenders to Americanism, look upon me as their banker . . . I am sick to death of my office—brutal captains and brutal sailors . . . calls of idleness or ceremony from my travelling fellow countrymen, who seldom know what they are in search of . . . beggars, cheats, simpletons, unfortunates, so mixed up that it is impossible to distinguish one from another, and so, in self defense, the Consul distrusts them all."[32]

Working conditions in Washington were less risky but equally tedious. Congress was always tightfisted. Until 1856 consuls did not receive regular salaries—they were supposed to survive on fees collected—and diplomats were quite poorly paid. The ceiling of $17,500 annually for heads of missions endured for 90 years, until 1946. Attractive posts were monopolized by ministers appointed by virtue of their personal wealth, past political services, and social position. If the cultured Charles Francis Adams worked effectively to prevent Britain from supporting the Confederacy in the Civil War, his counterpart in St. Petersburg, John Randolph, introduced himself to the Czar, according to legend, by saying, "Howya, Emperor? And how's the madam?"[33]

U.S. diplomacy in the Civil War was the best example of how effective

*Today, in contrast, over 3 million passports are annually issued, there are over 14.6 million in circulation, and 1100 Marines guard U.S. missions around the world.

the nineteenth-century State Department could be, given a talented, if inexperienced, secretary of state, William Seward, and an able envoy, Charles Francis Adams, in London.[34] Seward had a good working relationship with President Abraham Lincoln and was granted latitude in formulating and conducting policy. They successfully dealt with Washington's main diplomatic problem of preserving British neutrality and preventing any European aid or recognition for the Confederacy.

Adams, one of the most skillful diplomats in U.S. history, was the son and grandson of presidents. His keen understanding of the British and his sense of tact enabled him to modify Seward's instructions enough, toning down his superior's harsher criticisms of London, to achieve their commonly held objectives. Central in this effort were Adams's useful social ties with British leaders, which enabled him to deal quietly and effectively with issues arising out of U.S. actions against British ships as part of the Union blockade against the South.

When prospects for the North looked grim, Adams convinced the British government that the Union's victory was nevertheless inevitable and that any assistance for the rebels would not redound to Britain's long-run advantage. This was a successful example of classic diplomacy in a time when an ambassador's charm and skill could sway a royal court and avoid wars even if the two nations' interests were at odds. In that age of slow communications and small governing elites, an envoy was far more likely to operate as an independent force, personally determining the success or failure of his government's efforts. In the words of the British foreign minister, Lord Russell, Her Majesty's Government "had every reason to be satisfied with the language and conduct of Mr. Adams." Even an old antagonist, the *Times* of London, praised Adams for his "wide discretion and cool judgment."[35]

Thus, the Civil War was a high point in the otherwise lackluster performance of American statesmanship in the last half of the nineteenth-century. In 1869, President Ulysses Grant made his hometown friend Elihu Washburne secretary of state for 12 days so he could enjoy the honor before becoming minister to France. In contrast, the next secretary, Hamilton Fish, was by far the most able man in that corruption-ridden administration. When Gen. Orville Babcock, one of Grant's influence-selling cronies, tried to interfere in the department, Fish protected it by threatening to resign. Fish also organized the first geographical divisions at State.

Throughout the nineteenth century, the department was always shorthanded. In 1900, Sydney Smith, head of the diplomatic bureau, pointed out four aged clerks sitting in the four corners of his office, dividing the world among themselves "but not its goods," given the notoriously low salaries. Appropriately, the department was located in the former Washington Orphan Asylum from 1866 until 1875, when it moved to the State-War-Navy building next to the White House.[36]

The slow pace was overseen by the powerful chief clerk. Although Secretary of State Daniel Webster appointed his son to that job in 1841 and William Seward practiced similar nepotism 20 years later, authority was usually in the hands of long-established and fanatically dedicated veterans like Chief Clerk William Hunter Jr., whose legendary memory and linguistic skill were exercised during his 57 years in the department, from 1829 until his death in 1886; Alvey Adee, who joined the diplomatic service in 1869; and Wilbur Carr, who entered as a clerk in 1892.

Yet, choice posts remained prime targets for the spoils system. The 1883 Civil Service Act did not cover diplomats, and two years later, when the Democrats returned to the White House after a quarter-century in opposition, American diplomats across the world began packing their bags. Until well after the Civil War, most missions consisted only of a minister and his secretary, often a family member. A typical story was the attempt of Minister to Britain Rufus King to have his son appointed as secretary in the London embassy, a position then occupied by a man who had only recently vacated a secretaryship in Madrid to make way for Secretary of State John Quincy Adams's nephew. Adams finally agreed to find some new place for the twice-victimized official.

Certainly, diplomatic labors in Western Europe were light and pleasant. Missions and consulates opened from 10 A.M. to noon and from 2 to 4 P.M.; the evenings were devoted to parties. These were perfect jobs for rich and well-connected young men who, not wishing any more strenuous career, thought it pleasant to spend a few years in London, Paris, or Rome. The long-feared diplomatic aristocracy had been born.

After the Civil War, John Hay, later an able secretary of state, occupied his time as a diplomat in Vienna with sightseeing, the theater, and opera. One Kansas congressman remarked in 1889: "For months there was no minister at Vienna, I have never heard that there was any interruption of the ordinary course of affairs because of it. So I would be willing to take the chances of having no member at London or Berlin."[37]

In the 1870s and 1880s, political appointees heading U.S. missions abroad were still usually unqualified party hacks; career diplomats who held the small number of official positions overseas were largely dabblers; consuls were usually half-trained business agents (although a few used the positions to hone literary skills—for example, the writer James Russell Lowell, in Spain and Britain; Nathaniel Hawthorne, Liverpool; and Bret Harte, Geneva); and the Washington staff consisted mostly of clerks engaged in copying and filing dispatches. At this time, however, some especially able people began to develop as America's first career diplomats.

Henry White was typical of this group of talented young men. Born to a wealthy Baltimore family, White studied at Oxford, living for eight years in Britain and making many friends in political circles there. In 1879, having no vocation, he gained appointment to the diplomatic corps with a recommendation from a Maryland senator, a friend of his mother. White became a second secretary in London, knew job insecurity because his Republican allegiances made him grist for the spoils system, and felt that the U.S. service was inferior to Britain's because it lacked a professional merit-based corps. He later became a leading advocate of foreign service reform.

Alvey Adee, who became the department's key man at home, was representative of those who chose a Washington-based career in foreign policy rather than semipermanent service abroad. He started his career in Spain as private secretary to the flamboyant U.S. minister Daniel Sickles, a former Union general and congressman; Adee's duties included ordering the minister's fine wines and luxury goods from London and Paris. He returned to Washington as a clerk in 1876, rising to third assistant secretary in 1882 and to second assistant secretary shortly before his death in 1924. Adee drafted or approved almost all outgoing correspondence. He composed most of the treaties and was a stickler for style, writing critical comments in red ink on green paper.[38]

During a crisis, Adee slept on a cot in his office. Deaf, reclusive, and unmarried, he dedicated his life to the department. A Washingtonian commented as Adee passed, "There goes our State Department now." Once, in 1913, when in his seventies, Adee was hit by a car while riding his bicycle to work. Refusing all offers of help, he went to the department and put in a full day's work as if nothing had happened. Understandably, more than one secretary of state trembled at the prospect of Adee's illness.[39]

The center of Adee's life was the State-War-Navy building, whose many decorative porticos, mansard roofs, and pillars inspired its nickname, "the wedding cake." When it opened in 1875 it was considered the world's largest and finest office building, with 10 acres of floor space and 500 high-ceilinged offices. Saloon-style wooden swinging doors and overhead fans combatted, with little success, the city's hot and humid summers that caused the British Foreign Office to count Washington as a hardship post in the era before air conditioning. The State Department occupied the south wing, separated from the White House grounds by a narrow avenue. The secretary's office looked out on the Washington Monument and was adjoined by a reception room where treaties were signed and press conferences held. Leather chairs and sofas stood against walls lined with the portraits of past secretaries. A small diplomatic waiting room was designed for those who wanted a private audience.

State's routine was far more easygoing than it would be in later years. One historian described it in 1898 as "an antiquated feeble organization, enslaved by precedents and routine inherited from another century, remote from the public gaze and indifferent to it. The typewriter was seen as a necessary evil and the telephone was an instrument of last resort."[40]

Indeed, the department accepted typewriters only gradually for domestic correspondence in the 1890s. The last snobbish resistance was broken when a typewritten diplomatic note came from the British embassy. In 1881 there was just one telephone; a second was added 14 years later; transatlantic lines were first used in 1931.

While technology quickened the potential pace of diplomacy, domestic and international developments were changing the national outlook to favor active involvement in foreign affairs for the first time. As the continent was secured and the country industrialized, Americans increasingly felt their nation should take its place among the world powers.

By the 1890s, belief in an American global mission mingled with a desire to emulate the power politics and imperialism of the European powers, then engaged in dividing the world into spheres of influence. Within the United States, European-influenced conservatives wanted America to seek its own share of power and glory, while liberals sought to spread democracy and American practices abroad. The latter group also had a direct effect on the State Department through its advocacy of scientific management methods, professionalism in government, and a strengthened civil service.

These last ideals appealed particularly to the younger generation in the Foreign Service. To men like Henry White, William Rockhill, Francis Huntington-Wilson, among others, diplomacy was no job for amateurs. U.S. interests could only prosper if represented by a trained career service with a breadth of vision rather than by politicians and dabblers. Some of these reformers used their friendship with presidents Roosevelt and Wilson to lobby for an invigorated Foreign Service with better pay, allowances, and conditions, as well as a merit system and examinations for admission and promotions.

The consuls, looked down on by the diplomats, had a champion in their chief, Wilbur Carr, who served in almost every type of post and under 17 secretaries from 1892 to his retirement in 1939. Born on an Ohio farm in 1870, he studied business and bookkeeping at commercial schools and joined the civil service in 1892. He never served abroad and did not even visit Europe until 1916. Carr's background, like that of many consuls, sharply contrasted to that of the Ivy League diplomats, who opposed the consuls's desire to merge the two groups.

Still, a greater U.S. emphasis on foreign policy produced more presidential dependence on all parts of State. When President William McKinley declared war on Spain in 1898, he did not even consult his secretary of state, with whom he had met only once a month. In contrast, Teddy Roosevelt spoke to his secretary of state, the experienced John Hay, every day. Roosevelt began to fulfill hopes for reform in 1905 with an executive order establishing the merit system for all positions below minister or ambassador and by instituting the first exams for entering the service.

These were years of unprecedented U.S. diplomatic activism. America faced a world increasingly menaced by the European alliance system, an arms race, and the scramble for overseas colonies and bases that led to World War I.

Hay, who had so enjoyed his carefree years in Vienna, was unhappy under pressures of office so heavy that they broke his health. "It is impossible to exaggerate the petty worries and cares which, added to the really important matters, make the office of secretary of state almost intolerable," he wrote White. In particular he complained about the "unrestricted freedom of access" insisted on by members of Congress and "the venomous greed with which they demand and quarrel over every scrap of patronage." "A treaty entering the Senate," Hay continued, "is like a bull going into the arena. No one can say just how or when the final blow will fall—but one thing is certain—it will never leave that arena alive."[41] Hay did better with Congress than this pessimism suggests, and, discovering the newspapers as a new potential ally, he initiated background press briefings.

Roosevelt maintained a close relationship with Hay, his successor, Elihu Root, and with State's bureaucratic anchorman, Adee, who generally wrote the president's formal correspondence. "Why, there isn't a kitten born in a palace anywhere on earth," Roosevelt chuckled, "that I don't have to write a letter of congratulations to the peripatetic Tomcat that might have been its sire, and old Adee does that for me."

As today, this delegation of authority sometimes caused problems. When Peru's president sent a telegram suggesting a U.S. Navy visit, Adee prepared a vague but friendly response. The press had a field day, some trumpeting and others decrying a new extension of American military might. Roosevelt could hardly admit that he had not seen either the incoming or outgoing message.[42]

The department continued growing to deal with increased presidential foreign activism. During Roosevelt's term, divisions of Far Eastern,

Latin American, and Western Europe and Near Eastern Affairs were established, expressing Roosevelt's belief in European-style realpolitik. But President Woodrow Wilson, who came to office in 1913, distrusted State and his own secretaries, preferring to use agents, particularly the wealthy and talented, but uncharismatic, "Colonel" Edward M. House. Wilson made William Jennings Bryan secretary of state as a political reward and his many patronage appointments demoralized the department. "He is absolutely sincere," said Wilson of the pacifist and naive secretary after Bryan resigned. "That is what makes him dangerous."[43] Wilson wanted more pressure against Germany over violations of American neutrality during the early days of World War I, while Bryan sought more stringent neutrality. Wilson preferred to be his own secretary of state, typing out dispatches himself, and using Colonel House to bypass Bryan. The presidential aide secretly met foreign diplomats at an assistant's house to avoid publicity. Bryan attributed his 1915 resignation in part to House's—and White House—interference in all aspects of foreign policy.

Another feature common in future years was the presence of a White House confidante in a high department position, a role played for Roosevelt and Wilson by William Phillips. Phillips was a Harvard graduate from a prominent Boston family. A professor recommended him to the secretary of state, who suggested he study law instead, since the few available positions were usually awarded on political grounds. Phillips used his own contacts to become secretary to the new ambassador to Britain, where his main job was organizing the envoy's social schedule. But he also copied dispatches and taught himself diplomatic procedures until, through further connections, he received an appointment to the legation in China and a promotion to second secretary.

On returning to Washington, Phillips had to resign since the Diplomatic Service was completely separate from the home office—he could get into the State Department only as a messenger. Phillips quickly persuaded Secretary Root to establish a special office to handle dispatches from the Orient and Root made him head of this Division of Far East Affairs. Eventually, his hard work and aggressive approach gained him the job of counselor; he caught Teddy Roosevelt's eye, was invited to the White House, and formed a friendship with the president.

Phillips's comeback was disrupted in 1909, when he was again demoted because Senator Eugene Hale of Maine threatened to cut off State's appropriations unless Phillips's job was given to his son. Later, Colonel House had Phillips brought back to his old Washington post. There, Phillips examined Bryan's outgoing correspondence, managed the

Division of Western European Affairs, as well as diplomatic and consular appointments, protocol, and liaison with Colonel House.

Since Bryan spent much of his time on non-foreign policy activities, including his lecture tours, much of the day-to-day work was done by the staff of three assistant secretaries, a legal counsel, the chief clerk, and the director of the consular service. There were three regional divisions, a half-dozen administrators, two trade advisers, and about 125 clerks. The latter received civil service pay of $900 to $1800 a year; about a dozen of them were women. A clerk could still rise to head one of the bureaus. The Latin American section, largest of the regional divisions, had 10 clerks. Around 450 Americans served abroad.

Within the department there were few security rules. Dispatches could be read by virtually anyone and the public had free access to the building. Ambassadors' letters often went unanswered or replies were leaked to the press before they were received. Envoys abroad were not kept well informed of developments or diplomatic conversations in Washington. Few possessed any sophisticated understanding of the coming European conflict. The American vice-consul in Budapest, Frank Mallet, accurately predicted the coming war in an analysis drafted July 13, 1914, but considering the possible criticism for sending an expensive cable, put his dispatch in the regular mail. It arrived in Washington almost a month later, just as war was being declared all over Europe.[44]

The onset of World War I meant rapid growth and tighter security measures for the department, but, as in World War II, diplomats were often shoved aside by temporary agencies—the War Trade Board, the War Industries Board, the "Inquiry" on postwar planning—and by the War and Treasury departments. The department was so swamped with work and so short on funds that some ministers volunteered to stay on without pay to help evacuate Americans stranded in Europe. While Bryan's successor, Robert Lansing, and much of the department were more interventionist than Wilson himself, the reins remained firmly in the president's hands as he moved toward the April 1917 declaration of war.

After the war, against Lansing's advice, Wilson decided to attend the Versailles peace conference. He ignored his State Department advisers there and refused to accept an agreement drafted by lawyers—a direct swipe at his law-trained secretary of state. Lansing disapproved of Versailles but loyally supported it in public until its defeat in the Senate. When Wilson physically collapsed in the midst of his pro-treaty campaign and Lansing called cabinet meetings to conduct business, the president accused him of disloyalty. Lansing finally resigned with a sense of profound relief.

American policy toward the Russian Revolution was also characterized by Wilson's disregard of State Department channels. The U.S. ambassador in Petrograd, David Francis, a grain merchant with scant political knowledge, was told little about his superior's policies. Wilson preferred personal envoys. After the Bolshevik seizure of power in November 1917, members of the American Red Cross Commission and of some U.S. government missions in Russia—for example, the Committee on Public Information—performed what were in fact diplomatic functions. They dealt with Soviet authorities on a semiofficial basis, almost on the same footing as that of the American ambassador.[45]

For example, Arthur Bullard, a writer, was sent to Petrograd by his close friend Colonel House, ostensibly in a private capacity. He reported to Washington, participated in schemes to remove Ambassador Francis, and recommended recognition of the Bolsheviks. Edgar Sisson, a former editor, directed the work of the Committee on Public Information in Russia. Wilson gave him a personal note of instruction, and this fleeting contact induced Sisson to fancy himself as the president's personal representative. In effect, Sisson functioned as a political agent, also working to remove the hapless Francis, reporting on German-Soviet negotiations, and establishing contacts with the Soviet authorities.[46]

Raymond Robins, U.S. Red Cross representative, saw more of the early Bolshevik leaders than any other American. Until Washington relaxed its ban on official contacts, Robins's government-sanctioned liaison with Bolshevik leaders constituted virtually the only channel between the United States and the new government. Francis was told little about this; Robins acted as if he were the ambassador.[47]

While the war necessitated a more active diplomacy—one writer in 1915 called it "the first line of national defense," a phrase often cited by Foreign Service officers (FSOs) in later years—the experience seemed to reinforce traditional American distrust for both diplomats and foreign involvements. Isolationist Senator William Borah said in 1918, "The greatest war of all history was begun not to preserve liberty but to destroy it, and the scheme was hatched in the chancellories of Europe." That same year, liberal philosopher John Dewey commented, "Secret diplomacy . . . carries with it all the signs of a class so personally and professionally set apart that it moves in a high inaccessible realm whose doings are no concern of the vulgar mass."[48]

This attitude influenced American views of other secrets. Herbert Yardley, who joined the State Department in 1913 as a code clerk, was convinced that the department's ciphers were unsafe. He proved it by breaking all of them; but, failing to interest his superiors, he moved to the War Department, where he cracked foreign codes and devised safe ones for America. Yardley's greatest achievement was deciphering the Zimmerman note, in which Germany attempted to make an anti-American alliance with Mexico; the revelation helped turn neutral America against Berlin. After the war, Yardley broke the ciphers of twenty countries, including Britain, Japan, France, and the USSR. The department used his findings, but Henry Stimson, on becoming secretary of state in 1929, decreed, according to legend, "Gentlemen do not read each other's mail," and had Yardley's funds cut off.[49]

The 1920s was an isolationist era, but it also saw the success of the Foreign Service reform movement. President Warren Harding's secretary of state, Charles Evans Hughes, sympathized with the diplomats' campaign for better benefits to combat the high resignation rate and to encourage the service's professionalization. The resulting 1924 Rogers Act was a turning point in State's history, providing new travel and representional allowances, a pension plan, higher salaries, and standardized admission, promotion, and rankings.

Many of the reformers themselves did not realize that professionalization would challenge the aristocratic ethos that had dominated the Foreign Service from the 1880s. Although the diplomats resisted merger with the more numerous consuls into a single corps, unification was inevitable and was followed by further broadening of the Foreign Service. Amateur diplomacy characterized the nation's first century, and the aristocratic generalists ruled for the next 75 years, only gradually giving way to a professional-oriented service. The Rogers Act opened the door slightly through the establishment of admissions exams on language, law, international relations, and economics, and an oral exam before a board. The last-mentioned was still used, however, to sift out those who did not fit the service's predetermined social profile.

Many new department recruits now had university training in the growing field of international affairs. Georgetown University established the first School of Foreign Service in 1919. The Foreign Service's own institute opened in April 1925, though the curriculum consisted mostly of lectures by department officials.

What kind of people was the State Department seeking? In one lecture to recruits in the early 1920s, Allen Dulles, then chief of the Division of Near East Affairs, emphasized that each diplomat needed experience and intuition, sound and accurate judgment, and a solid grounding in the history and theory of international relations. He warned officers against receiving clandestine information, which might anger local authorities,*

*This is ironic since Dulles became a brilliant spymaster in World War II and director of the CIA in the 1950s.

and stressed the need to travel widely in the country where they served—but at their own expense.[50]

In a February 1926 broadcast, the experienced diplomat and future under-secretary of state Joseph Grew explained that American interests must be protected by "a new generation of red-blooded young Americans, straight-thinking, clear-speaking men, whose watchword is 'service' and whose high conceptions of integrity, sincerity and patriotism [are] steadily raising the standards of . . . the honorable profession they follow." Secretary of State Hughes commented that the "days of intrigue to support dynastic ambitions [and] to promote the immediate concerns of ruling houses" were gone. The new diplomacy was not based on "the divining of the intentions of monarchs" or "the mere discovery and thwarting of intrigues" but "on the understanding of peoples."[51]

It was ironic but understandable that Grew had to defend diplomacy as an "honorable profession." The State Department in 1929 was unhonored, underpaid, and understaffed. The Great Depression shattered many of the Rogers Act's promises. Pay was cut by 15 percent, the dollar's decreased purchasing power further sliced the income of those serving abroad, housing and promotional allowances were reduced; promotions, recruitment, and paid home leave were suspended. The number of officers fell; those remaining were demoralized.

Perhaps typical of the tedium at farflung posts was a plaintive letter from John MacMurray, U.S. ambassador to Turkey. Among other duties, he had been plagued by official cables ordering him to sell a piano belonging to an embassy couple who had been transferred home. "I suppose that I am getting the jitters," he wrote, "having had to live a better part of my three years in the intellectually and spiritually sterile climate of this cardboard capital in the wilds—having been allowed no home leave in almost four years and during that time a total of less than sixty days . . . of opportunity to breathe any other atmosphere—having had to send away first my children and more recently my wife, and to make out with the companionship of a couple of lovable but ungovernable native dogs—having pleaded almost in vain with the Department for more than a year to equip us with the personnel to meet a situation of predictably increasing difficulty, or at least to relieve us of certain liabilities—having received from the Department scarcely a word of guidance or of help . . . and having come to feel that we, who are sticking along here only because it would not be playing the game to resign or wrangle a transfer, are not a matter of interest to anybody in the Department except those who suspect us of some sculduggery [*sic*] about [furniture] exchange or about assignments of clerks." The secretary of state seemed to have as "his main preoccupation . . . Mrs. Gillespie's piano,'' but it couldn't be sold, ''So to hell with Mrs. Gillespie's piano.''[52]

Though some European posts retained more of the leisure and glamor of past days, work in Washington was hard. The secretary of state carried an enormous burden. One secretary's schedule for a single day included a meeting with the U.S. delegation to a conference on Chinese tariffs; a press conference; an appointment with the Spanish ambassador and a report from a Bolivian delegation. He then met a senator, a congressman, and a general. After lunch he signed official mail for one hour, then spent another hour with congressmen, more appointments, and a dinner in honor of the visiting Chilean finance minister. The press of meetings and decision making was relentless, whether they involved the seating arrangement at a state dinner or the U.S. response to Japan's invasion of China.[53]

Presidents Harding, Coolidge, and Hoover were not interested in foreign policy. When reporters spoke to the newly inaugurated Harding on this topic, he replied: ''You must ask Mr. Hughes. . . . From the beginning the secretary of state will speak for the State Department.'' Harding then withdrew, leaving his new cabinet member with the correspondents. Congress was another matter. The powerful Senator Boies Penrose of Pennsylvania explained, ''I do not think it matters much who is secretary of state. Congress—especially the Senate—will blaze the way in connection with our foreign policies.''[54]

The career personnel had been hostile to Wilson but were friendlier with his Republican successors. The secretaries of state and their own cliques protected Foreign Service interests and views, which included greater concern over conformity in character and intuition than with research and scholarship. The Foreign Service School's brief one-year course was shortened and more emphasis was placed on administration and visa work than on the principles and goals of American foreign policy.

Foreign Service officers generally shared 1920s nativist beliefs. They were suspicious of immigrants, prejudiced against southern and eastern European peoples, and skeptical about international organizations and disarmament, while believing in free trade, self-determination, and the efficacy of diplomacy in solving international disputes.

Since the United States was still a minor actor in the world, U.S. policy tended to be reactive. The State Department in Washington was still staffed mostly by clerks who handled embassy cables. Ambassadors received little guidance. Americans were bystanders to the tragic march of events in Europe that led to World War II.

During the 1929–1933 Hoover administration, Japanese expansionism in East Asia was a more immediate concern. The department was divided into pro-Japanese and pro-Chinese groups. The former believed that sanctions would only increase the likelihood of conflict with Japan, which preferred war to concessions. The latter thought that the United States should protect China from Japanese militarism and aggression. As Japan's advance continued, the China hands won out; but in the end their rivals were correct about which course would provoke war.

The views of diplomats were shaped by the places where they served. Those who worked in Moscow in the 1930s were critical of Stalin's dictatorship, while the U.S. ambassadors to Britain and France, Joseph Kennedy and William Bullitt, respectively, identified with those governments' appeasement policies toward Germany. U.S. diplomats gradually came to recognize the danger of Hitler in the 1930s but, like their British and French counterparts, desperately wanted to avoid conflict and believed that confronting Hitler would lead to war. James Dunn, conciliatory toward Mussolini's Italy, was not upset by Italy's invasion of Ethiopia and opposed sanctions against Rome's involvement in the Spanish Civil War during that decade. When World War II finally began, however, Foreign Service officers Anglo-French sympathies converted them into interventionists.

In contrast, the department quickly saw the USSR as a revolutionary force attempting to subvert other governments and opposed U.S. recognition of the Communist regime. This effort was led by Robert Kelley, chief of the Division of East European Affairs, who had studied Russian at Harvard and the Sorbonne and whose section was—in part, by necessity since it dealt with a closed country—more scholarly than its counterparts. By 1933, however, when the Roosevelt administration decided to reopen diplomatic relations with Moscow, most FSOs were ready to agree; the move was particularly opportune for the department's young group of Soviet specialists.

The training of language and area experts was an innovation during the early 1930s, opposed by some officials defending the traditional emphasis on generalists. Of the 14 men originally selected for studying Soviet affairs, most dropped out of the program or left the service entirely. But this effort showed admirable foresight, and the survivors, particularly Charles Bohlen and George Kennan, would play an important role in future American policymaking.

Along with FSO Loy Henderson, who had not received special training, Bohlen and Kennan served in Riga, where the State Department maintained a team of Kremlin-watchers just outside the USSR's borders. These men pored over available information and publications, using methods unthinkable to traditional officers who, with their direct access to foreign capitals and developments, preferred personal contacts and the famous Foreign Service intuition. While suspicious of Moscow's international intentions, as demonstrated by the revolutionary rhetoric and regimentation of foreign Communist parties, the young Soviet specialists were optimistic and eager when they arrived in 1934 to open the American embassy in Moscow.

These hopes were soon shattered. Harassed and isolated by the Soviet authorities, American diplomats witnessed the terror of the purges and show trials that decimated the skilled, educated, and leadership groups. Though the U.S. embassy knew little about Stalin's system of concentration camps, enough was glimpsed to evoke horror at the seemingly mad wave of self-destruction. Henderson likened the diplomats' position in Moscow "to the passengers from a ship which had been wrecked on a desert island surrounded by a shark-infested sea."[55]

Still, these and other experiences failed to free State Department thinking, still bound to the age of courts rather than to an era of expansionist, totalitarian dictatorship, and the department was unable to adjust quickly enough to a world in which states like Germany, Italy, Japan, and the USSR deviated so much from the past "rules" of international relations as understood by traditionalist diplomats.

Further, despite much talk about the need to analyze whole societies rather than merely the formal behavior of the government apparatus, State lacked the necessary skills and attitudes to perform this task. Despite their role as interpreters of the world to Americans, U.S. diplomats shared many of their fellow citizens' ethnocentric biases. The fault, of course, did not belong to the department alone. The American leaders and people were about to be called on for great psychological, political, and military efforts for which they had little preparation. What was most miraculous was the extent to which, in time, this test would be met and passed.

George Kennan, who combined an aristocratic mien with the analytical approach of the new FSOs, best expressed a side of traditional diplomacy still prized in the State Department: "The bland urbanity of word and conduct; the graciousness of manner; the wit; the good humor; the refinement of taste; the breadth of cultural interest; the largesse of perspective; the shrewd and skeptical view of men and governments; the appreciation for the values, in diplomacy, of elaborate indirection; the keen sensitivity to irony." This proud FSO elite believed that all would be well if only it were left alone to run American foreign policy. Often, their performance was not so remarkable as to merit such trust. The department's mentality was good at avoiding crisis due to misunderstanding but lacked the initiative necessary to cope with conflicts arising from clashing values and conflicting national ambitions.

Kennan, who in a real sense never felt comfortable in the changing American society of the mid-twentieth century, blamed the erosion of the traditional Foreign Service on "the great democratizers" whose attentions "encumbered" the department. This was only a further trial for "that honorable . . . company of men who have faithfully served successive American presidents and secretaries of state in a diplomatic capacity, often at considerable personal sacrifice, only to find themselves one day suddenly and mysteriously discarded."[56]

Despite the poor material rewards for loyalty and hard work, the traditional Foreign Service's characteristics of caste and snobbery, narrowness of vision, lack of sympathy for democracy, and conservative predictions ran counter to everything represented by the New Deal. President Franklin Roosevelt believed that diplomacy was too important to be left to the State Department. This was not an entirely novel view among presidents, but Roosevelt began a process of creating alternative agencies and channels that would permanently change the department's role in the policy-making process.

2

The Challenge of Global War: The Roosevelt Era

1933–1945

The crises of World War II and the Cold War brought the United States permanently to a central role in international affairs. Before 1941, America—and consequently the department—was usually a bystander. State reported world developments, but few events necessitated any U.S. response. The department's passivity, complacency, and aristocratic hauteur had made sense because its prime mission during that era of foreign policy quiescence was to avoid conflicts that were almost always eminently avoidable. Very much alive was the foreign-affairs philosophy developed at America's founding—sporadic diplomatic activity to maintain isolation and to avoid entanglement. Now the old, sleepy mechanisms and ideas no longer sufficed. Enemies were more implacable, threats were far more direct, geographic isolation less meaningful, and new responsibilities inescapable.

President Franklin Roosevelt tried to cope with the old guard's intractability by sprinkling his own men throughout the department, circumventing it with personal envoys, and dealing personally with the most important issues. Cordell Hull, secretary of state from 1933 to 1944 (longer than anyone else ever held the post), was the main victim of the Roosevelt system. Hull's predecessor, Henry Stimson, had suffered physically under a lesser strain, and worried whether the slim Hull would survive the job's rigors. Ironically, a decade later Stimson, by then secretary of war, was a daily listener to Hull's complaints on the difficulties of working with Roosevelt.

Born in a small Tennessee town, Hull studied law, served as a volunteer in the Spanish-American War, and became a judge. Elected to Congress in 1907 and to the Senate in 1930, he advocated free trade and opposed high tariffs, while supporting Wilson on the League of Nations issue. His excellent congressional connections were an asset for Roosevelt, and Hull even considered running for the White House himself; polls sometimes showed him more popular than the president.

This was not his only thwarted ambition. Hull often found himself "relied upon in public and ignored in private."[1] Roosevelt first developed a special link with Assistant Secretary Raymond Moley, but Hull quickly forced out this potential competitor. For many years thereafter, the president looked to Undersecretary Sumner Welles, a close personal friend, and to White House adviser Harry Hopkins.

Through a divide-and-rule approach, Roosevelt made subordinates compete in duplicate efforts. The contrasting personalities of Roosevelt and Hull—respectively, activist and cautious, energetic reformist and passive traditionalist—added to the problem. Hull's congressional work style was not good preparation for the far different world of executive-branch politics; he was neither decisive nor a good administrator.

Roosevelt never held a high opinion of Hull and thought State conservative, rigid, and unimaginative. "You should go through the experience of trying to get any changes in the thinking, policies, and action of the career diplomats," he complained, "and then you'd know what a real problem was." The department was still mired in the lethargy and complacency of earlier years. Roosevelt's complaints ranged from its passivity over Hitler's actions in Europe to State's occasional loss of documents and even the lack of ink in pens given him for signing international agreements. Hull was unable to bridge the wide gap between his own and his boss's personality. "If the president wishes to speak to me, all he has to do is pick up the telephone and I'll come running," Hull said. "It is not for me to bother the president of the United States." Yet he would also complain, with his pronounced speech impediment, of "that man acwoss the stweet who never tells me anything" and would periodically threaten to resign.[2]

Hull's grudges were, in Dean Acheson's words, "not hot hatreds, but long cold ones. In no hurry to 'get' his enemy, 'get' him he usually did." Unable to solve his problem with Roosevelt, he struck at the president's surrogates, particularly Welles, who Hull resented for going over his head to the White House and virtually usurping the role of secretary of state, albeit with Roosevelt's approval. Hull did not learn of Roosevelt's August 1941 conference with British Prime Minister Winston Churchill until Welles told him about it; the undersecretary also maintained his own correspondence with foreign diplomats and governments.

Welles, scion of a distinguished and wealthy family, was a good friend of Roosevelt and had been chief of the State Department's Latin American division in 1921 at the age of 28. During the 1920s, Welles was a diplomatic troubleshooter in the Dominican Republic and Honduras. Tall, slender, blond, and well dressed, a graduate of Groton and Harvard, and with field experience in Tokyo and Buenos Aires, Welles was a model member of the diplomatic aristocracy. But he was also a liberal who had pressed for the withdrawal of Marines from the Dominican Republic and preferred hemispheric cooperation to unilateral U.S. intervention. Roosevelt made him an assistant secretary, then ambassador to Cuba, and finally, in 1937, undersecretary, the number two post. "There is such a thing," wrote one of Hull's sympathizers, "as having a too capable Assistant from the point of view of the Chief."[3]

This rivalry was not the only headache for Hull and State. Roosevelt preferred using personal envoys, from the highly competent Averell Harriman to the unstable Patrick Hurley, over department channels. Other cabinet members, including Secretary of Agriculture (and later Vice-President) Henry Wallace, Secretary of Interior Harold Ickes, and Treasury Secretary Henry Morgenthau Jr., actively participated in making and implementing foreign policy.

The department's own growth during the Roosevelt years also posed problems. In 1936, the entire staff could stand on their building's steps for a group photograph; by 1946 the same gathering would have required a small stadium.

There were four assistant secretaries. One of them, Adolph Berle, had been picked by Roosevelt and Welles in order to put a New Dealer in the department. Son of a liberal Boston minister, he was graduated from Harvard at 18 and law school at 21. Berle was an adviser to Wilson at Versailles, a teacher, and author of an important book favoring government regulation of corporations. He was designated a one-man departmental think tank, but isolation from daily decision making—as often happened to would-be planners and thinkers in the future—left Berle with little power. Originally, Hull did not want him, sensing another Roosevelt-backed rival, but Berle tried hard to get along with his increasingly bitter boss. The career staff suspiciously viewed him as a White House spy.

Breckenridge Long, another assistant secretary, was an old-line conservative and narrow bigot as well as a Hull confidante. Originally from Missouri, he modeled himself a Southern gentleman, raising racehorses at his beautiful Maryland estate. He briefly served as an assistant secretary of state during World War I and twice ran unsuccessfully for the Senate. In 1933 he became ambassador to Italy, where his admiration for Mussolini wore off only gradually. As a rather virulent anti-Semite, Long used his control of immigration to bar American shores to the growing flood of refugees from Hitler's Germany.

Assistant Secretary G. Howland Shaw was responsible for administration, a job, said his former schoolmate, Dean Acheson, which "should be undertaken only by a saint or a fool"; Shaw tended toward the former; he was an ex-Foreign Service officer, a bachelor who kept much to himself, and a convert to Catholicism. Ultimately, Shaw became an expert on welfare problems and juvenile delinquency, retiring in 1944 to work on these issues.

Dean Acheson, the fourth of the assistant secretaries, was a highly skilled lawyer who later rose to be undersecretary and secretary under Truman. A conservative on economic matters, Acheson was a liberal internationalist and strong antifascist on foreign policy. He judged people sharply on his estimate of their intelligence and made many enemies, one of whom was Berle.

Acheson's sometimes abrasive self-confidence was revealed in one of his own anecdotes. At meetings, Hull would go over drafts of his speeches paragraph by paragraph, making any wider discussion impossible. Once, Acheson threw away an invitation to one of Hull's meetings. The next morning a second notice was discarded. Hull's messenger summoned Acheson to the annoyed secretary. "Are you refusing to come to my speech meeting?" Hull asked. Acheson said that he thought it was a waste of time. "I suppose," Hull replied sarcastically, "you think you could write a better speech." Acheson agreed to try. He obtained the division chiefs' support for his formulations, gained the president's endorsement, and the speech was well received. Naturally, this did not endear him to Hull.[4]

The assistant secretaries' duties were far-flung and often vague, forcing innumerable coordination meetings. "What was most often needed was not compromise but decision," Acheson later wrote. His tasks included responsibility for trade agreements, involving tedious and drawn-out negotiations—four years of discussion with Iceland taught him far more about sheepskins than he wanted to know—but the job of licensing military exports provided a chance to help Britain and apply pressure on Japan.[5]

Shaw, Acheson, and Berle sat on the Foreign Service Board that decided, among other things, whether officers could marry foreigners, meted out punishment for black market currency dealings (a constant temptation, given the often unrealistically low exchange rates), promotions, and retirements. "The small size of the Foreign Service," wrote Berle, "produced an intimacy that helped a great deal in personnel matters."[6]

Perhaps the single most important career official was James C. Dunn, who became the symbol of the conservative FSOs skeptical of New Deal reforms. Ironically, in contrast to the glittering background of so many of his colleagues, he had been a high school dropout and a bricklayer who studied at night, entered the State Department as a clerk, and later passed the Foreign Service exam. After overseas service and a stint as protocol chief, Dunn became Hull's special assistant in 1934. Meanwhile, he married the heiress to the Armour meatpacking fortune and bought a Washington estate adjoining the British embassy, signs of his socialization into the Foreign Service class.

From 1935 until he became Truman's ambassador to Italy a decade later, Dunn was head of the Division of West European Affairs. Roosevelt did not want FSOs to stay permanently in Washington, prompting Dunn and others to resign from the Foreign Service in order to extend their tenure as home-based division chiefs. These heads of regional divisions were like feudal barons, protecting their own favorites and constantly at odds with each other over jurisdiction. At the same time, the department provided no intelligence capacity, little economic or sociological analysis, and few broad policy guidelines.

This was less true for the department's research-oriented Soviet section. The Kremlin-watchers were, by necessity, more dependent on printed materials and scholarly attempts to piece together the hidden realities of Soviet politics, methods not well received by other bureaus, which preferred more intuitive approaches. In 1937, the section was merged into Dunn's Western European division; its files were given away to the Library of Congress and ultimately, a decade later, passed on to the newly formed CIA.

Individual Soviet specialists also suffered for their vocation. Bohlen, a Harvard graduate whose grandfather had been American ambassador to France, was relatively lucky. After joining the Foreign Service in 1929, he had two years of on-the-job training in Prague, where he was rotated through various consular duties—visas, invoices, passports, commercial reporting—before arriving in the political section.

He studied Russian in Paris for two and a half years and then joined the new U.S. embassy in Moscow, where he served from 1934 to 1940. After a tour in Tokyo, where he was interned for several months after Japan's December 1941 attack on Pearl Harbor, Bohlen became assistant chief of the Division of European Affairs' Soviet section, then headed by Loy

Henderson, which believed that Moscow's aims and ideology conflicted with U.S. interests. Whatever the specific approach of the Soviet specialists, they were mistrusted by many New Dealers. Bohlen's fluency in Russian enabled him to break this barrier during the war, when Roosevelt needed a career State Department man as translator (his previous one, a university professor, had reportedly regaled dinner parties with details of secret meetings).

While Bohlen found entry to the White House, his former boss, Loy Henderson, fell victim to the administration's mistrust of Soviet experts critical of America's wartime ally. Henderson's first visit to the USSR had been with a Red Cross relief mission in 1919, and two years later he had joined the Foreign Service. He monitored international Communist activities and in 1927 was sent to the Soviet-watch section in Riga before serving in Moscow between 1934 and 1938. After Soviet complaints about Henderson's work as head of the USSR section, the White House exiled him in 1943 by naming him minister to Iraq.

Hull tried to work closely with the career men. He met each Monday with the principal officials and often had staff discussions on Sunday as well. As a lawyer, he tried out various contradictory positions before settling on one, acting more like a senior partner than a commander. By the September 1938 Munich crisis, the department was working long hours and weekends of unpaid overtime. Hull took catnaps at a hotel near the office, receiving telephone calls and messengers even during these brief intervals. The pace of work was an ever-present reminder of the growing European crisis.

Fascism's rising power and aggression made liberal New Dealers even more concerned with a department they saw as a hotbed of reaction and appeasement, dominated by apologists for right-wing regimes and by bureaucratic caution. One liberal wrote: "The faded and moth-eaten tradition of Victorian diplomacy seeps out of every cranny. . . . It is a code of elegant cynicism, of tactical shrewdness that has small relevance beyond the horizons of the chessboard." Traditional routine was not acceptable in the face of world-threatening crisis. Rep. Emmanuel Celler rightly said that the department's indifference toward refugees from Hitler showed it "had a heartbeat muffled in protocol."[7]

Adolph Hitler's rise to power in 1933 and the consolidation of Nazi rule posed a difficult problem for U.S. policy, particularly given isolationist sentiment in Congress and in the country at large. While some U.S. diplomats in Germany quickly perceived the regime's nature and the threat it posed, State proved far better at reporting events than at understanding or responding to Berlin's foreign policy.

George Gordon and particularly George Messersmith, respectively

U.S. chargé d'affaires and consul-general in Berlin, were among the most astute analysts of Nazi Germany in any Western diplomatic service.[8] When Hitler took control, Gordon wrote, "The revolution has . . . transformed Germany into a completely centralized state. . . ."[9]

U.S. ambassador to Germany William Dodd at first believed that economic considerations and the Nazi leaders' inexperience would compel them to turn to traditional German conservatives who would stop the persecution of Jews and moderate the regime.[10] He urged Washington to give these figures a chance to exercise their supposed influence on Hitler. "A people has a right to govern itself [and] other peoples must exercise patience even when cruelties and injustices are done." Extreme reactions to Nazism by the United States might undermine the reasonable German officials who remained, he argued.[11] Dodd later changed his position and became a staunch advocate of collective security in Europe, peppering official dispatches and private letters to President Roosevelt with warnings.[12]

Unfortunately, despite accurate descriptions of German internal affairs, embassy analysis did not provide much useful guidance about the direction of Nazi foreign policy. Furthermore, Roosevelt's suspicions of State limited his willingness to hear any warnings provided, and the president was compelled to devote most of his attention to domestic economic recovery. Consequently, Roosevelt's early talk about blockade, boycott, and economic sanctions to pressure Germany was not translated into action.[13]

At one point, Roosevelt told Dodd that the German Jewish problem was not a U.S. government affair and that the United States could do nothing about the persecutions, though unofficial and personal influence of U.S. diplomatic representatives might be invoked to moderate depredations. Dodd's main task was to work for continued repayment of private U.S. loans and, if possible, for trade arrangements to increase German exports and help Berlin meet debt payments.[14]

While Hull was repelled by the Third Reich, the secretary of state was not inaccurately described by Germany's ambassador to Washington, Hans Dieckhoff, as "an idealist who lives somewhat in the clouds."[15] Hull could not see that his moral persuasion was useless in deterring or influencing Hitler from the road of aggression that eventually led to World War II.

Originally, then, the president and State maintained a policy of distant correctness toward the Nazi regime. American diplomats, except for Messersmith, seemed to have remained unaware of Hitler's unyielding expansionism. In 1936, as Hitler moved to annex Austria, Messersmith, now U.S. minister to Vienna, reported that if "one knows Mr. Hitler one must realize that his burning ambition is to impose his will on Europe by force of arms."[16]

In contrast, some department officials were prepared to appease Germany to obtain peace in Europe.[17] During the 1938 Munich crisis, the new American minister in Germany, Hugh Wilson, argued that Hitler's plans did not necessarily endanger the vital interests of Western powers. The Munich agreement, ceding western Czechoslovakia to Germany, opened the way to a better Europe, he said, opposing any U.S. action to hinder it. Wilson and other U.S. diplomats feared that Anglo-French resistance to Hitler's demands would bring war. Ambassador Joseph Kennedy in Britain believed that no price was too high to pay for continued peace and that, if war came, the United States must remain neutral.[18]

Most State Department officials did not understand that yielding to Hitler's demands only encouraged him toward more aggression and, ultimately, toward war. The dominant belief continued to be that Europe's problems did not impinge on America's vital interests. Hull's main efforts were applied to increasing U.S. pressure against Japanese aggression in China. Bypassed by the president on European questions, Hull was all the more upset by press reports of his absence when important decisions were made.

When Hitler seized the rest of Czechoslovakia in March 1939, Welles met with State's top Europeanists to decide the U.S. response. Messersmith advocated an immediate break of relations with Germany; another FSO, J. Pierrepont Moffat, and Dunn disagreed. Some representation, they argued, was better than none and could protect Americans and refugees. Berle feared that severing relations might stimulate Anglo-French hopes that the United States would join them in fighting Germany. A typical compromise was reached: The United States would neither break relations nor recognize Berlin's annexation. Hull resented Welles's leading role and falsely told the press that he himself had drafted the statement on the issue. When Roosevelt decided, without informing Hull, to order Welles on a European mission to see if peace was possible, the secretary was further antagonized toward both men.

By 1939, the inevitability of a new European conflict was clear. "While on the surface there is apparently a good deal of apathy," Berle wrote in his diary, "the State Department, at least, is rapidly getting to the boiling point."[19] On July 18, Hull and Roosevelt made a direct appeal to Senate leaders, grimly describing the likelihood of war and the need to repeal the U.S. arms embargo. Senator William Borah declared, "My feelings and belief is that we are not going to have a war. Germany isn't ready for it." Hull strongly disagreed, predicting war by summer's end and inviting Borah to read the embassy cables. "So far as the reports in your department are concerned,'' said the senator, ''I wouldn't be bound by them. I have my own sources of information . . . and on several occasions I've found them more reliable than the State Department.'' Hull was livid at another public humiliation.[20]

Despite Borah's optimism, the State Department was already planning for the conflict. Two officials were assigned to listen to the shortwave radio in seven-hour shifts and to send hourly bulletins to the president, Hull, and Welles. Roosevelt's pleasure with the system was the beginning of what in later years would become a worldwide radio monitoring operation.

The crisis finally broke on September 1, 1939, when the White House received a 3 A.M. telephone call from Ambassador Anthony Biddle in Warsaw reporting the German bombing of the city. Twenty minutes later, Ambassador William Bullitt called from Paris with the news that the war had begun. Roosevelt contacted Hull and Welles; by 4 A.M. the top State Department officials were meeting in Hull's office, listening to Hitler announce the attack over shortwave radio. The State Department, like the rest of the world, would never be the same.

Between 1939 and the war's end in 1945, the Foreign Service grew from 1000 to 3700 and the State Department expanded from 3700 to 7000 employees. America's new power and responsibility transformed it from a reluctant and secondary participant in international politics power into a global force, involved in every corner and issue of five continents. From technical backwardness—a dependence on officials with their ears against shortwave radios—State's communications, intelligence, and information-gathering capacities grew to the level of complexity necessary for survival in a modern and perilous world.

Yet, during World War II the department's influence remained limited. Hull and Welles led policy over Latin America and relations with neutrals, but the conduct of the war itself and policy in the battle zones and for the postwar world were controlled by the White House and the military. Everything was subordinated to the war effort. The United States generally took a backseat to Britain and France on the Middle East and in regard to their colonial empires, which still covered most of Africa and Asia. Roosevelt's determination to manage policy, the rise of new government agencies, and the foreign involvements of existing ones also curtailed State's jurisdiction.

Hull's vexation grew as he was not informed or consulted on important decisions. The secretary of state was excluded from the War Council and from the major Allied summit meetings in Casablanca, Tehran, and Yalta. Roosevelt even persuaded Churchill to keep his foreign minister, Anthony Eden, from going to Casablanca in order to prevent Hull from attending. Nor was the secretary of state informed about the effort to produce an atomic bomb. Hull could only threaten resignation, plot to eliminate Welles, and hope that Roosevelt would retire in 1940 so that he could become the Democratic candidate.

Slights against the entire department reflected those against Hull. The Board of Economic Warfare, the Treasury, the Office of War Information, the Coordinator of Inter-American Affairs, the Office of Lend-Lease Administration, the Office of Strategic Services, the Petroleum Coordinator, and others impinged on State's turf. These newcomers were resented, both for their liberal and multiethnic personnel and for their intrusions into foreign policy. One Foreign Service officer wrote in disgust, "Before long our foreign policy will be in the hands of everybody but the State Department."[21]

The war, however, only slightly eased State's naiveté on intelligence matters. Security from espionage was rarely taken seriously. No FBI clearance was required for government work in the 1930s. Canadian economist John Kenneth Galbraith joined the Agricultural Adjustment Administration and later State without even being asked if he was a U.S. citizen. One of Galbraith's friends, an anti-Nazi German refugee, recounted that he wandered into the State-War-Navy building in search of a public bathroom, just before America entered the war, and walked unhindered down the corridors, seeing one empty office after another with secret papers covering the desks. With Washington at the brink of war and Europe in flames, such carelessness was incredible.[22]

The unhappy relationship between President Roosevelt and the career Foreign Service was further undermined by the Tyler Kent affair. In May 1940, British counterintelligence arrested Kent, a young Princeton-educated diplomat and cipher clerk at the U.S. embassy, after the State Department waived his immunity. He was secretly tried and convicted of removing 1500 official documents from the embassy and passing them through British fascists to German intelligence.

Kent, an adamant isolationist, claimed he was only trying to uncover the president's activities to embroil the United States in the war. The case confirmed Roosevelt's belief that State contained rightists and isolationists bent on obstructing his foreign policy and even a few traitors working for the Axis powers. In addition, Kent's presumed compromise of U.S. diplomatic codes prompted Roosevelt thereafter to send the most sensitive communications with foreign leaders and his own personal envoys via the military communications system rather than through the State Department.[23]

Many New Dealers exaggerated State's rightist propensities. Secretary of Interior Harold Ickes, who never got along with State, wrote in his diary that it was "shot through with fascism."[24] Trusted presidential aide Harry Hopkins had particular contempt for the department, describing Foreign Service men as "cookie-pushers, pansies and usually isolationists to boot." Meeting Bohlen at a dinner in the autumn of 1942, Hopkins rudely asked him if he were a member of the anti-Soviet clique in the department.[25]

More serious subversion was coming from the opposite side. A seemingly minor incident, but one that eventually would devastate the department, came only three days after Germany's invasion of Poland. One Saturday evening, anti-Communist journalist Issac Don Levine brought Whittaker Chambers to meet Berle, whose duties at State included security against espionage. Chambers, a Communist party member and Soviet agent until 1935, told Berle that some department officials were also spies and mentioned several names, although he did not at that time single out Alger Hiss, the suspect in a later major spy case. Chambers offered no evidence and Berle made no serious investigation. The incident would be forgotten until it secretly exploded within State five years later and then surfaced publicly in 1948, a major cause for the department's problems during the McCarthy era.[26]

On a lighter note, the president used his wit to joke about his frustrations with State. In a draft for a 1940 speech, Roosevelt spoke of pro-Axis agents operating in the United States, and continued, "There are also American citizens, many of them in high places, who unwittingly in most cases, are aiding and abetting the work of these agents." When the State Department suggested deleting this phrase, Roosevelt quipped, "We'll change it to read—'There are also American citizens, many of them in high places—especially in the State Department. . . .'"[27]

Given its tradition of caution and the administration's commitment to action, the department was cut out of decision making. By 1941, many foreign missions in Washington were conducting important business directly with Hopkins, as did the U.S. embassy in London and Churchill himself. Hull received occasional polite notes from Hopkins enclosing copies of cables "for your information."[28]

Hull almost missed even the one conference he did attend, at Moscow in October 1943. Roosevelt preferred Harriman or Welles, complaining that Hull would just "mess things up." Hull was so angry at the prospect that, though he had never before been on an airplane—a remarkable fact given the far-flung travels of later secretaries—he told the president, "Wherever the conference might be held—anywhere between here and Chungking—I would be there myself."[29]

With the secretary of state stripped of any significant function, Undersecretary Welles too busy making policy to run the department, and communication between the two top men in disarray, feuds proliferated at the lower levels. Observing the disorder, Acheson learned lessons he would apply as secretary in later years. The department, he complained, "seemed to have been adrift, carried hither and yon by the currents of war or pushed about by collisions with more purposeful craft."[30]

All over Washington, combat over influence brought the department into collision with institutional rivals. Roosevelt never lost his sense of humor about these conflicts. The president joked that his dog Fala's food was nightly stolen by rats from the State Department across the street, but added laughingly that this was not a symbolic statement. Roosevelt explained seriously that he had wanted to reorganize the Foreign Service ever since taking office, and that the only way professional diplomats could ever understand America was to be sent for a year to Tennessee, Hull's home state.[31]

Even within State, the views of geographical divisions often conflicted. The Near East and Europe divisions, for example, differed on how strongly to support continued French control of its north African colonies. State's weakness and splits also did not prevent liberals from holding it responsible for positions actually dictated by the White House's pragmatic wartime policies, including recognition of regimes whose leaders formerly collaborated with Germany: Vichy France, the Darlan government in North Africa, Marshal Badoglio's cabinet of ex-Fascists in Italy, Franco Spain, and Latin American rightists.[32]

With regard to Latin America, the old debates about whether to embrace right-wing dictatorships or to press them toward democracy were heated, as were conflicts over how to handle states friendly to the Axis—Bolivia, Chile, and Argentina. Roosevelt's relative disinterest left the field open to quarreling among subordinates. Lawrence Duggan, the liberal head of the Latin American division, rejected British proposals to overthrow such regimes, while Welles also took a softer line than Hull against Latin American neutralism, arguing that hemispheric unity was more important than attempts to isolate Axis fellow travelers. Welles contemptuously considered his superior ignorant about the region.[33]

The showdown came when Welles, attending an inter-American conference in Rio de Janeiro, accepted a resolution allowing each country to determine its ties with the Axis, a compromise that let Argentina off the hook. Duggan and Berle tried to salvage the situation, while Hull, feeling his instructions had been deliberately flouted, reprimanded Welles in a heated long-distance telephone conversation, further embittering relations between the two men.

Other battles convulsed State's Far East section. Before the war, Ambassador Joseph Grew in Tokyo had warned that attempts to pressure

Japan over its aggression in China could lead to war and should be pursued only if the United States would back them up all the way. "If you can't find a rock to build your house on, but only sand," he wrote, "it's much safer not to build a house at all." But Hull, supported by Far East division director Stanley Hornbeck, ordered economic sanctions against Tokyo.[34]

Later, during the war, the Foreign Service specialists working at the U.S. embassy in China warned that Chiang Kai-shek's government was paralyzed by corruption and incompetence and that without reform the regime could neither fight Japan nor prevent a Communist takeover. FSOs like John Davies and John Service believed that the Chinese Communists were also nationalists who would eventually break with Stalin and prove useful in the fight against Japan. Still, State's main emphasis was on helping the non-Communist Chinese government to survive by encouraging it to strengthen its political base of support. Although the China hands' advice had almost no effect on wartime U.S. policy, conservatives would later hold them responsible for Chiang Kai-shek's fall, an event they had merely predicted and had tried to prevent.

Another controversy was a clash between the Far East and European divisions over decolonization. "Why should India defend a freedom she hasn't got?" Berle asked. "The age of imperialism is ended," asserted Sumner Welles. But America's need for European help in the war effort, and later in the Cold War, allowed the Europeanists to limit U.S. support for colonial independence.[35] While many Europeanists were stereotypical Foreign Service anglophiles, this predilection did not pervade the department. The Middle East section was split between sentiments of respect for the British as senior partners and mentors and preferences for decolonization. Many department officials believed America was popular in Asia and the Middle East as a supporter of development and independence, an image that could be a very important asset after the war when, they expected, the old imperial powers would lose control of those areas.

Roosevelt's policy had been heavily improvisational. "I have not the slightest objection to your trying your hand at an outline of the post-war picture," he told Berle in June 1941. "But for Heaven's sake don't ever let the columnists hear of it."[36] The most important goal was maintaining the wartime alliance in the years to come. Roosevelt believed that he could do this far more effectively than any professional diplomat. "I know you will not mind my being brutally frank," he wrote Churchill, "when I tell you that I think I can personally handle Stalin better than either your Foreign Office or my State Department."[37] The president was a strong believer in personal diplomacy and in keeping the reins in his own hands. Hull was not even allowed to keep copies of messages from

Stalin or Churchill, but had to read them in the presence of Roosevelt's military aide.

State Department officials, as well as Roosevelt, were very aware of Soviet defensive concerns and were willing to try to assuage them. The Middle East and Eastern European areas along Moscow's frontiers were viewed as testing grounds where one might see if Soviet demands were reasonable or expansionist. U.S. leaders would accept Soviet influence and neutral regimes in these areas, but not total Soviet domination. Berle's remark was typical: "I see no reason why we should object to" the small countries of Eastern Europe "being within the orbit of the Russians, provided we are assured that the USSR would not use this power to subvert the governments, and set up a regime of terror and cruelty among the peoples—in other words, deal with the situation as they dealt with the Baltic countries. . . . This is, indeed, the chief distinction which exists between a power which seeks world domination and a power which does not."[38]

Thus, the U.S. government hoped to avoid the frightening prospect of postwar conflict with the USSR; until the autumn of 1944, few thought such a confrontation inevitable. Meanwhile, the war was being fought and threats to the alliance had to be defused. From the beginning of World War II to the beginning of the Cold War, wrote Acheson, "We groped after interpretations, sometimes reversed the lines of action based on earlier views, and hesitated long before grasping what now seems obvious."[39]

The press of daily business and demands of bureaucratic warfare left little time for thought. Regardless of their views, everyone worked endless hours of unpaid overtime. Hull and others suffered health problems. Lack of space and equipment added to the burden; airgrams—shipped directly through the diplomatic pouch—replaced telegrams to avoid the time needed to encode less urgent reports. During the war, Acheson's overload included responsibility for economic warfare and raw materials, dealing with the Free French, setting future occupation zones in Germany, as well as his original job of administering foreign economic policy. He also chaired the department's Executive Committee on Commercial Policy, served as a member of the boards on Foreign Service personnel and the *Foreign Service Journal,* and supervised divisions dealing with his areas. Ironically, he had little contact with the high policy issues over which he would later be attacked by Congress and from the right.

Most FSOs were consciously contemptuous of traditional American idealism toward international affairs. They saw themselves as realists who thought that countries acted to serve national interests rather than to spread their ideas. White House appointees in the wartime department, however, projected New Deal appeals to freedoms, self-determination, and the struggle of democracy against dictatorship. Acting as a bridge between New Deal and traditionalist visions, Acheson rephrased the former's objectives in realpolitik terms. U.S. security required stability and order in the Moslem and Hindu worlds, from Morocco to India: "Certainly we favor the evolution of self-government for the diverse peoples of that area, as we favor the restoration of their liberties to the democratic peoples of France and Spain." U.S. interests entailed a stake in their political development: "If chaos prevails there, and the regions become a military vacuum, tempting adventure, we shall face the same danger of war which accompanied the collapse of the Turkish and Austro-Hungarian Empires."[40]

With the department itself so often bypassed, U.S. ambassadors were placed in a difficult position. Ambassador John Winant in London complained, "Nine-tenths of the information I receive comes from British sources." But it was clear that he had no authority, and Winant rarely saw the British leaders. The press even reported he was about to be fired. Winant wrote Hopkins, "I think the President . . . should know that no Ambassador can be an effective representative here in London unless he is given more information and more support than I am receiving."[41] Winant's position was made particularly difficult by Averell Harriman's presence, nominally as Lend-Lease representative, but in reality as the president's personal envoy. Even Harriman, with his direct line to the White House, had similar complaints when he later served as ambassador to the USSR.[42]

Friction within the department grew in the summer of 1943. Hull's slowness and Welles's activism led to constant quarrels that soon extended into the newspapers. Hull leaked his version of events to Arthur Krock of *The New York Times,* while Hull's liberal critics in the department and administration talked to columnist Drew Pearson. Krock wrote, "The State Department will function smoothly and effectively when the President permits the Secretary to be the undisputed head of a loyal staff."[43]

At that moment, Hull finally insisted that Roosevelt choose between him and Welles. The showdown involved personal rivalry as well as rumors, spread by the bitter and ambitious former ambassador to France, William Bullitt, that Welles was a homosexual. Since Hull's resignation would only aggravate the president's already serious problems with Congress, where Hull was both effective and popular, Roosevelt agreed to ease Welles out by sending him on a long mission abroad. When this plan leaked into the press, however, Hull described Welles as "disqualified for that job." Welles resigned in September 1943.[44]

If even Welles, after a decade of strong White House backing, could be so easily eliminated, it was understandable why the career staff wanted to avoid excessive identification with the New Deal. As one former official put it, they "knew that they would still be there when the Franklin Roosevelt Administration had been replaced by another one, which might well be reactionary and isolationist in accordance with the inexorable ebb and flow of American politics."[45]

In Adolph Berle's words, the career man, "has long since learned that if he stands up to a situation and gets into a row about it he is wrong, irrespective of the fact that the row might be a legitimate and honest defense of a legitimate and honest American interest. By consequence, when he sees trouble approaching, he slides away . . . having learned by long experience that it is the safest thing he can do."[46]

After four weeks of discussion, Hull and Roosevelt compromised on Edward Stettinius to replace Welles. Stettinius was popular but not respected. He worked well with people, but tended to be more concerned with personal relations than strong leadership. A former General Motors and U.S. Steel executive, Stettinius could satisfy conservatives but was, like Harriman, a liberal businessman. Heir of a wealthy family, he became U.S. Steel board chairman at the age of 37, then chairman of the War Resources Board, and later Lend-Lease administrator. His assignment was to reorganize State and resolve the administrative chaos arising from the wartime turmoil.

In January 1944 he regrouped geographical and functional divisions under assistant secretaries in a more logical fashion so that each had a clear area of responsibility on which he reported to the undersecretary. The secretary presided over a policy committee to mediate disputes between divisions and a staff committee to bring together the assistant secretaries.

Meanwhile, the health of the 73-year-old Hull continued to deteriorate. After so many past resignation threats made for tactical reasons, he finally quit in the fall of 1944. Hull's great contribution had been winning congressional support for Roosevelt's policies. He maintained good liaison with the leading Republican foreign policy specialist, John Foster Dulles, and ensured that Republican leaders were fully briefed and consulted. The administration remembered how President Wilson's failure with Congress and the opposition party had wrecked his plans for organizing the peace after World War I.

Nevertheless, by 1944 the relationship between the State Department and the White House had completely broken down. "The President cannot be Secretary of State; it is inherently impossible in the nature of both positions,'' Acheson later wrote. ''What he can do, and often had done with unhappy results, is to prevent anyone else from being Secretary of State.''[47]

Roosevelt chose Stettinius to succeed Hull as a handsome figurehead while the White House continued to control policy. Always conscious of the need for good employee relations, Stettinius even ordered a renovation of State's old building, with light-green paint, new plumbing, and modernistic furniture. During his brief tenure, however, he spent more than half his time attending meetings outside Washington, particularly in leading the U.S. delegation at the UN's founding conference. The day after that meeting ended in June 1945, the new president, Harry Truman, appointed James Byrnes to succeed Stettinius.

Stettinius was an unimpressive secretary, but he did put together an able team. For example, Acheson, now assistant secretary for congressional relations, worked effectively to build legislative support for administration policies. Breckenridge Long had handled the task on a personal basis, but Acheson's staff of 17 acknowledged congressional inquiries the day they were received, helped legislators arrange travel abroad, and offered to write speeches. The staff drafted legislation, arranged for its sponsorship, and marshaled votes and outside support. Acheson found lobbying a thankless job since it often entailed bringing sticky problems to Congress's attention. As soon as the war ended, Acheson seized the opportunity to escape back to private life, though he did not long enjoy his liberty.

Liberals criticized the appointment of Stettinius and his new team, regarding them as too conservative. The department, they argued, had been recaptured by a clique incapable of building a postwar world along the lines envisioned by New Dealers. ''Who won the election, anyway?'' they asked.[48] Eleanor Roosevelt wrote her husband that it made her ''rather nervous for you to say that you do not care what Jimmy Dunn thinks because he will do what you tell him to do and that for three years you have carried the State Department and you expect to go on doing it.''[49]

Still, the Cold War can hardly be attributed to a clique of State Department conservatives. Just as the atmosphere of wartime cooperation had led to high hopes of alliance, so the war's end and the emergence of political problems produced greater tension. Beginning in the fall of 1944, in response to Soviet behavior, the whole government began to shift its views.

While State was given some latitude on the Third World, the White House decided that relations with Moscow were too important to leave to the department. A single-minded emphasis on winning the war and fear that the Russians might collapse or make a deal with Germany encouraged U.S. open-handedness and seemed to necessitate trust in Stalin. For many nonprofessional policymakers, history began with the U.S.-Soviet alliance against Hitler—there was no memory of Communist international subversion, the 1930s purge trials, the Hitler-Stalin pact that carved up Poland in September 1939, nor any understanding of Soviet ideology, internal structure, and long-range objectives. (Institutionally, only State can be an administration's foreign policy memory, a duty the department has often failed but more often was given no opportunity to fulfill.)

"We are dealing with the Soviets on an emotional instead of a realistic basis," Bohlen later wrote, and it was "almost impossible to convince others that admiration for the extraordinary valor of the Russian troops and the unquestioned heroism of the Russian people was blinding Americans to the dangers of the Bolshevik leaders."[50] The State Department was more reserved than public opinion and other agencies in enthusiasm for Moscow's lasting friendship or good intentions.

Still, State and the government as a whole also had to keep in mind the importance of inter-Allied cooperation, the common U.S.-Soviet war goals, and the hopes for a new era of lasting international peace and harmony. The hardliners—who preferred rapprochement and the maintenance of the alliance, even while preserving their suspicions and stressing caution—were in retreat. Loy Henderson, a pessimist about Soviet intentions, lost his post as head of the East European section in 1943 because of his views. Veteran Kremlin-watchers Raymond Murphy and Raymond Atherton were also forced out. Bohlen, a more optimistic Soviet specialist, became special assistant to the secretary of state in December 1944 to act as liaison with the White House and to provide State with clues about the direction of a foreign policy over which it had little influence.[51]

Kennan also worried about an overly naive assessment of the Soviets, tracing it to a weakness that "causes Americans, once at war, to idealize their associates, to make inhuman demons of their opponents, and to become wholly oblivious to the long-term requirements of any balance of power."[52] Bohlen introduced Kennan to Harriman, the newly appointed ambassador to Moscow, and the three dined together one evening in Washington. Impressed by Kennan, Harriman requested him as an adviser, the post for which Kennan had been preparing his entire career.

Up to that point, Kennan had endured a frustrating war. Serving in Germany at the time of Pearl Harbor, he had been interned for several months. At his next post, Lisbon, instructions from Washington had been tardy and confused. The home office was almost laughably ignorant of conditions on the scene and had no idea what the War Department was doing. Dunn once told him to remember that State's policy role in wartime was secondary and that he should give advice only when asked.

Lonely and uninfluential in earlier years, Kennan felt himself engaged "in the curious art of writing for one's self alone." As late as September 1944, he wrote, "There will be much talk about the necessity for 'understanding Russia'; but there will be no place for the American who is really willing to undertake this disturbing task."[53] But in Moscow Kennan became Harriman's tutor; he was finally in the right place at the right time. American foreign policy was about to undergo its most important change since the foundation of the Republic.

The change began when the nearby Red Army refused to advance to help the August 1944 Warsaw uprising against the German occupation, condemning the Polish underground to a bloody annihilation. Subsequent Soviet treatment of Eastern Europe as conquered territory, imposing puppet governments and repressing independent resistance forces, also provoked mistrust. The United States knew that it did not have—and could not exercise, save at the cost of the alliance's breakdown and a new war—the strength to reverse these events, but this provided all the more reason to limit further Soviet gains.

One by one, Harriman, Bohlen, Berle, and others were converted. Those working on international organization were shaken by disputes over the projected United Nations. The first U.S. representatives in Eastern Europe gave Moscow the benefit of the doubt, but as time went on their disillusionment made for increasingly pessimistic reports to Washington.

U.S. diplomats had to determine which Soviet demands were provoked by legitimate defensive concerns and which gains were sought to improve the Soviet position for future offensive actions. The distinction was, to say the least, difficult to make: Poland's plains, for example, served both as Germany's invasion route to the east in 1941 and the USSR's road to the west in 1944–1945. Soviet claims that Turkey or Iran represented threats were met with greater skepticism; any increase in Moscow's influence in that region created the potential for later advances into the Mediterranean and the oil-producing areas of the Persian Gulf. State Department officials saw Soviet behavior as a continuation of traditional Czarist policy rather than as a product of Marxist ideology.

Kennan posed the key question in May 1944: "If initially successful will [the USSR] know where to stop? Will it not be inexorably carried forward, by its very nature, in a struggle to . . . attain complete mastery of the shores of the Atlantic and Pacific?" The USSR could only conceive of neighbors as either vassals or enemies, he wrote; "if they do not wish to be the one, they must reconcile themselves to being the other."[54]

Similarly, Harriman wrote Hull in September, "What frightens me . . . is that when a country begins to extend its influence by strong-arm methods beyond its borders under the guise of security it is difficult to see how a line can be drawn. If the policy is accepted that the Soviet Union has a right to penetrate her immediate neighbors for security, penetration of the next immediate neighbors becomes at a certain time equally logical."[55]

Berle made similar points: It was understandable that the USSR wanted to prevent threats from neighbors, but establishing puppet governments was a different matter. "The Soviet conception of 'security' does not appear cognizant of the similar need or rights of other countries and of Russia's obligation to accept the restraints as well as the benefits of an international security system."[56]

Henderson, who had suffered for his premature Cold War views, blamed the slowness of U.S. awakening on the department's internal structure. "There was," he later said, "an atmosphere of timidity in the geographic bureaus who had been pushed to one side by the up and coming New Dealers. The new people felt the 'old fogeys' should be pushed out. . . ."[56] There is some truth to this. With everyone busy planning for a postwar era of global cooperation, the department's activities shaped its views. The alternatives were unthinkable; memories were short.

While Stettinius, Bohlen, and other department officials accompanied Roosevelt to the Yalta summit conference with Churchill and Stalin in February 1945, where postwar issues were discussed—a reward Hull would never have won—Roosevelt still ignored State Department briefing papers. Records of the summit, including the Far East agreements over which State would be much criticized in the McCarthy era, were only available to Stettinius, Harriman, and Bohlen among department officials.[58]

The U.S. objective was to maintain some democracy in Eastern Europe, although Moscow interpreted the meeting as acknowledgment of its dominant role there. The USSR was given benefits in Asia in exchange for joining the war against Japan after Germany's surrender. These concessions would become highly controversial in the McCarthy era, but they were made on the basis of power realities: the Red Army's presence in Eastern Europe and the seemingly tough war ahead against Japan. "I didn't say the result was good," Roosevelt told Berle after the Yalta conference. "I said it was the best I could do."[59]

As it became clear that the Soviets were not living up to the agreements over Eastern Europe, Americans on the scene were increasingly angry. Bohlen recalled a meeting where General Lucius Clay, just appointed second-in-command to General Dwight Eisenhower, claimed that to get along with the Soviets you had to give trust to get trust. Bohlen assured Clay that he would soon become one of the American officials most opposed to Moscow. The more hope one placed in postwar conciliation, the more bitter the disappointment. Even Hopkins was beginning to have doubts, as did Roosevelt before his death in April 1945.[60]

Vice-President Harry Truman succeeded Roosevelt. He had only rarely met with the president and had received no foreign affairs briefings. Truman's advisers—Grew, Leahy, Stimson, Bohlen, and Harriman among them—were further along the road to Cold War, but they still preferred maintaining the alliance. This could only be done, they argued, if the United States stood firm.

When Harriman returned to Washington in April to meet the new president, both men still believed workable relations were possible between Washington and Moscow. The time had come, Harriman told a State Department staff conference, "to eliminate fear in our dealing with the Soviet Union and to show we are determined to maintain our position."[61] These two objectives—conciliation and strength—were seen as mutually reinforcing goals. Liberals and conservatives were reaching similar views of Soviet policy by different paths. The former rejected the abuse of human and national rights; the latter deeply mistrusted Moscow's Communist ideology.

The change between State's roles at the Yalta summit meeting of February 1945 under Roosevelt and at the July 1945 Potsdam summit under Truman illustrated the presidents' differing style of work. Truman preferred careful consultation and planning with his advisers, in contrast to Roosevelt's endless improvisations. While State generally supported Truman's decisions, some worried that the United States might have conceded too much. Bohlen thought it a mistake to ask Stalin to join the war against Japan; others disagreed with the proposed new boundary between Germany and Poland. Ironically, Dunn, who liberals saw as an arch-hardliner, was still relatively hopeful of Washington-Moscow consensus. Similarly, Kennan wanted to accept Europe's division into U.S. and Soviet spheres of influence and had to be persuaded not to resign when Washington refused to accept this view. Most State officials felt it wrong to abandon Eastern European peoples who still hoped for the application of wartime promises of freedom and independence.[62]

U.S. diplomats were also making decisions affecting the lives of millions in China. Originally, Washington tried to avoid involvement in that country's domestic politics. Chiang Kai-shek might want U.S. aid to fight the Communists, but the Japanese were the far more immediate threat. Those in the U.S. embassy and in the department who dealt with China policy were anti-Communists, but they were also realists. They urged that Chiang be supported and encouraged to reform in order to prevent a civil war likely to end in a Communist victory. But loans and public relations came easier than fundamental change. Throughout the war, Chiang's political base and military power deteriorated.

The U.S. ambassador to China, Patrick Hurley, one of the wildest presidential envoys ever to enter a Foreign Service nightmare, wrongly told Roosevelt that all was well. But while Hurley visited Washington in February 1945, his deputy sent reports suggesting that Chiang be pressured toward reform for his own good. Grew endorsed this approach, but Hurley was livid, viewing the FSO's message as sabotage.

The attitude of these China hands was in part based on their suspicion of Soviet ambitions to dominate China. They thought that Chiang could be strengthened by reform, but would lose if he sought a military confrontation with Mao Tse-tung's armies. The Chinese Communists might also be persuaded, in their own interests, to oppose Soviet imperialism. The embassy staff simply reported the facts about the growing popularity of the Communists and of Chiang's weakness and corruption. Harriman and Kennan shared this concern that Moscow might help Mao into power if there was no diplomatic settlement in China.

The old conflict between China and Japan specialists in State was also revived in a new form. "The China hands," said John Carter Vincent, then head of the department's China section, "could make a better case for China than China could for itself and the Japan hands could make a better case for Japan than Japan could for itself. No question about it, we were partisans. It got into your blood."[63]

There was also a revolt in the Far East division against its director, the autocratic Stanley Hornbeck. These complaints disturbed the consensus-minded Stettinius, and Hornbeck retired. Under Grew's aegis, Japan hand Joseph Ballantine became director of the Office of Far East Affairs in autumn 1944. The Inter-Divisional Committee on the Far East, also controlled by Japan specialists, recommended letting Japan retain the emperor to encourage Tokyo's surrender, a highly controversial issue at the time.[64]

The leak of department documents to the magazine *Amerasia* seemed to be another part of this conflict. *Amerasia,* associated with the prestigious Institute of Pacific Affairs, had links to the Communists, and the FBI discovered it was receiving many classified reports. Grew pressed Truman to arrest six low-level officials, including John Stewart Service, who had lent some of his dispatches to the magazine's editor in June 1945. But the affair was hushed up for two disparate reasons: The FBI had used illegal methods and it was feared that accusations of Soviet espionage might damage the alliance. Service was cleared, but the affair would resurface.

In April 1945, Hurley was foolishly optimistic about Soviet willingness to accept a non-Communist government in China. From Moscow, Kennan and John Paton Davies were far more critical than Ambassador Hurley of Stalin's attempts to maximize his influence in China. Ironically, Davies tried to convince Hurley of this danger, attempting to moderate the naiveté of the man who would later have him persecuted as naive and pro-Communist.

Hurley resigned on November 26, 1945, accusing seven Foreign Service officers, including Service, Vincent, and Davies, of favoring the Chinese Communists and imperialists (meaning Great Britain) who wanted to keep China divided. This was the first shot in the bloody "Who lost China" debate and the accusations of State Department treason that dominated much of the next decade.

As this turbulent era began, Truman felt new leadership was needed at State. Stettinius himself told Truman of the great disorder in White House–State Department relations and of State's liaison problems with other agencies.[65] On July 3, 1945, only three days before leaving for the Potsdam summit conference, Truman named James Byrnes secretary of state. Byrnes had little experience in foreign affairs, but, like Hull, he had served in both houses of Congress. Given the troubles between Truman and Byrnes, it is understandable why later presidents avoided choosing any secretary of state with his own political base. Byrnes's belief that he would make a better president than Truman was a major cause of friction.*

As powerful head of the Office of War Mobilization, Byrnes once told Hopkins to "keep the hell out of my business." Obviously, he was not a man who would be satisfied as a mere figurehead.[66] Roosevelt had been certain of this: "Jimmy had always been on his own in the Senate and elsewhere and I am not sure that he and I could act harmoniously as a team," Roosevelt told Stettinius in November 1944. "In other words," Stettinius asked, "Jimmy might question who was boss?" The president replied, "That's exactly it."[67]

As secretary, Byrnes quickly ended Bohlen's liaison job, refusing to allow anyone to broker his relations with the White House, and he largely cut himself off from the career staff. By weakening the department, this unilateralism also undermined Byrnes's own effectiveness and power.

*Except for John Foster Dulles's brief Senate appointment and the short tenures of former congressmen Christian Herter and Senator Edmund Muskie, Byrnes was the last secretary of state with Capitol Hill experience.

The State Department, which almost doubled in size between 1939 and 1945, and the Foreign Service, which quadrupled in that same period, were profoundly transformed by the war and by their dramatic new tasks. Many things happened in the summer of 1945—the death of Roosevelt and elevation of Truman; the replacement of Stettinius by Byrnes; the war's end; the dropping of the atomic bomb; and attempts to patch up the alliance for final victory, coupled with the beginning of its collapse. Those fateful months pushed the State Department to the center of the international scene and brought it into its modern period of global responsibility.

3

State Takes Command: The Truman Terms

1945–1952

Since perilous times produce heroic actions, the dangers and conflicts the United States faced after World War II helped create a high level of State Department performance. The nature of the Cold War, in contrast to the shooting war just concluded, put diplomacy in command. This responsibility strained State's capacities to the fullest and gave rise to both the McCarthy attacks on the department and the need for a full reorganization of the U.S. foreign policymaking process.

The range of new foreign policy instruments, including economic and military aid, cultural exchanges, public and media relations, and covert operations, also added to the State Department's burden. America's position as the world's strongest country made the department's decisions, once so esoteric, a matter of great moment in every world capital. These changes were symbolized by the increasing use of the term *national security* to indicate the complex mix of diplomacy, military strength, and intelligence gathering that furnished U.S. foreign policy's boundaries and tools.

President Truman, conscious of these new perils and complexities, was willing to delegate authority while retaining key decisions and an access to information which Roosevelt had denied him as vice-president. Truman's administration began the daily two-page summary of important diplomatic developments for the president, supplemented by the secretary of state's verbal reports, as well as daily intelligence summaries and a weekly CIA briefing.[1]

Much of today's foreign policy machinery dates from this era. The 1947 National Security Act created the National Security Council (NSC) and the CIA to help manage the flow of information and options. The merger of the War and Navy departments into a Department of Defense was completed in 1949. These changes created new competitors for State as the dominant foreign policy influence.[2]

During and immediately after World War II, State and the military departments held regular meetings to work out mutual problems and common strategies.[3] The NSC was established to institutionalize such cooperation, providing a forum in which agencies might reconcile their views and give the president unified conclusions, proposals on current issues, and warnings of future problems. NSC meetings were attended by the president, vice-president, secretaries of state and defense, the director of the CIA, the chairman of the military Joint Chiefs of Staff, and other high officials brought in at the president's request.

A small secretarial staff handled paperwork and informed departments of NSC decisions, but most of the work was handled by delegates from existing departments. As long as the NSC staff acted as a committee of the whole and the secretary of state was the president's main foreign policy adviser, the State Department dominated. Truman's White House staffers, as veterans of the bureaucracy, saw themselves as implementers rather than as policymakers and rarely challenged State's prerogatives.

State, however, did not control new foreign aid and information programs. Technical assistance to underdeveloped countries was first developed by a junior State Department official, Benjamin F. Hardy Jr., who proposed it to the White House staff after his own superiors had twice rejected the idea as too costly. When aid programs began State changed its mind, but the White House, Congress, and even many FSOs preferred an independent institution to run them. Otherwise, as one official explained, State might be dragged away from its main responsibility, becoming "a general store, where surplus property sales, publishing ventures and other extraneous commodities displace diplomacy on the shelves."[4] The European Cooperation Administration, which administered the Marshall Plan to help rebuild Europe, and the Office of Mutual Security, for other parts of the world, were established as separate agencies. They worked fairly well with State and Defense, although this arrangement did not prevent constant tinkering in later decades.

As always, the character and relationships of high officials were central in determining State's structure and influence. Secretary of State James Byrnes was personally powerful during his 1945–1947 tenure, but his aloofness from both the White House and the departmental bureaucracy limited his effectiveness. Truman replaced Byrnes with George Marshall

(1947–1948), succeeded by Dean Acheson (1948–1953), both of whom, in contrast, maintained good ties up and down the line. Marshall delegated authority well, letting his undersecretary, Acheson, run the department on a day-to-day basis and reserving only major decisions for himself.

Such an arrangement was needed since the secretary was so often abroad, negotiating over the postwar international order. Byrnes left for the 1945 Potsdam summit conference only three days after taking over and spent 62 percent of his time in office, 350 out of 562 days, away from Washington. Since Marshall was absent about one-third of the time, Acheson was in charge during much of his tenure as undersecretary.[5]

Even aside from travel, the pressures of office left the secretary with little time to think beyond immediate issues. He needed to confer constantly with the president, with leading members of Congress, assistant secretaries, and division chiefs. No wonder these groups often felt themselves ignored by the secretive Byrnes. "The State Department fiddles while Byrnes roams," went one department witticism. The secretary's reluctance to share information and responsibility contributed to his deteriorating relations with the president, Capitol Hill, and his own department. Byrnes did not effectively use Undersecretary Acheson to administer the department in his absence; Acheson and Truman, both bypassed by Byrnes, tended to band together.[6]

"If Truman had chosen anyone else for secretary of state, it would have been astonishing," exclaimed *Newsweek,* calling Byrnes "unquestionably the No. 2 Democrat of today."[7] This was precisely the problem. Truman could only be uncomfortable with a man who wanted to be president, though he picked Byrnes partly because the secretary of state was then—in the absence of a vice-president—the post next in line for the presidency. In an inaugural speech, Byrnes carefully noted that his job was "to carry out" policies "determined by the President and the Congress," but he often seemed to forget this in practice.[8]

At first, Byrnes seemed set for a successful tenure. His popularity in the Senate gained him unanimous confirmation without a hearing or debate. In 1945, the triumph over the Axis, the afterglow of the UN's founding, optimism over maintaining the wartime alliance with the USSR, and Byrnes's own self-confidence made the future seem bright.

Some U.S. diplomats worried that the old legislative compromiser might prove too yielding to Moscow. Byrnes's willingness to recognize Soviet puppet governments in Romania and Bulgaria in exchange for promised elections was not well received by Truman or by the Republicans. But the real problem was that American sentiments were changing;

Soviet good faith was increasingly under question. If officials at State took a tougher line than did U.S. public opinion, it was not due to some conservative cabal but to a belief in a Soviet threat stemming from observation of Moscow's tightening hold and repression in Eastern Europe and elsewhere, Stalin's territorial demands, and his breaking of earlier agreements.[9] The more the two countries got down to details over postwar boundaries and governments in areas controlled by the Soviets, the more conflicts emerged from month to month throughout 1945 and 1946.

Byrnes's style also produced friction with Truman. In Moscow for a December 1945 foreign ministers meeting, Byrnes told Ambassador Harriman that he would not telegram the White House on the discussions: "The President has given me complete authority. I can't trust the White House to prevent leaks."

Byrnes's attitude toward State was similar. He only sent a sketchy report after the conference was over. Truman was angry, and powerful senators complained about being left out of decisions. Senator Arthur Vandenberg remarked, "We didn't know how lucky we were to have Stettinius until we got Brynes."[10]

Given these policy and personal conflicts, Truman decided to replace Byrnes with General George Marshall. Determined to establish his control over State, Truman told a press conference, "The State Department doesn't have a policy unless I support it."[11]

Policy was already in disarray from the avalanche of events, wartime pressures, and expansion producing State's disorganization. A "tangled legacy from all sorts of conflicting policies and personalities," said *The New York Tribune,* was "being administered by an equal tangle of conflicting agencies and authorities. . . . The United States cannot indefinitely leave its foreign policy to the accidental interplay of the brilliant amateur, the opinionated eccentric, and the bureaucratic intriguer."[12]

After all, vital international issues—what Truman, in a 1945 message to State, called the "increased responsibilities" arising out of the postwar world's "difficult and complex international problems"—were affected by intradepartmental struggles. For example, Grew and the Japanese hands unsuccessfully advocated revising the demand for Japan's unconditional surrender by pledging to allow the emperor to retain his throne. Tokyo ignored U.S. peace proposals that omitted such an explicit promise, leading to hundreds of thousands of deaths when U.S. planes dropped two atomic bombs in August 1945 to bring about Japan's surrender.[13] After the war, the Japanese were allowed to keep their emperor anyway.

In another instance, Moscow saw Washington's sudden termination of Lend-Lease in September 1945, shortly after the war's end, as a deliberate provocation. Actually, it was a routine bureaucratic decision made by Grew while other high officials were out of the country.

The times were conducive to such errors. Complex and fast-moving events brought a confusion intensified by many transfers and retirements. Liberal China expert John Carter Vincent became office director for East Asia. A new bureau, Special Political Affairs, staffed with liberal idealists and headed by Alger Hiss, was created to deal with the freshly founded United Nations. Infusions from the wartime agencies diluted the department's traditional tone and brought new bureaucratic struggles and security problems.[14] Transferees "floated in limbo," wrote H. Stuart Hughes, then director of research on Europe. "We felt most of the time as though we were firing our memoranda off into a void. The atmosphere was that of Kafka's Castle, in which one never knew who would answer the telephone or whether it would be answered at all."[15]

The department faced the immediate transfer of more than 10,000 employees from such temporary bureaus as the Office of War Information, Office of Inter-American Affairs, and the Office of Strategic Services (OSS). State's staff grew from 4000 in 1939 to 11,000 in 1946. Altogether 40 percent of the personnel were new. Embassies were also expanding. In 1934, the U.S. mission in Ecuador had been staffed only by a minister, a second secretary and two clerks, all of them underworked. By 1946 there was an ambassador, a counselor, 10 officers, and 30 clerks, plus military and naval attachés. That year there were 300 U.S. embassies, legations, and consulates around the world. Overall, in Washington and in the field, 18 percent of Foreign Service officers were engaged in political work, 22 percent were on economic issues, 13 percent worked on trade and commercial matters, 24 percent labored in consular work, 12 percent were in information and cultural activities, and 11 percent were administrators.[16]

The handling of intelligence became a point of special controversy. The OSS, responsible for that task in the war, was dissolved in September 1945. Col. Alfred McCormack, director of military intelligence, became a special assistant to the secretary of state and 1600 OSS employees were transferred to form a new group "whose business," McCormack explained, was "to turn information into intelligence." Acheson wanted this work in State Department hands, while OSS director "Wild Bill" Donovan and Secretary of the Navy James Forrestal wanted an independent intelligence agency. Truman decided to give State a chance to run the show.[17]

Unfortunately, Congress did not yet understand the need for professional intelligence work, and the military was jealous of State's victory.

There was also opposition in the department led by Spruille Braden and Loy Henderson, heads of the Latin America and Near East bureaus, respectively. They thought McCormack's operation would duplicate their offices' labors and distrusted the OSS analysts' liberalism and "amateurism." Byrnes gave in, Truman also concluded that an independent agency would be best, and McCormack resigned. Eventually, about half the OSS transferees remained to form State's Bureau of Intelligence and Research (INR), but State had thrown away the opportunity to dominate intelligence analysis. The CIA was soon created for that purpose.[18]

While Acheson sought a greater intelligence role for State, its own reliability was being attacked in Congress and from the right. During the previous decade, most criticism of the department had come from the left. Now, in the wake of the *Amerasia* espionage case and Hurley's resignation, conservatives began to make charges of security leaks and even treason because of allegedly insufficient support for Chiang Kai-shek. White House adviser Admiral Leahy echoed these sentiments. "The President is all right—he's behind Chiang," he told another officer. "But those pinkies in the State Department can't be trusted."[19]

Right-wing Senator Kenneth Wherry of Nebraska condemned Acheson, for example, for the routine comment that Washington and not Gen. Douglas MacArthur's military government would determine U.S. policy on Japan. Wherry said of Acheson: "He will be the main leader of the new group of liberals that will draw up the State Department policy." Senator Connally replied for the Democrats, "Anyone with a morsel of sense knows the President and Jimmy Byrnes are going to set our foreign policy."[20]

The Senate Foreign Relations Committee heard both Hurley's charges of disloyalty and rebuttals from Byrnes and Acheson. "Men who have rendered loyal service to the Government," said Secretary of State Byrnes, "cannot be dismissed and their reputations ought not to be destroyed on the basis of suspicions entertained by an individual." Hurley told the committee that disloyal subordinates sabotaged administration policy supporting Chiang Kai-shek, but the examples he provided involved only petty bureaucratic infighting. Senator Connally accused Hurley of merely seeking headlines; the committee dropped the investigation.[21]

While Hurley's case was trumped-up, there were some real security problems within the department. The very ferocity of the politically motivated attacks made State all the more eager to cover up difficulties. Security risks, often recent transferees from wartime agencies, usually held minor posts. For example, ex-OSS agent Carl Marzani was later convicted of falsely denying Communist party membership under oath.

Further, cases of subversion brought out in the 1950s dated from before World War II. H. Julian Wadleigh, an expert on trade issues, admitted the truth of Whittaker Chambers's accusation that he gave information to Soviet agents in the 1930s. On July 26, 1946, Byrnes announced that preliminary screening of 3000 employees transferred from wartime agencies led to recommendations against the permanent employment of 285, of whom 79 had been terminated by that time. A March 1948 report by a House of Representatives committee found possible security problems among 108 applicants or employees, of whom just 57 were currently employed at State (only 40 remained by 1950), all cleared by full FBI investigations. These studies were later distorted by Senator Joe McCarthy in his accusations of widespread subversion in the department.[22]

In addition to Wadleigh, Noel Field, an official in the West European division, was probably a Soviet agent before he left the department in 1938 to join the League of Nations. In later years he went to Eastern Europe, was put on show trial as an American agent in Czechoslovakia, and settled in Hungary. Some former Soviet agents claimed that Laurence Duggan, head of the Latin America division in the 1930s, who committed suicide during the McCarthy era, refused recruitment as a spy but gave them general background briefings.

The most important firsthand account of subversion at State came from Whittaker Chambers, who had been a Soviet intelligence agent in the 1930s. Chambers's accusations against Alger Hiss in 1948 set off a sensation that would drag the department over many reefs in the following years. Yet Chambers stressed that Moscow's objective had been to steal reports rather than to influence policy, which he called "a magnificent waste of time."[23]

There was never any evidence that U.S. policy had been altered, certainly not on China, by subversive efforts, nor was there anything to show that State's higher-ups had conspired to protect spies or leftist attempts to influence decision making. There can be no doubt that most of the controversy was caused by partisan efforts to discredit the Democratic administration, but the problem was worsened by State's own lax security before 1945 and by its attempts to avoid adverse publicity, giving some basis to charges of a cover-up.

Chambers's warning to State on the eve of World War II about Soviet infiltration had not been taken seriously. Many people were hired during the rapid wartime expansion without proper precautions; postwar transfers allowed others into State on the erroneous presumption that they had already been fully checked. By 1945, however, State was conducting its own investigations and looked into Chambers's story, particularly as it related to a rising career officer, Alger Hiss. French intelligence had also warned about Hiss, who had been an aide to Asia division director Stanley Hornbeck in the early 1940s. The FBI questioned Hornbeck, who spoke highly of his assistant. In May 1944, Hiss moved to the Office of Special Political Affairs, dealing with UN and postwar planning problems, and became its director in March 1945. He enjoyed access to the highest officials and to a broad range of documents. More and more questions were raised about his reliability; the FBI intensified the investigation and, in May 1945, interviewed Chambers.[24]

Igor Gouzenko, a Soviet embassy code clerk who defected in September 1945, spoke of a highly placed Soviet agent at State, which the FBI concluded was Hiss. The FBI tapped Hiss's telephone, opened his mail, and kept him under surveillance. Little new evidence was found, but in early 1946 FBI director J. Edgar Hoover told Byrnes that he thought Hiss was a Soviet operative; State's security staff concurred. Hiss's promotion and assignment were suspended and his access to documents restricted.

Hiss denied the accusations when he was informed of them in March. Internal reports recommended that he be permitted to resign or, if he refused, be fired. But dismissal would involve a hearing, publicity, and a legal decision that might rule the department's evidence insufficient.

State dawdled in forcing the issue and Hoover began leaking information to friends in the press and Congress. To the rescue came Republican foreign policy adviser John Foster Dulles, who renewed an offer for Hiss to become president of the Carnegie Endowment for International Peace. Hiss resigned in December 1946 and Byrnes wrote a polite note of regret at his departure. Hiss's replacement at Special Political Affairs, Dean Rusk, ordered a thorough security check of the bureau's personnel.[25] Two years later, after Chambers publicly accused Hiss of espionage, the department's reputation plummeted. Those who defended Hiss, including Acheson, were strongly attacked in the press and in Congress.

State suffered for keeping its own investigation and removal of Hiss a secret. State's fear—that exposure of the real, though limited, extent of infiltration would raise congressional and public criticism—was realized in full. Poor management of security made State even more vulnerable. A congressional study that examined Hiss's file before the case became public called him "the greatest security risk the Department has had," but the security office's records did not even mention his resignation.[26]

State's desire to keep the affair quiet was based partly on the fact that its evidence against Hiss might not have met judicial standards. Yet the department needed to prevent infiltration by Soviet agents who might copy documents or report on secret decisions, recruit more spies, or influence policy decisions. It was also necessary to ferret out corrupt or unreliable applicants and employees as well as those who might be subject to blackmail. As a result, in 1946 Congress passed a new law empowering the secretary of state to dismiss any employee without reason or defense if deemed necessary for national security.

A congressional staff study surveying 108 security files found most of them concerned transferees in relatively minor positions. Derogatory information was often questionable, based on hearsay or the accusations of personal enemies. In one January 1947 case, concerning an employee who had signed several petitions to allow Communists on election ballots, the assistant secretary for administration ruled that the grounds for dismissal as a security risk were "substantial evidence of Communistic affiliations past or present . . . without equally substantial refutation or a substantial evidence of a change of heart." The Security Committee in this case decided that the woman was not a security risk; nonetheless she was designated an undesirable employee and resigned.[27]

Many of the cases studied were not security risks but rather those of applicants with criminal records or psychological problems, and most of these had been turned down by the department. People were, however, sometimes given responsible positions on the basis of limited information. Some employees against whom there were no direct complaints had friends who were Communists or even possible agents. One case, for example, concerned a man described by acquaintances as interested in Communism but also as a very ardent New Dealer. One friend's father was a Communist party member, and the employee himself had suffered at least one mental breakdown. A January 1947 internal department memo suggested that he be brought "before the [Security] Committee as a security hazard—possible break and embarrassment if Congress gets this." He quit to take a job at the United Nations, where a number of officials asked to leave State found employment.[28]

A staffer in one assistant secretary's office was observed by investigators with a member of an espionage group in August 1946. In December, he successfully solicited his boss's intervention to help someone with Communist party connections, who had transferred to the Foreign Service from another agency, obtain a position. The new employee was quickly dismissed but had continued access to classified material a week after he was supposed to be out of the office.[29]

The large-scale personnel transfers, State's rapid expansion, and the shift of the USSR from ally to enemy swamped the small security apparatus. In February 1947, Secretary of State Marshall chose John Peurifoy to coordinate security measures. The FBI made recommendations, a new Personnel Security Board rescreened employees, and the security bureau was reorganized. The department issued a twenty-five-page manual cautioning employees about being "chatterbugs or know-alls" and promising that security consciousness would help bring promotions.[30] These reforms strengthened internal security but were too late to stem political attacks.

Traditional criticisms of State Department cosmopolitanism were now mixed with criticisms of real administrative problems and alleged subversion. Throughout 1946, salvos came from all quarters. A former high official in wartime economic operations complained of State that "the people doing the clerical end of the work there don't have the faintest idea of the standards prevailing in the well-run agencies." One congressman called the department's cultural relations program "a hotbed of Reds" and claimed Americans "are tired of this cultural relations stuff." Berle, retiring as ambassador to Brazil, said efficient administration was impossible because of all the "watertight compartments."[31]

The House Military Affairs Committee claimed that State's intelligence section was full of pro-Soviet employees. Alfred McCormack, the bureau's director, demanded the charges be investigated or dropped. "It is no answer to say your committee lacks jurisdiction to make a fair and an adequate investigation of the charges," he wrote. "If that were so, you should not have published the charges in the first place." Acheson said that accusations of Communism had only been sustained against one employee, and that many were hounded simply for being New Dealers. "I have not considered that to be a crime," he told the House Foreign Affairs Committee.[32]

Given these pressures, Foreign Service careers became less attractive. FSO recruits, told that the U.S. government was "reposing special trust and confidence in your integrity, prudence and ability," earned only $57 a week in 1946, about the same as skilled blue-collar workers. There were problems finding and retaining good people on salaries ranging from $2900 to $10,000 a year. One well-regarded career officer turned down the ambassadorship to Argentina because he could not bear the personal expenses. Ambassador to Moscow Charles Bohlen had no funds to replace his ancient auto. It broke down on a trip to the Kremlin as a Soviet limousine carrying the ambassador from Outer Mongolia passed by.[33]

Another FSO complained that a lack of clerical staff forced officials to spend the "bulk of their time entertaining applications for . . . visas, passports, consular invoices, replying to postage stamp inquiries from school children. . . . [The] fault of the Foreign Service lies not in the type of officer but in the work required of him."

One result of these problems was the 1946 Foreign Service Act, which raised salaries, provided for "selection out" of substandard officers, and improved home leave. It also created a Foreign Service Reserve for temporary use of outside specialists.[34]

But further reorganization was necessary both within State and to regulate relations among the chief foreign policy agencies. Formulation of U.S. positions on atomic energy, Germany, trusteeship for European colonies, hemispheric defense, and relations with the USSR were delayed by bureaucratic shortcomings. The United States, wrote James Reston in *The New York Times,* was "trying to play a new and vastly different role in the world with an old Government machine that is neither geared nor staffed for the job."[35]

Byrnes's frequent absences during his tenure underlined these problems. The old State-War-Navy committee could not function without the secretary's presence, and the department's staff committee rarely met.[36] Contradictory policy statements resulted from a lack of planning and a decline in discipline. The continuing division between a Foreign Service corps serving abroad and a civil service staff in Washington caused personnel problems. State-White House and State-congressional relations had deteriorated while assistant secretaries lacked sufficient power to run their own bureaus.[37]

Two examples of this era's problems with policy coordination were the public split between Secretary of State Byrnes and Secretary of Commerce Henry Wallace and the dispute over U.S.-Argentine relations. Wallace, formerly Roosevelt's third-term vice-president, was out of step with the emerging Cold War consensus. After Byrnes criticized the USSR for violating agreements in September 1946, Wallace made a speech at a New York rally condemning the tougher line. He claimed the United States was too friendly to Britain and too hostile to the USSR. Truman had carelessly approved the statement without understanding its implications.[38]

An outraged Byrnes demanded Wallace's resignation. "You and I spent 15 months building a bipartisan policy upon which the world could rely," Byrnes wrote the president. "Wallace destroyed it in a day." Truman fired Wallace. "The Government of the United States must stand as a unit in its relations with the rest of the world," the president told a press conference. "No member of the executive branch . . . will make any public statement . . . in conflict with our established foreign policy." Everything would have to be cleared with the State Department.[39]

The Argentina battle was equally contentious. Spruille Braden grew up in South America and worked there as a mining engineer before joining the Foreign Service. As U.S. ambassador to Argentina during World War II, he fought German influence and collaborators. Braden saw Argentine dictator Juan Peron as one of the latter, and Peron made him *persona non grata* for his criticisms. Braden's tenacity was rewarded with the post of assistant secretary for Latin America in August 1945. His predecessor, Nelson Rockefeller, had taken a relatively soft line toward Buenos Aires and the same could be said of the then new U.S. ambassador to Argentina, George Messersmith.

Peron was, in Acheson's words, "detested by all good men—except Argentinians," and U.S. criticisms strengthened the dictator's nationalistic appeal at home. He even ran on the election slogan of "Braden or Peron." Messersmith, known at State as "Forty-Page George" for his long cables, went over Braden's head, writing Truman and Byrnes to urge bilateral conciliation. Braden accused Messersmith of insubordination. State ended the quarrel by forcing both Braden and Messersmith to resign in June 1947. That action, wrote Acheson, "had the powerful effect of transforming an instruction from the Department from an invitation to debate to an order to act."[40]

A more important issue was the need to formulate a proper U.S. stance toward the USSR in the postwar era. It is most ironic, in view of McCarthyist charges, that the State Department did more than any other government agency to warn of the emerging dangers in Stalin's policy. Here, Soviet specialist George Kennan made a major contribution. Bedridden by illness at the U.S. embassy in Moscow, Kennan received a pessimistic cable from the Treasury, the department which had held the greatest hope for postwar U.S.-Soviet collaboration. If even that bureau confessed itself baffled by Moscow's actions, Kennan reasoned, perhaps Washington was ready to listen to his long-neglected views. He dictated an 8000-word telegram in February 1946 that explained Soviet behavior as springing from an internal need for expansion. Kennan's ideas quickly dominated the debate within the U.S. government.

His telegram came at precisely the right moment, Kennan later reflected, since "more important than the observable nature of external reality" in forming the government's view of the world "is the subjective state of readiness on the part of Washington officialdom." This raises the question of "whether a government so constituted should deceive itself into believing that it is capable of conducting a mature, consistent, and discriminating foreign policy." Kennan concluded it was not, and the same problem—and the same sad answer—would mark the aftermath of many future foreign policy crises.[41]

Kennan was not a good bureaucratic operator, being an outsider and critic by disposition, but he gave articulate expression to ideas already held by many at State as a result of their firsthand experiences. Kennan saw Soviet suspicion as unresolvable through U.S. concessions, since survival of the Moscow regime was dependent on generating foreign threats. Only U.S. strength could discourage Soviet expansion and encourage enough confidence in Western Europe to permit reconstruction there. Eventually a policy blocking the spread of Soviet power—to become known as *containment*—would force Moscow into a more reasonable bargaining position and permit diplomacy to take over. A parallel policy of "patience with firmness," to use historian John Gaddis's phrase, was already reflected in Byrnes's speeches. Despite popular pro-Soviet feelings during the period of wartime alliance, public opinion polls by 1946 were showing a favorable response to tougher U.S. policies.[42]

A second, related priority was to devote American attention and resources toward a long-term international leadership role rather than retreat into the nation's traditional isolationist mood. As Acheson said in one speech, the task was in "focussing the will of 140 million people on problems beyond our shores" at a time when they "are focussing on 140 million other things."

Liberals feared the new policy was heading toward a U.S.-Soviet conflict; conservatives were not sure that they wanted to spend the money needed for foreign commitments. By 1947, however, the country began to unite behind accepting the new responsibilities. Byrnes's replacement by George Marshall in January 1947 was ecstatically hailed at State as a major step toward rebuilding morale and discipline. Marshall took the job out of a sense of duty, though he preferred, Joseph Grew told a Red Cross women's meeting in a memorable slip of the tongue, "to retire to Leesburg and spend the rest of his life with Mrs. Eisenhower."[43]

Truman had great confidence in Marshall, while Undersecretary Acheson, wrote one of his friends, "spends a good deal of time bubbling over with his enthusiasm, rapture almost, about General Marshall." Acheson wrote former Secretary of State Henry Stimson that Marshall "has taken hold of this baffling institution with the calmness, orderliness and vigor with which you are familiar. We are all very happy and very lucky to have him here." This attitude was partly due to Marshall's willingness to trust, and delegate authority to, his staff. One of Marshall's first acts was to call in leading State officials to ask their opinions on current problems, something unthinkable under Byrnes.[44]

Marshall believed in the effectiveness of a clearly defined chain of command. "Gentlemen," he once said, "don't fight the problem, decide it." While some liberals thought having a general as secretary of state set a bad precedent, Marshall seemed more like, as one reporter put it, "a statesman who happened to be a general, not a general trying to be a statesman."[45]

The break with the past was also symbolized by State's move to a building in Washington's "Foggy Bottom" section that remains the department's offices today. This was a spartan headquarters, almost irreversibly drab, with cramped offices. The effect came close to being demoralizing. Veteran FSO Henry Villard nostalgically recalled "when policy-makers met informally over a pipe or a cigar . . . in a semicircle" around the secretary of state, "unhampered by squads of technicians and specialists. But in the dehydrated air of New State, jobs were further compartmentalized, policy papers were composed and 'staffed' along military lines" and decisions were made by checking a "yes" or "no" box on documents.[46] This new age marked State's transformation, at least in theory, from an aristocratic to a bureaucratic institution.

Despite traditionalist complaints, enhanced efficiency came just in time. In January 1947, the British embassy informed State that London could no longer afford to help Greece and Turkey in combatting, respectively, Communist guerrillas and Soviet military pressure. By the following day, State's staff produced a proposal for responses and delivered it to Acheson's home. Within 48 hours the conclusions were on Marshall's desk. The following day the plans were also endorsed by the president and by the secretaries of war and of the navy. Truman, Marshall, and Acheson briefed and won over congressional leaders to their view of growing Soviet strength, European weakness, and U.S. responsibility. State then drew up a detailed aid program and drafted a presidential message to Congress.

This effort "unleashed for the first time the creative effort" of State's staff, one participant later wrote. Only 19 days after the crisis began, the president made the Truman Doctrine speech pledging U.S. aid for countries threatened by foreign aggression. Preparation of this initiative, which laid the basis for containment of Soviet expansionism and for foreign aid efforts, was a triumph of teamwork. The Near East and European bureaus developed ideas, others added comments and criticisms (Kennan thought the commitment too open-ended), State's top economic affairs official, William Clayton, used his prestige with business and Congress, and the information staff worked to explain the new policy to the public.[47]

Other initiatives quickly followed. In May, Acheson tested the water with a speech suggesting U.S. aid for European recovery. The following month Marshall made his famous Harvard commencement address proposing the European Recovery Program, better known as the Marshall Plan. The new Policy Planning Staff and the Intelligence and Research Bureau played major roles, while Bohlen wrote the first draft of the speech. The pressure of time was relentless. Robert Lovett, who replaced Acheson as undersecretary in June 1947, was exhausted. "It has been ghastly,'' Kennan commented in October. ''I'm afraid if he ever goes to bed, he'll never get up.''[48]

As Lovett was well aware, however, organizational efficiency was closely related to policymaking and the nation's security. For example, he discovered that State's analysts correctly predicted that Moscow would complete its takeover of Hungary in 1947, but their reports had never reached the top. He started a briefing book, kept in the secretary's right-hand drawer, with the latest data on potential crises and proposed responses.[49]

A more complex world and government required more sophisticated methods. Marshall's reforms dealt with many problems revealed by the Byrnes era. The undersecretary was made State's chief operating officer, ensuring smooth performance during the secretary's frequent absences. A newly established Secretariat directed reports to decision makers and monitored the bureaucracy's compliance.

The State Department was dubious about the idea of a National Security Council to improve interagency communication. Secretary of Defense Forrestal agreed that NSC would not be a policymaking agency but only ''a place to identify for the President those things upon which policy needs to be made.'' The NSC staff would be temporarily assigned there by existing agencies. State would set U.S. diplomatic objectives, the Joint Chiefs of Staff would prepare plans for national security, and the NSC would ensure that political goals were matched with military capacities. This sensitivity dictated that NCS's first executive secretary, Admiral Sidney Souers, at first delivered NSC papers to the president through the State Department. But this awkward arrangement was soon revised so that Souers would have direct access to Truman.[50]

The Policy Planning Staff faced a similar maze of channels. It was created by Marshall after a meeting with Soviet leaders convinced him of Moscow's expectations for Western Europe's imminent collapse. The group, under Kennan's leadership, worked at top speed, debating well into the night, and developed the Marshall Plan to avoid this potential calamity.[51] Between 1947 and 1949, some 60 of the staff's briefing papers formed the basis for NSC decisions shaping the new U.S. foreign policy. Its papers for NSC consideration would first be submitted to lower-level officers in relevant departments, revised for consideration by senior officials, and only then passed to the NSC's members. If necessary, each agency would present its own position, with Souers acting as referee, before the final product was given to the president.

The lengthy process was best geared to preparing for distant problems. The NSC's first study—whether a U.S. military withdrawal from Italy would increase the likelihood of a Communist takeover—took eight months. Still, it was better to debate issues in advance than to allow a last-minute crisis to force decisions. State was also satisfied with its continued influence on the NSC. This arrangement continued even after Truman decided during the Korean War to create a permanent NSC senior staff.[52] The NSC and the CIA only gradually began to challenge State's hegemony.[53]

When Secretary of State Marshall retired in January 1949, Dean Acheson returned to government to replace him. He was the fifth man to hold the office in five years, but morale was high and State's relations with both the White House and Defense Department were excellent. "For the first time in the memory of living man," wrote one reporter, "the American foreign office comes somewhere near being adequate to the needs of the country."[54]

Dean Acheson evaluated people on the basis of their "intelligence," explained Archibald MacLeish. "He did not shrink at making enemies and had almost no tolerance for what he felt was inferior intellect or stupid questioning."[55] Such attitudes were barriers to success in the Foreign Service, but they may well be necessary in prodding complacent, self-serving bureaucracies into action. The irony is worth underlining: The personality of Acheson, one of State's most successful leaders in building institutional power and effective policies, was the antithesis of that expected from a diplomat.

Acheson turned his traits to advantage by becoming a unique figure—a "career appointee"—during a decade's service in high positions. Thus, he was able to attain the career officer's experience while retaining a policymaker's decisiveness and clarity of vision. Other secretaries made themselves strong through personal influence in the White House; Acheson, in his own right and under Secretary of State George Marshall, was the only one to utilize fully the department's resources.

At the top level, the mutual respect among Truman, Marshall, and Acheson was an essential element in their cooperation. Policy was built on partnership between a strong president and a strong secretary of state. Acheson did not love State; frustration inspired his repeated attempts to resign and escape back to private life. He well understood the department's weaknesses, and while not overawed by the mystery with which professionals try to surround diplomacy, he was able to respect their abilities. Consequently, Acheson tried to make State work rather than bypass the staff.[56]

Acheson personally mellowed as secretary of state, showing, said one colleague, "none of the strain and terrible irritation (almost to the fly-off-the-handle kind) I had seen him exhibit a few times, when he was working with Byrnes."[57] He avoided travel, believing the secretary should stay in Washington as much as possible. Acheson's success lay in good bureaucratic practices: He participated in policy formation at the earliest possible point, listened to subordinates in staff meetings, gave credit and rewards for excellence and refused to tolerate mediocrity, maintained high standards, backed subordinates in fights with other departments, and energetically tried to win policy battles.[58]

Acheson's effort to dominate overall U.S. policy was helped by his control of the NSC. Ineffectual Secretary of Defense Louis Johnson could not compete with him. In 1950, when Johnson's incompetence and petty anti-Acheson sabotage led to his replacement by Marshall, the close cooperation between the two departments was renewed. Thus, Acheson was not merely the government's chief diplomat but its strategic coordinator as well.

"A man's stature is measurable by the time he has had the Soviet's aggressive number," editorialized *The Washington Post* on Acheson's appointment.[59] Some thought that the new secretary had not been a cold warrior long enough, unfairly blaming him for Roosevelt's alleged appeasement of Moscow. In fact, both Acheson and State tried to make Americans recognize a Soviet threat long before most of their later critics were ever aware of it, but this did not save them from a humiliating rout over security matters and the Hiss case.

The case exploded into the headlines when Chambers finally made public his accusations against Hiss. State was also accused of failing to uncover spies and even of being controlled by Stalin's agents. Although State had forced Hiss to quit, the press and public thought the department had been blind to the danger. Acheson's old rival, Berle, falsely portrayed him as Hiss's protector, incorrectly identifying Alger Hiss—rather than his brother Donald—as a former Acheson assistant. Found guilty of perjury and widely assumed to have been guilty of espionage, Alger Hiss became, for conservatives, a symbol of the hated Eastern Establishment, New Deal liberals, and the State Department.

The flames of contention were fed in January 1950, after Hiss's conviction for perjury, when Acheson cited a Biblical injunction on charity in refusing to "turn his back" on Hiss. Senator Richard Nixon called this statement "disgusting," and Acheson's invocation of pity—his father was an Episcopal bishop—damaged both himself and State. "After a while you get tired of the curs yapping, and have to have your say," Acheson told a friend, but a sympathetic Senator Vandenberg suggested he make a stronger statement about "the appalling dangers of national security in the State Department leaks." Acheson even offered to resign, but President Truman refused the offer. In this heated emotional atmosphere, Senator Joe McCarthy began his witch-hunting career.[60]

Meanwhile, on the international scene, State had invented and implemented the concrete programs needed to limit Soviet expansion: the Marshall Plan, military aid programs, and the North Atlantic Treaty Organization (NATO). It worked closely and successfully with Congress, particularly with Republican leader Senator Vandenberg, on these issues. It was still a period of bipartisan foreign policy, built on a consensus over the need to combat Soviet power.[61]

Kennan's overoptimistic hopes for U.S.-Soviet compromise and German reunification were not well received in Washington. His influence was reduced when Acheson became secretary of state and ordered Policy Planning to clear its papers through the assistant secretaries instead of sending them directly to him. Kennan saw this as a contradiction of his staff's whole purpose and resigned in 1949. Acheson replaced him with Paul Nitze, who organized a comprehensive review of U.S. foreign policy for the NSC.[62] State remained dominant over the military and NSC throughout the 1950s.[63]

It was easier to reach agreement over strategic and European issues than on some of the developments ending colonial empires and creating a whole new diplomatic front, the *Third World*. One of these events—a U.S.-supported UN decision in November 1947 to partitition British-ruled Palestine—produced Israel's May 1948 declaration of independence and an unsuccessful Arab invasion of the new state.[64]

While department officials criticized the White House for allegedly playing domestic politics, supporters of the Zionist cause accused State's Arabists of anti-Semitism.[65] Service in the Arab world and responsibility for relations with those states caused State's area specialists to worry about the effect of the creation of Israel on U.S. interests in the region. At the same time, White House policy on Palestine was shaped not merely by domestic considerations, but also by a different view of strategic interest and humanitarian responsibility.

State's bureaucracy opposed permitting Jewish refugees from the European Holocaust to go to Palestine. Assistant Secretary for Near East Affairs Loy Henderson saw the Arabs as a barrier to Soviet expansion and feared U.S. policy would drive them into Moscow's arms. The department's views were opposed and defeated by President Truman, his aides Clark Clifford and David Niles, and Congress. Despite State's opposition, Truman supported the UN recommendation to partition Palestine into Arab and Jewish states. When Undersecretary of State Lovett ordered the U.S. UN delegation to vote for giving southern Palestine to the projected Arab rather than Jewish state, Truman reversed the decision as conflicting with his own commitments.

After the UN voted for partition, State argued that Arab opposition made that plan unenforceable. Without outside help, Policy Planning reported in January 1948, Israel could not survive. It recommended abandoning partition and seeking a new trusteeship arrangement. After State held secret negotiations with Egypt and Saudi Arabia in an attempt to prevent war, U.S. UN Ambassador Warren Austin made a dramatic February speech reversing Washington's position on partition.[66]

Truman, shocked by Austin's statements, accused State of sabotaging his policy. Certainly Marshall, Lovett, and Henderson were doing everything possible to reverse the U.S. stand, but sloppy communications apparently caused the unapproved step rather than any deliberate derogation. Marshall had left for a Latin America trip having agreed to support trusteeship if partition could not be implemented, but no specific finding was ever made that partition had in fact failed. Truman knew a policy switch was being discussed, but had no idea it was going to be put into effect. The eagerness of many at State to do away with partition swayed their judgment on the issue.[67]

While State's efforts on behalf of trusteeship continued down to the last moment, neither Arabs nor Jews would agree on any new U.S. proposal. Another round of embarrassing confusion followed in May 1948 when the White House quickly decided to recognize Israel while the department was still trying to avoid partition.[68] By his interventions on the issue, President Truman reminded the State Department of White House command over U.S. foreign policy, but a better system of coordination would have avoided much confusion and demoralization.

While Britain's exit from Palestine provoked difficulties, the refusal of European powers to leave other colonies was equally troublesome. State's Asia and Near East bureaus, and later the Africa section as well, tended to side with local nationalists; Europeanists supported their own clients and usually triumphed. The Soviet factor added another consideration. Marshall professed himself unwilling to see "colonial empires . . . supplanted by philosophies and political organizations emanating from and controlled by the Kremlin." During the debate over the future of Holland's colony in Indonesia, Lovett and the Europeanists opposed the Asia and UN bureaus. In this case, however, Marshall supported independence, believing the Dutch could not win a military victory and accepting his analysts' view that a protracted rebellion would strengthen Communist forces. Pressure was put on Holland to grant independence to the nationalists.[69]

Indochina was a different story. In the urgent pressure of the early postwar era, the Third World was usually a low priority. Acheson recalls believing that the United States opposed France's return to Indochina. "The next thing I knew about it . . . I suddenly found myself carrying on arbitration with the French back in Saigon." Roosevelt's musing about temporary trusteeship, leading to independence for colonies, had been opposed by the military and European bureau as well as by the British.[70] After Roosevelt's death and the onset of the Cold War, Washington adopted a largely passive position. Ho Chi Minh's 1945 declaration of independence was ignored. In May 1945, Washington had decided to accept reimposition of French colonialism in exchange for Paris's support on international issues. Identification with France by influential FSOs who loved that country and culture paralleled the Near East hands' empathy for the Arabs. By February 1950, Washington's need for French cooperation in Europe and awareness of Communist control over the Vietnamese nationalist movement brought U.S. support for France's war there. The Asia hands, headed by John Carter Vincent, warned of future dangers, but immediate and East-West problems proved more persuasive for policymakers.[71]

A few department officials suggested that the Vietnamese Communists might not retain control—or might want reasonable relations with Washington—if they came to power. The risks against taking such a course were tremendous, and China was not a promising precedent. "It may not be certain," Woodruff Wallmer of European Affairs wrote, "that Ho and Co. will succeed in setting up a Communist State if they get rid of the French, but let me suggest that from the standpoint of the security of the United States, it is one hell of a chance to take."[72]

This was a major reason for the Europeanists' frequent victories. The effect of Western European imperialism in spreading anti-Americanism and strengthening Communist forces in the Third World was understood, particularly in the regional bureaus. Still, the most immediate U.S. need was for French support in Europe; the most obvious danger was of Soviet expansionism. Longer-run considerations would be left for future administrations. Furthermore, though State wisely supported the Yugoslav Communists when they broke with Stalin, the original emphasis on struggle against the USSR would become in the 1950s an undifferentiated opposition to any left-of-center or even nationalist forces that might seek Cold War neutrality.[73]

China was the cautionary example that frightened policymakers. In 1945, Secretary of the Navy James Forrestal, a relative hard-liner, opposed any public declaration of conflict with Moscow to avoid escalating the Cold War. Three years later Marshall rejected, with Truman's support, any explanation to the public about Chiang Kai-shek's incompetence and the deteriorating military situation in China. Such criticism, Marshall argued, would produce Chiang's collapse. Only in August 1949, two months before the final Communist victory and Chiang's withdrawal from the mainland, did the State Department issue a "White Paper" to explain why Washington had not been able to save Chiang.[74]

Of course, Chiang's fall was due neither to U.S. actions nor to conspiracies in the U.S. government. Ironically, the Washington dilemma over China and the ensuing attack on the State Department was largely caused by excessive foresight and intelligence success rather than by any shortcomings. State Department analysts understood that only major internal reforms or compromise could save China from a Maoist victory. Chiang lacked the flexibility and the United States lacked the leverage to undertake either effort. China was too large and distant for direct U.S. military intervention to be successful even at a terrible price.

Thus the department knew that if Chiang persisted in his course, he would be defeated by Communist leader Mao Tse-tung. State also believed, however, that even a Communist Chinese regime would not long accept Soviet hegemony. Eventually, Peking would seek its own path and open the possibility of U.S.-China accommodation. In fact, the Sino-Soviet split did occur a decade later, followed eventually by U.S.-China détente.

Still, the fall of an important, historic U.S. ally to a foe widely seen at the time as part of a Kremlin-controlled monolithic empire became a matter of heated public debate. The difference between the conspiratorial and pragmatic views of events in the Far East is clearly revealed by comparing Senator Joseph McCarthy's February 1950 speech at Wheeling, West Virginia, which launched his career as a red-hunter, with one by Secretary Acheson at Washington's National Press Club the previous month.

"At war's end we were physically the strongest nation on earth," said McCarthy. "The reason why we find ourselves in a position of impotency is not because our only powerful potential enemy has sent men to invade our shores but rather because of the traitorous actions of those who have been treated so well by this Nation. . . . This is glaringly true in the State Department. There the bright young men who are born with silver spoons in their mouths are the ones who have been most traitorous."

McCarthy continued, "How can we account for our present situation unless we believe that men high in the government are conspiring to deliver us to disaster? This must be the product of a great conspiracy on a scale so immense as to dwarf any previous venture in the history of man." After attacking several individuals, including FSO China specialist John Service who, the senator falsely claimed, "had previously urged that communism was the only hope of China,'' McCarthy concluded, ''I have in my hand a list of 205, a list of names that were made known to the Secretary of State as being members of the Communist Party and who nevertheless are still working and shaping policy in the State Department.''[75]

This was merely an old list of employees being investigated for a variety of reasons, most not even accused of Communism, and only 46 of whom—all cleared by the FBI—still worked at State. But McCarthy's inaccuracy on this point was only one area of distortion. Equally important was the broader question: Were international developments detrimental to U.S. interests caused by Moscow's conspiracies and its agents in Washington or by more complex and largely indigenous forces?

Acheson, in contrast, tried to explain Asian revolutions as stemming from ''a revulsion against the acceptance of misery and poverty as the normal condition of life'' and ''foreign domination.'' Much of the ''bewilderment . . . about recent developments in China,'' he explained, ''comes from a failure to understand this basic revolutionary force which is loose in Asia.'' Chiang's fall was not due to American bungling, but to his own frittering away of military power, U.S. backing, and popular support. The Chinese abandoned their own government. ''Added to the grossest incompetence ever experienced by any military command was this total lack of support both in the armies and in the country. . . . The Communists did not create this [revolutionary spirit] but they were shrewd and cunning enough to . . . ride this thing into victory and power.''

The United States wanted to stop the spread of Communism as an instrument of Soviet foreign policy and imperialism, Acheson continued, and supported the desire of countries for independence and economic progress. ''If we can help that development . . . then we have brought about the best way that anyone knows of stopping this spread of communism.'' Soviet domination would bring on the Chinese people's wrath. Any ''foolish adventures'' on the part of the United States would make it patriotic for Chinese to ally with Moscow. The first rule of U.S. policy in Asia remained opposition to any violation of China's unity and integrity.[76]

Panic over treason easily supplanted such longer-range strategies in the public mind. The Soviets controlled Eastern Europe as well as powerful Communist parties in Western Europe and elsewhere, and China had joined the Soviet camp. Hopes for a new era of peace and harmony had quickly collapsed. As always, it was easier to seek scapegoats than to try to find real causes.

The department's immediate reaction to McCarthy's speech was optimistic: "Now, he will have to prove it. Then we will have an end to the matter," was a comment often heard among its officials. But McCarthy was only one of a number of Republican politicians seeking partisan gain through the security issue. Senator Robert Taft said in January 1950 that the department had "been guided by a left-wing group who obviously have wanted to get rid of Chiang and were willing at least to turn China over to the Communists for that purpose." After Hiss's conviction on January 25, the chorus grew. Congressman Richard Nixon claimed the following day, "We are not just dealing with espionage agents who get 30 pieces of silver to obtain the blueprint of a new weapon . . . but this is a far more sinister type of activity, because it permits the enemy to guide and shape our policy."[77]

Ironically, by this time the security problems existing immediately after the war had been largely resolved. Executive Order 9835, of March 1947, provided that any official could be removed if "reasonable grounds exist for the belief that the person involved is disloyal to the Government of the United States." In August 1950, Congress enacted Public Law 733, which allowed the secretary of state, at his discretion, to suspend an employee in the interests of national security. In April 1951, "reasonable doubt" about loyalty became grounds for removal.

This system erred on the side of careless strictness rather than of excessive leniency. The Loyalty Security Board confronted employees with gossip and rumors, which they were then challenged to disprove. Suspicions, negative publicity, and congressional pressure for quick action tended to overwhelm evidence to the contrary, ruining several careers on the flimsiest scraps of evidence. Some used the security system to strike at personal rivals or for self-aggrandizement. One top Far East specialist was charged before the Loyalty Review Board for having associated with the Japanese Communist party's leader. He had done so on department orders to discover what the politician was planning. Even when charges were dropped, the incident haunted the FSO; when he was to be appointed an ambassador, the Senate refused confirmation.[78] By February 1952, the department had handled 604 loyalty-security cases, nearly half of which ended without interrogation or charges. After 54 hearings, 11 employees were separated as security risks, though none were judged disloyal.[79]

At first, State strongly combatted McCarthy's charges—Acheson, after all, was the witch-hunters' main target—and generally tried to protect its employees. Still, morale plummeted. Employees and spouses dodged questions about their place of work at cocktail parties, prospective recruits decided not to apply at State, officials were afraid to make recommendations lest their views be used against them in years to come.

Acheson, columnist Drew Pearson wrote in his diary, was "too harassed, too tired, and too numb" to mobilize support in Congress.[80]

Deputy Undersecretary Peurifoy, in charge of security, tried to refute McCarthy point by point. The senator, he noted, provided no proof and constantly changed his numbers and claims. Service, for example, was "able, conscientious and . . . demonstrably loyal." He had passed five loyalty checks in five years although they disrupted his assignments and personal life. "It's a shame and a disgrace that he and his family should have to face, once again such humiliation, embarrassment and inconvenience; and I'd like to say that the sympathy and good wishes of the entire Department go to them."[81]

Ambassador Phillip Jessup, one of McCarthy's main targets, spoke of his own anti-Communist statements and Soviet press attacks on him. He pointed out, for example, that he was not a sponsor of the American-Russian Institute, as McCarthy complained, but only of a 1946 dinner where that group honored President Roosevelt. While McCarthy noted the few Communist supporters of the affair, Jessup pointed out that the great majority of the 100 endorsers were distinguished citizens. The institute was not even listed as subversive until three years later.[82]

It was hard to counter the simplistic headlines, but at first State seemed successful. In March 1950 *The New York Times* correspondent Arthur Krock wrote that the score was nine to nothing for the administration against McCarthy as the senator's charges were disproved. Yet this had no effect on the public mood. The American Society of Newspaper Editors applauded after McCarthy called Acheson "incompetent" and attacked Marshall. "His hearers apparently concluded that though he is a barroom fighter who pays no attention to the rules designed to make fighting fair, he has something," one observer reported.[83] The hysterical atmosphere and intimidation created by McCarthy reached into every aspect of American political life. Dozens of speeches and articles portrayed State as full of traitors.

Before the same audience of editors, Acheson tried to counter the department's Eastern Establishment image by pointing to officials who came from apparently respectable states like the Carolinas, Wisconsin, Texas, and Georgia. The "wrong way" to combat Communism, he warned, is to "destroy the institution that you are trying to protect." Acheson did not ask for sympathy: "I and my associates are only the intended victims . . . But you . . . are participants."[84]

North Korea's invasion of South Korea in June 1950 intensified the attack on the department while, ironically, showing the inaccuracy of State's image. The Pentagon had wanted to withdraw U.S. troops from Korea as fast as possible in 1948, while the much-maligned Office of Far

East Affairs insisted on awaiting the creation of adequate South Korean defense forces. On another key regional issue, the Defense Departmnt had warned that any treaty with Japan without Soviet participation would be risky while State had insisted an agreement was needed to put Japan back on its feet.[85]

The conservatives' hero, General MacArthur, like Acheson, had left South Korea out of a U.S. defense perimeter based on Japan and the Philippines. Congress had refused to approve the Korea Aid Bill of 1949 without which, Acheson warned, South Korea might not survive as a free nation. When the war began, Acheson had acted quickly to secure U.S. intervention and to gain UN approval for this policy. After U.S. forces were forced to retreat in the early days of the war, MacArthur and other generals panicked and called for a pullout, but Dean Rusk rallied Defense Secretary Marshall and others to hold on, comparing the moment to the Battle of Britain in World War II.[86]

Behind the scenes, the Joint Chiefs of Staff were as dubious as State about U.S. field commander Gen. Douglas MacArthur's call to widen the war by attacking China. Truman's firing of MacArthur, which brought a firestorm of vituperation against the president and Acheson, was supported by the top generals. In the Korean armistice negotiations, when the Peking government insisted that defecting Chinese prisoners be forcibly returned, the Defense Department advocated meeting this condition to gain the repatriation of American POWs, but the State Department successfully opposed the concession.[87]

Far from being unpatriotic or incompetent, the State Department—even during the peak of attacks against it—skillfully defended the nation. Kennan, who did not get along well with Acheson, praised his accomplishment: "Here he was: a gentleman, the soul of honor, attempting to serve the interests of the country against the background of a Washington seething with anger, confusion, and misunderstanding, bearing the greatest possible burden of responsibility for a dreadful situation he had not created, yet having daily to endure the most vicious and unjust of personal attacks from the very men—the congressional claque and other admirers of General MacArthur—who . . . had created it." Almost every Democrat running for office demanded Acheson's removal as a political liability. In August 1950, Republican Senator Wherry said, "The blood of our boys in Korea is on [Acheson's] shoulders, and no one else."[88]

Incredibly, this wave of defamation peaked at the same time State was taking the lead in creating a new militant American strategic conception, NSC-68, characterized by a hard-line maximization of Soviet capacities and intentions; supported the building of the hydrogen bomb; and advocated a major buildup of U.S. conventional military capability.[89]

Throughout 1951, however, the insanity mounted. McCarthy accused State of plotting to give Western Europe to the Soviets. Wherry called for a no-confidence motion against Acheson; another senator proposed State's abolition and replacement by a new institution. Taft accused Acheson of favoring "appeasement" in Korea. Acheson must be removed, said the Republican leader, and sympathy for Communism must be eliminated at State. Others proposed cutting off the department's funds; Democrats urged Acheson's resignation.[90]

Those opposing Acheson, wrote James Reston, were no longer listening; those sympathetic were no longer enthusiastic. "The great experiment in keeping foreign policy out of domestic politics has failed." As McCarthy and his allies produced new charges faster than they could be refuted, the pressure grew to punish someone with the department. On July 1, 1950, in the midst of congressional investigations, two China hands, John Paton Davies and O. Edmund Clubb, director of the Office of Chinese Affairs, were suspended. Both men were cleared, though Clubb soon resigned and Davies was later forced out by the Eisenhower administration. John Carter Vincent, also suspended, was cleared by the State Department but terminated by a Civil Service Committee Loyalty Review Board. Five hundred security cases were reopened. "We didn't want it said that we whitewashed them," explained one official.[91]

How could department officials perform their duties while McCarthy attacked Jessup for being "found at the time and place when disaster hit Americans and success hits Russians" or when congressmen sought to eliminate Acheson's salary from State's appropriation? Such measures failed, but the budget was cut sharply. Republicans claimed high positions were filled with political hacks, though Truman appointed more Republicans and independents—including Marshall, Lovett, Dulles, and many ambassadors—than virtually any other president; 62 percent of Truman's chiefs of mission were career people.[92]

These were sad times indeed for the State Department, particularly since the avalanche of hatred and calumny had come so soon after its greatest achievements. As the *Foreign Service Journal* editorialized, "Shall [the FSO] report only what will harmonize the temper of the times. . . . Knowing the dangers of honesty and risk to his career and reputation?"[93]

Dwight Eisenhower won the presidency in the November 1952 elections and on his inauguration, the following January, the Republicans regained the White House for the first time in 20 years. Acheson's farewell message to State was grim: "Yours is not an easy task, nor one which is much appreciated. You don't ask much of your fellow citizens, because you are dealing with matters which, though they affect the life of every citizen of this country intimately, do it in ways which it is not easy for every citizen to understand . . . One thing I think you are entitled to ask—that you should not be vilified; that your loyalty should not be brought in doubt; that slanders and libel should not be made against you."[94]

4

The Horseless Rider: The Eisenhower Administration

1953–1961

By January 1953, the State Department was seriously discredited and, ironically, many of those now coming into power had helped undermine its reputation. John Foster Dulles, the new secretary of state, distrusted the institution he inherited, considering it the product of 20 years of Democratic rule and, as such, unpopular with his allies in the White House and Congress. Consequently, Dulles was an extremely powerful figure in the Eisenhower administration, but cut himself off from State's career staff. While Dulles personally dominated policymaking, the department's institutional primacy was further weakened. Morale remained low and the organizational changes implemented during the 1950s, though solving some problems, initially produced a great deal of confusion.

In short, the Eisenhower foreign policy system was characterized by a strong secretary of state, with wide powers delegated by the president, alongside a weak State Department. Recent studies, showing that Eisenhower was more active than contemporary observers thought, only partly modify this picture. After all, the White House staff and NSC did not yet have the size or structure necessary to seize the foreign policy reins, although the former could institute initiatives and challenge or restrain Dulles on specific issues.

Eisenhower's creation of the post of special assistant for national security affairs caused few problems for Dulles, since this official and the NSC staff he supervised were still more expeditors than decision makers. "I shall regard the secretary of state as the Cabinet officer responsible for advising and assisting me in the formulation and control of foreign policy," said Eisenhower in June 1953, and as "my channel of authority within the Executive Branch on foreign policy."[1] All presidents make similar statements, but Eisenhower actually followed this principle.

White House foreign policy advisers were subject to Dulles's approval, and three were forced out after policy disputes with the secretary of state. One of them, Nelson Rockefeller, tried to hide his staff far out in Virginia to avoid Dulles's oversight while developing a peace plan for the 1955 U.S.-Soviet Geneva summit.[2] Even from his deathbed, Dulles telephoned State to investigate whether one of the president's aides had overstepped his own authority.[3]

Although Eisenhower was not mesmerized by Dulles, as many early accounts claimed, neither did he use his secretary as a pliant tool, as some recent historians portray the relationship. The president intervened on several occasions to temper Dulles's "brinksmanship," interposing his own ideas for easing tensions with Moscow. Eisenhower pressed for the Geneva summit meeting, refused military intervention to avoid French defeat in Indochina, and suggested the United States and USSR open their skies for mutual inspection.

On matters of policy, the secretary's rigidity sometimes frustrated even the president. "I'm tired—and I think everyone is tired—of just plain indictments of the Soviet regime," Eisenhower told an aide. "Instead, just *one* thing matters: what have *we* got to offer the world? What are we willing to do, to improve the chances of peace?" After Stalin's death in 1953, Eisenhower went against Dulles's advice to make a conciliatory speech written by White House aide Emmet John Hughes and Policy Planning Staff Director Nitze.[4]

When the president told a press conference that neutralism was generally a good thing, Dulles "clarified" this by adding that "except under very exceptional circumstances, it is an immoral and shortsighted conception."[5] Thus, while Eisenhower sometimes altered the direction of policy, Dulles often persuaded the president to accept his own interpretations.

Still, Dulles was as powerful as any secretary of state had ever been, with influence on a par with that of Marshall and Acheson. But Dulles acted more like a vice-president for foreign affairs than as a leader of a large government department. He carefully nurtured links with congressional Republicans while his relationship with the career department staff deteriorated. In the McCarthy era, as Acheson had discovered, the secretary of state seemed forced to choose between alienating either legislators or diplomats.

Given Dulles's power, challengeable only by the president's direct intervention, clear lines of authority allowed for a relatively smooth policy process that handled some difficult crises while also creating several "time bombs" for future conflicts. Dulles worked with a few personal advisers, including his brother Allen, who was CIA director, devising policies. On his frequent travels, he often supplanted ambassadors and assistant secretaries in policy implementation as well. These absences abroad, like those of Byrnes, also made it harder for State to function effectively. When too much emphasis is placed on a single individual, the quality of policy becomes greatly dependent on his health, attention span, and idiosyncrasies.[6]

Dulles's growing illness and eventual resignation in April 1959 revealed the flaws in a system so dependent on one man. Consequently, by the end of Eisenhower's second term, growing dissatisfaction with the foreign policy machinery led to proposed reforms that further challenged State's position. While McCarthyism's crusade against State and Dulles's centralization of authority without an adequate institutional basis contributed to this decline, there was also the older criticism that, in an increasingly fast-paced and complex world, State was too unwieldy, slow, cautious, and unimaginative.

As in the Truman administration, the special relationship between president and secretary of state was a vital factor in policymaking. Dulles's grandfather, John Foster, and his uncle, Robert Lansing, had been secretaries of state, and Dulles had served as his grandfather's secretary at international conferences. Dulles "has been in training for this job all his life," said a respectful Eisenhower. For his part, remembering how President Woodrow Wilson's adviser, Colonel House, had come between the chief executive and uncle Robert Lansing, Dulles was always careful to maintain close coordination with Eisenhower. "The passing of Dulles," wrote Arthur Krock, expressing the contemporary, if exaggerated, perception, "was Eisenhower's heaviest burden."[7]

Despite the secretary's lineage, the Eisenhower-Dulles team, like other administrations, was surprised to find policymaking far more difficult than expected. Hughes describes the confusion over their first foreign aid package in May 1953. State and the Budget Bureau saw the White House's plan only at the last minute. State wanted more money, the Budget Bureau demanded less. Since congressional hearings began the next morning, the dispute had to be quickly resolved. Back and forth Hughes ran between various institutional opponents, carrying the draft message until after sundown. Finally, he bent the wording in Dulles's favor and the president accepted it. "How reassuring it would be to . . . our allies around the world," he noted sarcastically, "if they could see the disciplined and dedicated way we plan and provide our economic assistance.''[8]

At State itself, many staffers were hardly reassured by Dulles's maiden speech to them in January 1953. What was needed, the new secretary said, was ''competence, discipline, and positive loyalty to the policies that our President and the Congress may prescribe . . . less than that is not tolerable at this time.'' In McCarthyism's shadow, these words challenged the audience's professional integrity. Even worse, Dulles's qualified compliment that ''those who comprise the Department of State and Foreign Service are, as a whole, a group of loyal Americans'' was deemed more accusative than supportive.[9]

Decades later, department veterans still recall those phrases. State's besieged employees desperately needed loyalty from the top, and they did not think Dulles gave it to them.

Security was an obsession during those years. In February 1953, books, music, and paintings by Communists or fellow-travelers were banned from State's overseas libraries. The department was ordered to cooperate with a ludicrous junket by McCarthy aides investigating the volumes stocked by its information program and to administer a restrictive passport and visa policy that sometimes denied foreign travel to those of leftist political views.

Even some White House aides charged that failure to criticize Senator McCarthy encouraged continuing assaults against officials. But they could convince neither Eisenhower, who felt ignoring McCarthy would undermine the senator, nor Dulles, who did not want to risk congressional hostility.[10] One early test was the controversial nomination of veteran Soviet specialist Charles Bohlen as U.S. ambassador to Moscow. Bohlen was surprised by the appointment, since he had differed with Dulles at a CIA briefing shortly before the election on the practicality of ''unleashing'' Chiang Kai-shek against Communist China. When Kennan had disagreed with such a position, albeit before it became official policy, he was forced to resign.

The nomination went before Congress in March 1953 and was opposed by the McCarthyites. McCarthy himself attacked Bohlen as ''worse than a security risk'' because of his association with past Democratic administrations. To complicate matters, McCarthy received leaks from State's secret files, through security director Scott McLeod, a former FBI agent and congressional aide. A nervous Dulles asked Bohlen if anything in his past might embarrass the administration, and when Bohlen said no, the secretary replied, ''Well, I'm glad of that because I couldn't stand another Alger Hiss.'' Dulles tried to avoid being photographed with Bohlen at the hearings, but requested confirmation as "an acid test of the orderly process of our government" and asked whether charges lacking even the "substance of rumor would prevail." Bohlen was confirmed because the administration stood firm, Eisenhower defended his appointee, and congressional Republicans rallied to their party. Senator Robert Taft warned, however, that there should be "No more Bohlens."[11]

Paul Nitze, Kennan's talented successor at Policy Planning, did not fare so well. Early in the administration, Nitze brought prepared notes for congressional testimony to one of Dulles's personal assistants, "What's this?" the man asked. "The Secretary didn't tell you to do this." Nitze explained it was routine. "Well, don't let it happen again," the assistant responded. The impolite appointee was soon gone, but so was Nitze. Charles Wilson, the conservative secretary of defense, asked Nitze to work for him, but congressional Republicans would not approve. There were already too many holdovers from past administrations, they complained.[12]

McLeod at first controlled both personnel and security, although his embarrassing performance soon led to his divestiture of the former responsibility. But this administration, which had promised so earnestly to improve government and end treason in Washington, had to produce some dramatic changes. "The question is frequently asked," said McLeod, "'Has the State Department changed? Has the mess been cleaned up?'"[13] There had been only 2 or 3 dismissals of employees in 1949, 12 in 1950, 35 in 1951 and 70 in 1952. Eisenhower era investigations quickly removed 425 employees, more for homosexuality and other personal considerations than for any direct security problems. The higher statistics were misleading, as employees could no longer resign voluntarily before administrative charges were brought against them. Between 1947 and 1954, only about 1.3 percent of applications for employment were turned down due to security reasons, a fairly consistent proportion. The main personnel problems were resignations due to low morale, the disinterest of talented young people in such a maligned profession, and funding cuts that suspended the recruitment of any new FSOs between 1952 and 1954.[14]

Dulles complained that he was receiving "security" files involving accusations of drunkenness, pacifist relatives, or membership in the World Federalists. Eisenhower agreed that such sloppiness showed a new head of security was needed, but McLeod survived for over three more years. Dulles's objections to trivial and unsupported allegations did succeed in easing the burden on those accused—instead of being suspended without benefits, employees were transferred to nonsensitive work pending the outcome of their cases—but officials still daily faced the threat of denunciation on flimsy grounds followed by a damaged career, dismissal, or resignation. Vice-President Nixon, particularly strident in demanding purges, argued for an investigation of the U.S. Information Agency because not enough security risks had been found in proportion to its size.[15]

McLeod ordered a full field investigation of all 11,000 employees, from ambassadors to clerks, and security teams fanned out across the globe looking for possible leftist connections, drinking problems, and sexual escapades. As rumors of secret dossiers and denunciations spread suspicion, officers began to write blander reports, torn between conscience and career on whether they should risk reporting their observations. Even to study the USSR, Russian, or Marxism might make one suspect.

This atmosphere ran counter to the Foreign Service's whole purpose. One diplomat complained, "If I had a son, I would do everything in my power to suppress any desire he might have to enter the Foreign Service." In a January 1954 letter to *The New York Times*, five distinguished former diplomats, including Grew, Phillips, and Shaw, warned that the obsession with security was destroying "accuracy and initiative. . . . The ultimate result is a threat to national security."[16]

Former President Truman commented, "What Eisenhower doesn't seem to realize is that when a man doesn't back his subordinates, the whole morale of government is shot to pieces." Acheson noted, "A great institution ought to command respect from anyone who is given the responsibility of command." What was happening at State "is very bad and gives one a contempt for the cowardly fellows who are doing it. . . . Dulles's people seem to me like Cossacks quartered in a grand old city hall, burning the panelling to cook with."[17]

The China hands remained particular targets. In November 1954, Dulles dismissed John Paton Davies, one of the last survivors. Dulles did not find Davies guilty of disloyalty, but accepted the Security Hearing Board's questionable finding that his continued employment was not clearly consistent with national security interests. When Dulles visited Pakistan shortly thereafter, he complained over dinner at the house of U.S. diplomat John Emmerson of spending an entire weekend reading Davies's file. Emmerson, another former East Asia hand, was not sympathetic since he himself had undergone seven months of hearings and was kept out of the region for 16 years. As late as 1962, Emmerson was denied an ambassadorship because of old grudges from Senate rightists.[18]

When Dulles, in accepting John Carter Vincent's retirement, added that the China expert had failed "to meet the standard which is demanded of a Foreign Service Officer," FSOs considered this a further insult. They felt that colleagues were being punished precisely because they had lived up to the highest standard: reporting the facts as they saw them. Asked a letter to the *Foreign Service Journal,* "Are we all subject to being labeled security risks if ten or fifteen years after we've observed and evaluated . . . a foreign political situation, it's decided that our observations were wrong—or right?" The Foreign Service Association's board of directors commented that the Loyalty Review Board's position on Vincent apparently meant "that any Foreign Service Officer reporting confidentially to his superiors may cast a doubt on his own loyalty if his reports contain criticisms of a friendly government."[19]

Corruption of the reporting process and failure to obtain accurate information on Vietnam and other controversial issues can be partly traced to intimidation, but State's internal processes awarding conformity and discouraging criticism or warnings about policy shortcomings also contributed to this atmosphere. Links between career advancement and a willingness to echo internally a current administration's mood were not restricted to the Eisenhower period. Still, the situation became so bad under Dulles that one writer suggested political appointees were superior, "since professional diplomats are now afraid of giving frank reports."[20]

Among the career staff, the era's traumas produced much soul searching. Earlier reforms had fallen short of expectations. Not only did State's employees have declining influence, but professionalism itself was under assault. Politicians and political appointees mistrusted the department, other bureaus provided increasing competition, and Congress sniped and cut budgets. The Foreign Service, Kennan concluded gloomily, "was weakened beyond hope of recovery"; it was "an administrative ruin." Another writer commented, "Being a diplomat during these last few years has been like being a soldier caught between the fire of friends and foe."[21]

Low morale and the department's poor functioning were due to far more than security harassment or the willfulness of politicians alone. Almost everyone agreed on the desperate need for reorganization and more effective management to cope with the continuing Cold War and intensive U.S. involvement in the world. The heroic, free-wheeling days of the late 1940s, when a small group created new ideas and major programs overnight, were replaced by complex bureaucratic problems.

"Managerial control is next to impossible," complained a congressional report. "There are now five Departments of State instead of one," due to the chronic difficulties of internal coordination and jurisdictional disputes with other departments, including the new aid and information agencies. State's handling of funds was "not only complex but incredible." Between 1944 and 1956, there were eight different chief administrators and an equal number of reorganizations. Faced with administrative breakdowns and lingering McCarthyist suspicions, Congress constantly made deep budget cuts. Even Dulles's pleas to grant money for entertainment allowances ("the booze fund," in one congressman's words), better living conditions abroad, and a larger staff were only partly successful.[22]

Other problems at State were due to outdated thinking. The pre-World War II department was small enough to be run through personal relationships and managed by FSO amateurs. During the war, a new breed of professional managers was introduced into State, to the resentment of FSOs who saw them as outsiders intruding into their private networks. But this growing presence was made necessary by the refusal of FSOs to become better administrators. Senior officers bragged, "I don't know a thing about administration . . . nor do I wish to learn," or claimed that administrators "seem to forget that they are essentially valets. Instead of pressing our pants they are trying to wear them." It was true that most of the professional managers did not understand diplomatic and personnel requirements, but if FSOs wanted to do better, they would have to provide these skills themselves.[23]

To deal with these and other problems, Henry Wriston, president of Brown University, was picked in March 1954 to lead a special Committee on Personnel which quickly gained presidential endorsement for long-discussed, dramatic reforms. The committee's main proposal was to merge most of the civil service positions, which dominated the Washington slots, into the Foreign Service, which staffed overseas missions. If the FSOs would not become administrators or specialists in congressional issues, intelligence, labor affairs, and other areas of work, the managers and specialists would be made into FSOs. Civil service posts would be converted into FSO positions, and civil servants were urged to join the Foreign Service. This would allow FSOs to spend more time in Washington—of 197 officers with over 20 years' service, only 45 had held assignments in the United States—and give the home guard more field experience. Midcareer entry of other nongovernment specialists into the Service and an energetic recruitment program would strengthen and expand the corps. Theoretically, the Foreign Service would gain a group of experts on economics, labor, intelligence, public affairs, and other areas.[24]

Wriston tried to end the Service's old elitist and aristocratic ethos and suit it to a modern world of mass movements, revolutions, and ideological struggle. "The theory or philosophy that the Corps should be made up of 'generalists' only," he explained, "was far better adapted to . . . a second class power with a tradition of isolation than it is to the leader of the Free World. . . ." He called for "a genuinely representative, democratically oriented service."[25]

The Wriston committee initiated major changes in the department's structure. The Foreign Service's membership was tripled; 1400 people transferred from the Civil Service rolls, uprooting their settled lives in Washington to go overseas. Some became successful diplomats, others dropped out. Professional managers and permanent employees in Washington were replaced by career people, a decision symbolized by the appointment of veteran FSO Loy Henderson as chief administrator.

Yet, while "Wristonization" permitted expansion, it actually reduced specialization. The expertise of transplanted civil servants was rarely used in their new embassy jobs while continuity was lost as they joined the merry-go-round of shorter-term assignments. The department's memory was damaged and officers had less incentive to learn about a given country or functional specialization. As one FSO put it, Wristonization meant "the break-up of experienced teams of specialists and the failure adequately to rebuild them within a personnel system which has now become much too fluid." Contemporary studies found only 15.8 percent of FSOs had spent more than six years in any region and that only 32 percent were in any one part of the world for more than three years.[26] Many of these problems still plague the department today.

State's control of overseas diplomacy also continued to dwindle as other agencies exported their own representatives. State employees comprised fewer than 15 percent of government civilian employees abroad during the Eisenhower administration; even excluding Defense Department workers, State still had less than half the total. The FBI insisted, for example, on its own attaché in Japan, although army intelligence, CIA personnel, and a legal attaché were already serving there. Soon, the FBI man had his own assistant. Harold Stassen demanded a Tokyo office for his Mutual Security Agency to assist Japanese industry, asking, "How will we be able to live with ourselves five years from now if Japan's economy has gone to pieces and we have done nothing to help it?" He was persuaded to send only 25 people, but somehow the Japanese economy survived. Embassies were often lost in this tangle, relegated to servicing other agencies.[27]

Ambassadors themselves also came under criticism. Scandal has long followed the use of these posts to compensate political allies or campaign contributors. "It has always been possible to purchase an embassy for cash on the barrel head," complained one FSO. This practice was highlighted in the Eisenhower era by a Senate report showing 19 appointees among large donors to the Republicans' 1956 effort. The case of Ambassador-designate Maxwell Gluck became notorious when he could not name the prime minister of Ceylon (now Sri Lanka), his destined post, and knew little about the region.

"Surely," wrote *The New York Times*'s C. L. Sulzberger, "it is possible to find rich men who comprehend the better-known foreign tongues"; but he agreed that wealth was needed to pay entertainment bills at prize European embassies, given the relatively low pay and expense money. The British ambassador to Washington received $103,000, compared to $60,000 for his American counterpart in 1957. (Ironically, U.S. embassy funds were often expended to entertain visiting congressmen who had voted to cut State's allowances.)[28]

Every president promises to do better in his appointments, but the Foreign Service is always disappointed. The FSO's disdain for appointees is no mere snobbery: While some do well, the selection of these outsiders destroys the hopes of officers who have labored decades to become ambassadors. The growing number of Third World countries has meant an increase in the number of career appointments, but the plum European posts have usually remained in the hands of appointees.[29]

Some of these organizational problems were resolved, but many remained after the Eisenhower administration's end. The Senate approved a bill in 1957 to encourage ambassadors and FSOs to know the language of the country in which they served. After an extensive training program, there was marked improvement: 25 percent of FSOs were bilingual, 35 percent had professional proficiency, and 26 percent had a working knowledge of another language. In particular, there was recognition of the need to train more people for work in Third World countries and languages. The establishment of an Africa bureau in 1957 showed an appreciation of the coming wave of independent states. The fading of McCarthyist excesses also encouraged recruitment by the end of the 1950s.[30]

However, the funding situation remained frustrating. As James Reston wrote in 1958: "Let Secretary of Defense Neil McElroy ask for another billion to be ready to wage a nuclear war and there would be scarcely a vote against him. But let Mr. Dulles ask for a million to replace poorly qualified pork-barrel ambassadors with well-qualified professionals or . . . for a few hundred thousand to . . . train his top FSOs and he is in for a protracted debate. It is an odd defect especially since the purpose of diplomacy, like the purpose of missiles, is not primarily to win wars but to prevent them."[31]

Despite the lowered morale, sliced budgets, the confusion attending reorganization, and Dulles's aloofness, the State Department still carried much of the burden of daily diplomacy. On longer-range and major decisions, however, it was at an increasing disadvantage. Policy Planning, the department's own think tank, which had dominated the NSC, was downgraded by Dulles. State lost another source of ideas by reducing contact with outside academic specialists, who had been one of McCarthy's main targets. By the mid-1960s, when the department again sought to use social science research and outside consultants, opposition to Vietnam made professors unwilling to help. The volume of department paperwork, the need for endless compromise between bureaus over policies and instructions for embassies, the bureaucracy's slowness, and other agencies' demands to have their viewpoints heard required new means of coordination.

President Eisenhower considered having a second vice-president or a secretary for international coordination, with a subordinate secretary of state for overseas travel, administration, and congressional testimony. In the meantime, he strengthened the NSC, dividing its work between a Planning Board to explore future issues and an Operations Control Board (OCB) to monitor the bureaucracy's implementation of White House decisions. Mid-level officials from different departments served on these committees. The NSC still had no life of its own; it was the staff for a foreign affairs cabinet where the president could hear different viewpoints and make decisions.

The OCB met every Wednesday to distribute work assignments for each agency and to issue progress reports, but this process could take six months for a given issue. It could only advise, not order compliance. The NSC itself, consisting of the president and leaders of relevant departments, met on Thursdays in the cabinet room for two or three hours. While the president always had the final say, Dulles's primacy and the low profile of national security advisers generally avoided major conflicts.[32]

The CIA's growing role resulted from the Cold War and the militancy of the Eisenhower administration in waging it. Successful covert operations in Iran and Guatemala, close collaboration between the Dulles brothers, and the lack of congressional oversight or media criticism made the Agency an attractive tool for the 1950s. Undersecretary of State Walter Bedell Smith, Eisenhower's wartime chief of staff and the previous CIA director, provided another link between the two organizations. The Special Group, an OCB subcommittee, met weekly with a CIA representative presiding and with Smith representing State, to supervise covert operations.

The Truman administration had hesitated to overthrow the nationalist regimes of Mohammed Mossadegh in Iran and Jacobo Arbenz in Guatemala, but the Eisenhower appointees eagerly implemented coups against both in 1953 and 1954, respectively. Far from the CIA acting independently, John Foster Dulles was enthusiastic in pressing covert operations. CIA Director Allen Dulles used his brother, as well as direct briefings, to put his views before Eisenhower. The two brothers rarely disagreed and usually acted as a team, conducting a daily telephone dialogue on issue after issue. This level of State-CIA coordination has not been equaled since.

Nonetheless, many high-ranking State officials and ambassadors were not informed of CIA operations, even those affecting their regions or countries. Sometimes State and CIA were on opposite sides—as in Indonesia and Costa Rica, where the Agency was trying to overthrow governments with which State sought to build good relations. Some ambassadors and Agency station chiefs worked well together, but in many cases there were conflicts and the diplomatic mission had little control or knowledge of CIA activities.[33]

The 1954 coup in Guatemala illustrates some of these problems. Ten years earlier, the military had installed a progressive regime that implemented land reform and other efforts to help the country's poor peasant majority, actions that adversely affected United Fruit Company interests. While a small number of Communists held state positions, by no stretch of the imagination could they be described as controlling the regime.

United Fruit, however, portrayed Jacobo Arbenz's government as a Red menace to the hemisphere. "The Arbenz government, beyond any question, was controlled and dominated by Communists," said the U.S. ambassador to Guatemala on remarkably flimsy evidence. When Guatemala suggested a nonaggression pact with the United States, this was deemed "a giveaway of the inspiration whence this maneuver came. It is a Soviet term."[34] The company had good access to the State Department: Dulles was a former company counsel, Smith became a board member after he left government, and John Moors Cabot, the assistant secretary of state for the region, was a major stockholder while his brother, head of State's Office of International Security Affairs, was a former director and president of a company-related bank.

Acheson, discovering a CIA plan to overthrow Arbenz during Truman's presidency, talked the president out of the idea; but there would be no such roadblocks in the Eisenhower administration. Smith and the Dulles brothers agreed on a coup plan and gained the president's approval, contrary to the advice of Assistant Secretary Cabot.[35]

When Cabot had became assistant secretary for Latin America, Dulles had told him, "I want you to devise an imaginative policy for Latin America—but don't spend any money." After Cabot's dissent on Guatemala, Dulles replaced him with someone he hoped would be more pliant; but the new assistant secretary, Texas lawyer Henry Holland, and Deputy

Assistant Secretary Robert Woodward also had reservations. Holland opposed U.S. military supplies for Guatemalan rebels, warning that intervention would be unpopular in Latin America and might lead to a bloody civil war.[36]

But Allen Dulles deemed the aid vital for the coup's success. "Mr. President," the victorious CIA director told Eisenhower, "when I saw [Holland] walking into your office with three large lawbooks under his arm, I knew he had lost his case already." This illustrated State's reputation for excessive legalism and its inability to compete with the CIA. The White House saw the former as stuffy, passive, effete, always telling the president why he should do nothing. The CIA, on the other hand, was activist, in a "can do" American spirit, solving problems quickly, cheaply, and decisively. Faced with alleged imminent Communist takeovers, presidents found it easy to choose between the two. A similar process occurred in Iran, where the CIA helped restore the shah to power after two years of State's frustrating attempts to get along with nationalist premier Mohammed Mossadegh.[37]

Washington also had difficulties dealing with Egypt's charismatic leader, Gamal Abdel Nasser. The State Department generally favored working with the new Arab nationalism to build Middle East defenses against Soviet intrusion. At the same time, it wanted to maintain the important alliance with Britain and France and to promote an Arab-Israeli settlement. As often happens, conflicting though equally vital goals were reflected by differences within the department.

The U.S. embassy in Cairo knew relatively little beforehand about Nasser's July 1952 coup, but the conspirators kept the CIA informed. On the scene was one of America's ablest diplomats, Ambassador Jefferson Caffery, watching the steady decline of King Farouk's corrupt, incompetent, and unpopular regime. Scion of a wealthy Louisiana plantation family, Caffery seemed an ambassador from central casting—one colleague called him "Mr. Diplomat." He wore suits tailored in London's Savile Row, custom-made English shoes, and a Homburg hat. Caffery was a shy man, as are many FSOs who find the introspection and caution expected in their official role to fit well with personal preferences. Also common among department officials was Caffery's suaveness, egoism, and cold manner.[38]

Caffery entered the Foreign Service in 1911 at age 24 and spent most of the next 44 years abroad, in Greece, Japan, Spain, Sweden, El Salvador, Colombia, Cuba, Brazil, France, and finally Egypt. He served five years in Paris as the first postwar U.S. ambassador and the first career man ever to hold that position, though Caffery remarked that almost every week he heard rumors of his imminent replacement by a political appointee. In

Cairo, Caffery's practice of cultivating close personal relations with leading figures worked extremely well among the royalist, and later the republican, elite. His skill and contacts made him personally popular and effective.

Soon after the coup, Caffery met with the new leaders and quickly sized them up as good from the American viewpoint, eager to cooperate in regional defense but lacking experience and organization. U.S. aid would be needed to preserve the influence of the junta's "more cautious and pro-Western" figurehead, General Mohammed Naguib, Caffery warned. Dulles, however, decided to take his time reviewing the matter.

Cairo's most immediate concern was the British military installations in the Suez Canal zone, which gave London great leverage in Egyptian affairs. State Department briefing papers told Dulles that Britain's "colonial and imperialistic policies are millstones around our neck." Once the Suez problem was solved, they suggested, the United States could move on to seek an Arab-Israeli peace accord. Influenced by these arguments, Dulles asked Britain for concessions, and in July 1954 an Anglo-Egyptian treaty was signed to remove British troops.

As a next step, Assistant Secretary of State Henry Byroade pressed for quick military and economic aid to Egypt, but nothing happened. Dulles held back the aid originally offered in exchange for the Suez deal as an incentive for Cairo to reach an Egypt-Israel peace settlement. Byroade, formerly the United States' youngest World War II general, was chosen to replace Caffery when he retired because Dulles hoped that shared youth and military background would ensure felicity between Byroade and Nasser.

But growing Egypt-Israel tension and Nasser's interest in neutralism, closer relations with the USSR, and Arab leadership all antagonized Washington. Nasser attended the first nonaligned conference in 1955, where he secretly explored the possibility of obtaining Communist arms. Dulles ordered Byroade to boycott Nasser's ceremonial arrival home, leaving the Soviet ambassador alone as the airport reception committee. Unlike his brother, the secretary of state thought Nasser's threat to buy weapons from the Communists was a bluff; he was surprised when a huge arms deal was announced in August 1955.

Overestimating American leverage, Dulles was determined to show Nasser he could not play off the great powers. The main U.S. asset was possible funding for a Nile River dam at Aswan, a project the Egyptians believed to be the key to developing their economy. On March 28, 1956, Dulles approved Operation Omega, a plan to teach Nasser a lesson by denying him arms and aid, strengthening anti-Nasser elements in the Arab world, and initiating covert operations to remove unfriendly Arab governments. Instead of caving in, as Dulles and most high State Department officials expected, Nasser dramatically nationalized the Suez Canal Company, leading to a retaliatory invasion by Israel, Britain, and France. Dulles opposed military intervention and convinced Eisenhower to force the invaders' withdrawal despite the president's hesitation at breaking with such close allies. Nasser emerged from all this as the leading Arab hero for the next 15 years.

Dulles understood the Third World well enough to see that such open attacks would alienate those peoples and perhaps push them toward Moscow, but he was equally quick to condemn would-be neutrals for abandoning the Free World cause. By late 1956, Dulles and the State Department hierarchy regarded Nasser as little more than a Soviet tool. Other Third World leaders were branded as such for far less.

In a rare battle between ambassador and secretary of state, Byroade's cables criticized Dulles's new policy and argued that Washington should try harder to reconcile with Nasser. He thought Egypt's acceptance of Soviet aid did not imply sympathy for Moscow and felt the Dulles approach "suggests we continue to judge Egypt solely by whether—measured by our own criteria—she is for us or for the Soviets." Washington seemed to expect Middle East states to be totally in the Western camp, but "neutralism exists over a large portion of this part of [the] world. If we fail to develop means of fruitful cooperation with this large body of people and continue to consider them as being either in enemy camp or as 'fellow travellers' I fear that before too long we will begin to appear in [their] eyes . . . as being the unreasonable member of East-West struggle."[39]

Dulles complained that despite U.S. technical aid and good-faith negotiations over arms sales, Nasser supported leftist elements in Syria, attacked the collective security Baghdad Pact, intervened in northern Africa, and attempted to undermine U.S. influence in Saudi Arabia. The real reason for the break, he said, was the gradual unveiling of Egyptian policy as "maintaining U.S. expectations of future Egyptian cooperation while demanding immediate U.S. assistance and in fact pursuing policies detrimental to U.S. objectives." The administration had sought cooperation with Egypt, "often at considerable political cost," but further efforts at conciliation would appease, and likely strengthen, a hostile regime. The secretary of state, of course, had his way and Byroade was soon transferred to a less sensitive post.

Within the department, the switch from pro- to anti-Nasser positions met with dissent. Like the earlier anticolonialism of Asian specialists, there was now an emerging Third World lobby at State favoring greater concentration on the problems of Asia, Africa, and Latin America, and more sympathy with the nationalism, nonalignment, and processes of change in those regions. It fought usually frustrating and losing battles, although such ideas won some influence in the Kennedy administration.

Dulles did not find it easy to take advice. Of his three successive undersecretaries, only Herbert Hoover Jr. was influential. Although Walter Bedell Smith fulfilled an important administrative and intelligence liaison role, Dulles later prevented his installation in the White House as a presidential consultant.

Dulles's one-man show provided relatively little work for his final undersecretary, Christian Herter. At their first meeting, Dulles forgot what Herter had come to discuss and offered him only an assistant secretaryship.[40] Installed finally in the number-two position, Herter complained, "It is hard to know what use I am around here. I have been given no authority, and no area of work. . . . Everyone finds it difficult to know what Foster is either doing or thinking. . . . It is discouraging. I have left this office, many nights, thinking quite clearly that I should go home and pack my bags."[41]

Given the centralization of power in Dulles's hands, his resignation in April 1959, forced by the cancer that was ending his life, necessitated major revisions in the policy process. The change gave an opening to Eisenhower's more flexible proclivities on the Cold War through Dulles's successor, the previously neglected Herter. Well liked in the department, Herter had begun his career as an FSO before switching to electoral politics. Characteristically, Dulles warned his successor, "I have always had the greatest reverence for the office of Secretary of State. I have fought every effort to interpose anyone between the Secretary of State and the President. I would advise you to do likewise, Chris."[42]

Foreign policy leadership in the administration's last 18 months became more of a team effort. Deputy Undersecretary Douglas Dillon, an investment banker and former ambassador to France, was influential, as was veteran FSO and Soviet specialist Bohlen, though the continued hostility of congressional Republicans prevented his heading a proposed Office of Soviet Union Affairs. The post-Dulles line was illustrated by State's support for a nuclear testing moratorium and by plans for what later became the Alliance for Progress in Latin America.[43]

With Dulles removed as the keystone of his policymaking system, Eisenhower again turned his thoughts to reorganization, contemplating an above-cabinet official to handle diplomacy. With his own health in question after two heart attacks, Eisenhower wanted to limit the burden of travel and negotiations. He relied considerably on Gordon Gray, his assistant for national security, a foretaste of that job's growing importance in later years.[44]

But the administration did not do well in the two major crises it faced in the post-Dulles era, Cuba and the U-2 incident. The United States had a long, complex relationship with nearby Cuba, frequently intervening in its affairs and dominating a great deal of its economy. In the 1950s, U.S. ambassadors got along well with dictator Fulgencio Batista; others, aware of Cuba's problems, saw the likelihood of political change there. This group included Assistant Secretary of State Roy Rubottom and William Wieland, director of the Office of Caribbean and Mexican Affairs, along with some CIA officials. When Fidel Castro began his guerrilla struggle in the mountains of eastern Cuba, these people hoped he would emerge as a dynamic, reform-minded leader.

On the other side, Ambassador to Havana Earl Smith and Admiral Arleigh Burke, chief of naval operations, argued that Castro was a Communist. State and CIA reports, while noting some Communist influence in the movement, denied this. Convinced that Batista was doomed, the State Department began to hold up arms sales to him in autumn 1957. Finally, an emissary was sent to suggest that Batista leave Cuba. The revolutionary forces entered Havana in early January 1959. Only a few days before Castro actually took power did the CIA begin to suggest that his triumph might be inimical to U.S. interests.[45]

Washington believed that moderate political, business, and union leaders could play a major role in the new regime. The weakness of these forces and their habitual dependence on the United States, as well as Castro's determination and charisma, prevented this. The new ambassador, Philip Bonsal, one of the more reform-minded FSOs working in Latin America, tried his best but it was clear by December that Castro's suspicion of the United States made good relations impossible. The final straw was a Treasury Department decision, made without consulting State officials responsible for Cuban relations, to support the refusal of U.S. oil companies to refine Soviet oil Castro had purchased. Havana responded by nationalizing the refineries. Washington decided that Castro must be overthrown and ended his sugar quota; Cuba forced the closure of the U.S. embassy.[46]

While presidential candidate John Kennedy would criticize both earlier U.S. overidentification with Batista and the State Department's failure to heed anti-Communist warnings, it is surprising that there was no major McCarthyist-type reaction or "Who Lost Cuba?" controversy comparable to earlier debates over China. Still, Earl Smith, Spruille Braden, and several other ambassadors darkly hinted at State Department conspiracies in congressional hearings. Caribbean affairs director Wieland became a particular target. The Senate Internal Security Subcommittee questioned his "integrity and general suitability" for diplomatic work and his security clearance was held up for months. President Kennedy's strong defense of Wieland, by then an administrative analyst, showed FSOs that the new chief executive would protect them, in contrast to the Eisenhower administration's dalliance with McCarthyism.[47]

If the Cuba crisis showed poor preparation and decision making, the response to Moscow's downing of a U-2 reconnaissance plane and capture of its pilot in 1960 demonstrated poor coordination. When Eisenhower, acting Secretary of State Dillon, Allen Dulles, and military aide Gen. Andrew Goodpastor met to deal with the crisis, Goodpastor did not inform them of a change in previous plans which had put the National Aeronautics and Space Administration (NASA) in charge. NASA sent out an obviously faulty press release claiming the U-2 was an off-course weather research plane. Soviet Premier Nikita Khrushchev had clear evidence to the contrary.

The State Department hotly debated whether to admit that the U-2 was an American spy plane. Bohlen suggested a "no comment" position, but others wanted to justify overflights of the USSR, arguing that the United States had a right and duty to carry out such reconnaissance. The department finally decided to acknowledge the aerial spying program, a step also favored by Eisenhower. While honesty is usually commendable—and given the wreckage and pilot in Soviet hands, prevarication would have been hopeless—it can often create more diplomatic problems than it solves. The statement of responsibility forced or rationalized Khrushchev's cancellation of a summit meeting with Eisenhower that State had hoped would create a mood of détente.[48]

By the end of Eisenhower's second term, the whole policy process, with or without Dulles, was coming under increasing criticism from Congress and independent foreign affairs groups concerning the overcentralization of power in Dulles's hands, managerial knots unsolved by Wristonization, and the proliferation of competing agencies. Senator Henry Jackson's Subcommittee on National Policy Machinery, which took testimony from many experienced policymakers between 1959 and 1962, and several government and private studies suggested major changes for State and the policymaking system. The fact that most of the structural problems described in the Jackson hearings still exist today is impressive proof of their intractability.

"Faulty machinery is rarely the real culprit when our policies are inconsistent or when they lack sustained forward momentum," Jackson concluded. Lack of direction and policy coherence at the top is usually responsible. Nevertheless, the conduct of policymaking determines the accuracy of information on which policy is based, the quality of implementation, and the competence of those who make decisions. The problem, Jackson noted, "is not reorganization—it is getting our best people in key foreign policy and defense posts."

While critical of State, Jackson defended its indispensability at the center of the policy process. "No task is more urgent than improving the effectiveness of the Department of State," he wrote. "In our system, there can be no satisfactory substitute for a Secretary of State willing and able to exercise his leadership across the full range of national security matters, as they relate to foreign policy." But "State is not doing enough in asserting its leadership." There was insufficient planning, inadequate training, duplication and overlapping of functions, excessive staff, too many committees and levels of bureaucracy ("layering") between the department's top and bottom. Responsibility was diffused to the point where the structure became a roadblock to new ideas, rapid action, and effective performance.[49]

The Jackson committee wanted to rationalize the flow of information and decisions. The secretary of state would be the president's main lieutenant on foreign affairs and the central authority in the policymaking process. The State Department would plan for future contingencies and coordinate effectively with other agencies; the NSC would act as a filter by discussing and resolving dissonant views, thereby providing the president with a forum for the discussion of major problems.

The committee's diagnosis and prescriptions are as good as any ever produced, but the battle of individual and institutional wills, the confrontation of ambitions, and the speed and complexity of events made any such neatness difficult. Roughly every four years a new group descends on Washington, convinced that it can impose order on the government and even on the world at large. Within a short time, even though it may have some successes, the unsolvable pressing problems of external crisis and internal process leave it weary, sore, and somewhat disillusioned.

For the secretary of state, the burden had grown immensely. The job required somewhat contradictory and time-consuming roles. He must be personal adviser to the president; ranking diplomat dealing with foreign governments; administration spokesman to Congress, the public, and the world; and institutional leader. Dulles himself characterized the position as "The 'impossible' job." "The question remained, and still remains," wrote Paul Nitze, "whether it is possible for a Secretary of State simultaneously to maintain the good will of Congress and of the press, and also maintain the respect and confidence of our allies and friends abroad."[50]

Consequently, the Jackson committee, as did Eisenhower, considered dividing the job into two parts, creating either a "supersecretary" for foreign affairs, above the secretary of state, to coordinate all foreign policy or a minister of foreign affairs below him to travel abroad and to negotiate. Nelson Rockefeller favored the former option; Robert Lovett endorsed the latter, suggesting, "The Department will not run without a Secretary of State on the job." But there were objections to both proposals. Foreign countries would not want to talk only with a number-two man, while a supersecretary might challenge the president's authority.[51]

The organization of the department itself was also subjected to much questioning. Kennan was one of the most concise critics. "The present system is based, throughout," he wrote, "on what appears to be a conscious striving for maximum fragmentation and diffusion of power." Administrative functions were divorced from the chain of command and placed under control of "special independent hierarchies of administrators, managers and security officials." This separation from substantive considerations in the name of an efficient personnel system convinced officers that their "fate will . . . be determined by members of an invisible fraternity of administrators and security agents whose identity he does not know, who do not know him, and for whom he is only a card from the business machine. . . ."[52]

State's increasing size and complexity enhanced this alienation. The department grew from fewer than 6200 people when America entered World War II to 24,000 in 1964 (7000 at home and 17,000 abroad, including about 10,000 foreign nationals) in 274 posts. There were now two undersecretaries, two deputy undersecretaries, 13 assistant secretaries, over 30 deputy assistant secretaries, more than 60 area and other office directors, and over 90 country desk officers. They received 1300 incoming cables a day and sent out another 1000, of which the secretary would see 20 to 30 of the former and 6 of the latter. Commented Dean Rusk, "Junior officers in the Department today . . . have to deal with matters which before World War II would have come to the secretary of state."[53]

This apparent authority was misleading since junior officers had to clear actions with seven or eight higher levels and a variety of bureaus. Consequently, while it was possible to make a decision quickly, the trip through channels could take seven to ten days, resulting in the same, often obvious, answer. "There are those who think that the heart of a bureaucracy is a struggle for power," noted Rusk. "This is not the case at all. The heart of the bureaucratic problem is the inclination to avoid responsibility."[54]

The result was a proliferation of committees, which protected individuals by spreading responsibility among many. "If you want to see anybody in Defense or State, or any other department I know of," said veteran diplomat David Bruce, "they seem to be perpetually off in committee meetings." Consultations and discussion have many advantages, of course, but the committee system also produced endless compromises, watered-down decisions, busywork, lowest-common-denominator solutions, and a fear of creativity.[55]

"The system of diffused authority spreads outwards into a thousand branches and twigs of the governmental tree," wrote George Kennan. At every level, decision making was made by consensus among bureaus and agencies, any of which could veto or delay action. The operative principle frequently voiced by officials becomes, "Anything you fellows can agree on is all right with me." Such methods, in Kennan's words, produce "a hodgepodge inferior to any of the individual views out of which it is brewed" and require enormous amounts of wasted time and paperwork.[56]

Lacking trust in the State Department, the presidents following Eisenhower reposed more confidence and responsibility in smaller and seemingly more efficient groups: the NSC and interagency task forces. Few of the participants in the pre-1961 debate, however, advocated raising the NSC's status. After all, the Eisenhower-era NSC had taken on all the department's weaknesses—bogged down in detail and by interagency compromises—without having its operational strengths. The Jackson committee thought the special assistant for national security affairs should merely "Keep the President informed" and "staffed on issues that he takes into his own hands." It noted with approval that during the NSC's early days it was chaired by the secretary of state in the president's absence.[57]

So strong was this commitment to strengthen the State Department and to limit the NSC that the Kennedy administration quickly abolished OCB to improve State's position. National security adviser McGeorge Bundy wrote Jackson in September 1961 that the president did not "wish any question to arise as to the clear authority of the Secretary of State, not only in his own Department, and not only in such large-scale related areas as foreign aid and information policy, but also as the agent of coordination in all our major policies toward other nations." In addition, Kennedy issued a May 1961 order designating ambassadors as leaders of the "country team with authority over other agencies represented in the embassy."[58]

But could State exercise such leadership? Given the policy machine's inertia and inflexibility, as well as all the vested interests involved, structural reform did not satisfy the White House. "We must expect that the regular apparatus of the government will become, with time, of less value to the President and the Secretary of State," concluded Kennan. In critical areas, "American statesmen will have to take refuge in bypassing the regular machinery and in the creation of ad hoc devices—kitchen cabinets, personal envoys, foreign offices within foreign offices, and personal diplomacy.'' This is precisely what happened during the next two decades.[59]

Posterity would harshly judge Dulles for his tough talk and rigid policy formulations. Yet rhetoric was usually different from practice: Chiang Kai-shek was not ''unleashed,'' the Suez conflict was defused by U.S. mediation, and the Soviets were not pushed over the brink.[60] The real damage to U.S. interests during this period came in two other areas—the wasting of opportunities in the emerging Third World and, more immediately, the demoralization and atrophy of the policymaking apparatus itself.

5

On The Team: The Kennedy-Johnson Years 1961–1968

In 1982, former secretary of state Dean Rusk gave some advice to George Shultz, who had just been named to the office, suggestions that told much about Rusk's, and State's, experiences in the 1960s. "Domestic issues can only lose elections," he wrote Shultz, quoting President Kennedy, "but foreign policy issues can kill us all." Although State's tasks were of the greatest importance, the White House was always the master. "There is only one president," Rusk continued. "A secretary of state serves at the pleasure of the president; his resignation is implicitly always on the president's desk." As a veteran of the military and State Department bureaucracies, Rusk had fully absorbed the civil service philosophy. To a far greater extent than did his predecessors, Acheson and Dulles, or his successors, Kissinger and Haig, Rusk saw his role as one among several advisers to the president, rather than as a viceroy.[1]

This attitude reflected presidents Kennedy's and Johnson's determination to be strong leaders on the diplomatic front. The State Department functioned as part of a government team. While Kennedy asked State to help build a creative and energetic foreign policy, his irritation and impatience with it encouraged White House officials to step into the breach, assuming more authority over foreign policy than did their counterparts in previous administrations. State's yielding of turf and authority showed loyalty to the president but weakened it for the future by seeming to confirm that the department was incapable of decisive action.

"Fortunately," Rusk told Shultz, "the secretary is backed up by a professional diplomatic service that is second to none. . . . It seems to be fashionable for new boys surrounding a new president to approach the foreign service with a mixture of suspicion and derision. After all, the foreign service does not share their view that the world was created at the last presidential election or that a world of more than 160 nations will somehow be very different because we elected one man rather than another as president.''

This is a classic statement of the career service's attitude toward political policymakers. With every change of government, the incoming officeholders are optimistic that a rapid revision of policies will produce dramatic victories on the international scene. They expect the bureaucracy to oppose and delay these innovations and, consequently, have hostile feelings toward the people and machinery they find in place. Most secretaries of state see themselves as aligned with the politicals against the careerists, but Rusk was an exception, reflecting his experiences as one of the latter in the 1940s and early 1950s.

Rusk's advice to Shultz also reflected his own friction with the young New Frontiersmen: ''Members of the White House staff do not and cannot share [the secretary of state's] responsibilities. Their job is to assist the president, not to substitute for him. It is one thing for a member of the White House staff to transmit to a Cabinet officer an instruction from the president; it is quite another thing for such a staff officer to try to issue his own directives.'' If the president allows this group to penetrate the department's chain of command, ''he is asking for a lot of trouble.''

Despite his problems with the White House staff, Rusk proudly noted that there were relatively few top-level feuds in the Kennedy-Johnson years. He got along well with Secretary of Defense Robert McNamara, who brought the Defense Department to one of its peaks of influence, and with national security advisers McGeorge Bundy and Walt Rostow. In light of interagency rivalries in later administrations, Rusk stated, ''Guerrilla war among those at the top of the government is simply too dangerous in the kind of world in which we live.''

Although this was an appropriate attitude, State's abdication of leadership damaged U.S. policy by denying alternative points of view. The department took a backseat on Vietnam—a war directed mainly from the White House and carried out by the Defense Department—and played a secondary and not always helpful part in the Bay of Pigs and the Dominican Republic interventions. Its record was better in handling conflicts with European allies and in several other contemporary crises. Nevertheless, the prevalent view at the end of the Eisenhower era—that State must play the central policymaking role—was replaced by the time of Richard

Nixon's 1968 election by his belief that the NSC staff must shoulder this task.

At the beginning of his administration, Kennedy asked the department to imitate his youthful, dynamic leadership, contemptuous of John Foster Dulles's glacial views, and orient itself toward the emerging Third World. "Instead of becoming merely experts in diplomatic history . . . or in current clippings from *The New York Times*," Kennedy told FSOs, "now you have to involve yourselves in every element of foreign life—labor, the class struggle, cultural affairs and all the rest—attempting to predict in what direction the forces will move."[2]

Diplomacy, wrote Kennedy aide Ralph Dungan, can "no longer be only a matter of making policy by cables, of playing it by ear, day-to-day," but required anticipation of events, not just reaction to them.[3] The Peace Corps, Alliance for Progress, and, within State, the semiautonomous Agency for International Development (AID) and Arms Control and Disarmament Agency (ACDA) were formed to increase the breadth and flexibility of U.S. policy. Kennedy's designation of each ambassador as head of a "country team" and Johnson's 1966 statement making the secretary of state responsible for overall direction and coordination of government activities overseas were also attempts to give leadership to the department.

Still, Kennedy never trusted State. "These fellows really object to my being president," he said a few months after his inauguration. The department was seen as sluggish and transfixed by the status quo. Kennedy called State a "bowl of jelly" and, according to Arthur Schlesinger, dreamed "of establishing a secret office of thirty people or so to run foreign policy while maintaining the State Department as a facade in which people might contentedly carry papers from bureau to bureau."[4]

The professionals had their own view of the matter. When Kennedy asked veteran FSO Bohlen, "What is wrong with that God-damned department of yours?" Bohlen suggested that Kennedy's constant intervention in the process was part of the problem. The president was not convinced.[5]

One limitation on Rusk's autonomy was that Kennedy had personally filled many of State's top positions, choosing Chester Bowles as undersecretary, Averell Harriman as ambassador-at-large, former governor of Michigan G. Mennen Williams as assistant secretary for African affairs, and Adlai Stevenson as UN ambassador. Rusk was able to appoint George McGhee, a former colleague at State, as head of Policy Planning over Kennedy's candidate, Walt Rostow, who later in that post became one of Rusk's closest allies after McGhee was promoted. In contrast to

Rusk's predicament, Secretary of Defense McNamara was able to choose his own assistants.[6]

Kennedy's personal activism further challenged traditional procedures. He frequently telephoned or met with lower-ranking officials rather than go through the chain of command; his White House aides constantly intervened in policy. Presidential assistants Arthur Schlesinger, Richard Goodwin, and Ralph Dungan were heavily involved in foreign issues, as were General Maxwell Taylor, the president's military adviser, Theodore Sorensen, perhaps Kennedy's closest adviser, and Attorney-General Robert Kennedy.

As his special assistant for national security affairs, the president chose McGeorge Bundy, a scion of the Establishment (Bundy's brother William was married to Dean Acheson's daughter) and a Harvard dean. Bundy and his deputy, Walt Rostow, eliminated Eisenhower's committee system, making the NSC a more compact body—a presidential discussion group rather than the top of a bureaucratic pyramid. Bundy's 10 assistants included Carl Kaysen on arms and European issues; Michael Forrestal, son of the first secretary of defense, on the Far East; and ex-CIA man Robert Komer, later Johnson's key Southeast Asia specialist, on the Middle East. While this group formed the White House's "little State Department," in Kennedy's phrase, Bundy and Rusk maintained good relations due to the efficient channel between the NSC staff and State run by Rusk's staff. Through this link, State cleared papers and decisions to a far greater extent than it would with Johnson.[7]

The national security assistant managed the flow of information, intelligence, and decision-papers to the president, and, replacing the OCB, monitored the operations of other agencies. Rather than acting as a neutral clearinghouse for independent departments and top policymakers, the NSC staff began to lobby for its own policy preferences, particularly in the Kennedy years. Although its members avoided press attention or direct negotiations with foreign governments, they did make bureaucratic alliances—or engage in struggles—with other agencies. The White House started its own Situation Room and installed equipment giving the staff direct access to State, Defense Department, and CIA cables. The energy, small size, and bureaucratic compactness of the NSC staff allowed it to run rings around State in the competition for influence.[8]

Rusk's crew could not match this pace. In the words of former FSO John Davies, the New Frontier sought "bold new ideas and quick decisions . . . of men who had learned from long, disillusioning experience that there were few new ideas, bold or otherwise, that would solidly produce dramatic breakthroughs and whose experience for a decade had been that bold ideas and actions were personally dangerous and could lead to congressional investigations and public disgrace."[9]

Rusk's reluctance to express firm opinions sometimes meant that no one clearly spoke for State. Assistant secretaries, lacking guidance or consensus, had to seek the undersecretary's mediation or tell Rusk, "Unless you disagree, Mr. Secretary, I propose to do such and such." Consequently, subordinates would deal directly with the president or with NSC staff counterparts without fully informing Rusk. In this manner, cross-agency teams organized themselves to lobby for a particular position on Vietnam, nuclear weapons, and African issues.[10]

As a secretary whose ideas were closer to the career rather than the political model, Rusk was not in tune with the administration's operational style. Cautious and reserved by nature, Rusk apparently tried to model himself on George Marshall. But, Marshall had Lovett and Acheson to provide energy and policy advice, a president highly attuned to State, and little competition from other agencies; Rusk faced a different situation in all three respects. Consequently, his refusal to give more leadership to subordinates or to be more assertive with the president weakened State, ironically in line with Rusk's own dictum: "Power gravitates to those who are willing to make decisions and live with the results, simply because there are so many who readily yield to the intrepid few" who seek authority.[11]

Rusk was born in 1909 of a poor Georgia family and became a Rhodes scholar and teacher. He joined the army during World War II and later served in the Pentagon, switching to State in March 1947 as director of special political affairs to replace the soon-to-be-disgraced Hiss. Rusk was eager to strengthen the UN and was also sympathetic to building good relations with the emerging Third World rather than supporting European colonial powers. Like Acheson, he believed that Chiang Kai-shek could not be saved but that Taiwan must be protected from a Communist takeover. In the spring of 1950, at the peak of the attacks on State by McCarthyites and the pro-Chiang lobby, Rusk courageously volunteered to take the tough post of assistant secretary for Far East affairs, a sacrifice that made Acheson an enthusiastic supporter of Rusk's selection as secretary of state a decade later. When North Korea invaded South Korea, Rusk pushed for a defensive force under UN auspices, a brilliantly successful move. In December 1951, he resigned to become president of the prestigious Rockefeller Foundation.[12]

All this government experience helped prepare Rusk for the secretary's high-pressure job. During his first 19 months in office, he traveled 161,000 miles abroad, attended 15 major international conferences, conducted numerous bilateral talks, testified before congressional committees 47 times, held 40 press conferences, and appeared on 23 television and radio programs.[13] Rusk had help from a group of talented and experienced colleagues, some of whom had especially good relations with the president. "I hope," Rusk pointedly commented, "no one expects that only Presidential appointees are looked upon as sources of ideas."[14]

A problem for Kennedy appointees was the often breakneck pace of job rotation. In six years, including holdovers from the Eisenhower administration, there were five assistant secretaries for Latin American Affairs; four each at Congressional Relations, Public Affairs, Educational and Cultural Affairs, and Far East Affairs; and three in charge of Economic Affairs. Kennedy's ambassadorial appointments generally received high marks, but Senate Foreign Relations Committee Chairman William Fulbright criticized the "idiotic policy" of shifting envoys every two years on the average. Without continuity it was impossible for an assistant secretary or ambassador to know his bureau or country very well.[15]

Chester Bowles, the senior undersecretary of state and a veteran politician, was the first to lose the bureaucratic "musical chairs" game. In theory, Bowles was expected to run State while Rusk made policy. As a leader of the Democratic party's liberal wing, however, Bowles also saw himself as the New Frontier's representative at State. He favored bringing into top positions more people "who understand the Kennedy policies and believe in them," but opposed the Bay of Pigs invasion of Cuba and worried that the administration was too eager to use military options.[16]

"Most of the political assumptions under which we now operate," Bowles wrote in one memorandum, "are hand-me-downs from a period in which the balance of power was vastly more favorable than it is now." In arguing that the department was unnecessarily archaic in its thinking, Bowles startled the president by telling him that under the Foreign Service promotion system, Kennedy's age would make him only an FSO-3, a little more than halfway up the ladder.[17]

Bowles found himself held fully responsible for the very management problems he was trying to combat as the White House concluded that State was not operating up to speed. But it must also be said that he was not well suited to the job, by temperament or experience. Like many outsiders to the policy process, Bowles was more used to discussion and idea making than to decision making and bureaucratic warfare. To make matters worse, Rusk thought Bowles went over his head to the White House too often; in turn, Rusk rarely consulted with Bowles, preferring counsel from one of his own selections, veteran FSO, Deputy Undersecretary of State U. Alexis Johnson.[18]

This conflict, like many internal government battles, was fought out in the media. Bowles was unhappy about critical leaks from his enemies, but when Kennedy tried to press a Latin American ambassadorship on him, Bowles's friends staged their own press campaign, one of them commenting, "It will be a curious result if the first head to roll after the Cuban affair is the head of the man who opposed it." Of one article, Kennedy said, "You can tell how that story was written. . . . One paragraph is from Bowles or his people. The next paragraph is from someone at State trying to make a case against Bowles."[19] Meanwhile Bowles's fellow undersecretary George Ball took on more of the responsibility for running State. In November 1961, the inevitable shake-up occurred. Bowles was first given a face-saving position as a special presidential adviser on Third World issues, and later left Washington as ambassador to India.

The reorganization made Ball the number-two man at State. Ball was energetic, outspoken, liked to play devil's advocate, and strongly opposed the escalating U.S. role in Vietnam. His competitiveness made him seek ways to circumvent the NSC staff, at one point using an FSO working in the White House as a secret channel to President Johnson. When National Security Adviser Bundy found out, Ball's contact was quickly transferred to Pakistan.[20]

Other changes strengthened White House influence in the department. Averell Harriman, Roosevelt's wartime ambassador to Moscow, became assistant secretary for East Asia, a post beneath his experience and seniority but one gaining increased importance as the Vietnam War became a priority. Harriman also handled policy on Laos.

Frederick Dutton, a White House aide, reinforced Rusk's considerable political skills on Capitol Hill as assistant secretary for congressional relations. Richard Goodwin, another White House deputy, represented presidential concern on Latin America as deputy assistant secretary for that region, though the bureaucracy quickly isolated and forced him out. Rostow moved from deputy national security adviser to bring renewed life to the Policy Planning Staff, producing a report recommending responses to 44 potential crises and the emerging Sino-Soviet split. He became increasingly preoccupied with Vietnam, convinced that the United States must teach Communists a lesson there to discourage aggression elsewhere, ideas Rostow reinforced in President Johnson's mind when he returned to the White House as national security adviser.

Nevertheless, State's problems continued and even intensified throughout the 1960s. Lack of efficiency was a constant issue. White House aides criticized its demands for them to endorse quickly cables that had taken State three weeks to approve. Kennedy frequently told such stories, including one involving State's tardiness in complying with his request for ideas to update the Monroe Doctrine; once given to Bundy, the job was quickly done.[21]

Asked to respond to an important Khrushchev note over Berlin, State took 43 days to produce a draft the White House deemed too long, negative, and conservative in tone. A requested survey of possible future Cuba policies after the Bay of Pigs invasion resulted in a 30-page "laundry-list" of all possible moves, with little analysis and a strongly interventionist tone not to the administration's liking. Kennedy repeatedly turned to people outside State for advice and to interagency working groups for policy direction. He chose Paul Nitze, then an assistant secretary of defense, to head a Cuba task force, and Deputy Secretary of Defense Roswell Gilpatric to chair another on Vietnam. Rusk had to talk with Kennedy to bring the two groups under State's control.[22]

The department's shortcomings arose from a range of bureaucratic, personnel, and management problems. William Attwood, a journalist and one of the energetic outsiders Kennedy appointed as ambassador in 1961 (to Guinea and, later, to Kenya), provided an excellent definition of the attitudinal problem at State. He found too many "people for whom . . . a satisfying week's work consists in initialing as many reams of paper and deferring as many decisions as possible; with whom you can talk of 'action' only in terms of setting up a committee, hopefully one that will spawn subcommittees. The chief considerations of a bureaucrat are to abide by the letter of the regulations, whatever the consequences, to keep a clean desk, and never to 'make waves.'" Attwood wrote about his ambassadorial post, "While Guinea lacked people who knew what to do, Washington seemed to have too many." His officers were tied to desks, writing and signing papers instead of making contacts and investigating events, although most of their reports received little attention.[23]

Meetings and paperwork took time from other tasks, including the need to think creatively about existing and future problems. One FSO suggested a corollary to Parkinson's Law: "Every producer of paper added to the government roster creates the need for an additional consumer of paper." The consumer, in turn, also becomes a producer. Travel and expense vouchers typified the pettifogging nature of State buraucracy. In one often-told story, an FSO returns to Washington traveling with his mother. On his last item—the cost of a taxi from Union Station to his hotel—he carefully separated out his share of the fare. "Did your mother ride in the cab with you," asked the accounting office. "No," he replied, "my mother walked and carried the bags."[24]

Conferences also ate up the workday. Rusk held a staff meeting at 9:15 A.M.; Ball chaired a discussion on operations shortly after 10. At 10:45, assistant secretaries briefed their staffs on the first two sessions until about noon. "All energies are employed in arranging for so many people to live together," Ambassador John Kenneth Galbraith wrote in his diary.[25]

Averell Harriman told the Jackson subcommittee in 1963, "Men with a spark and independence of expression are at times held down, whereas caution is rewarded." Arthur Schlesinger, one of the department's harshest critics, wrote, "One almost concluded that the definition of a Foreign Service officer was a man for whom the risks always outweighed the opportunities." He called the promotion system "a conspiracy of the conventional against the unconventional" and considered State a "benevolent society, taking care of its worst as well as—sometimes better than—its best." Whether or not these comments were accurate, they certainly reflected the suspicions between the activist, change-minded appointees and the career people.[26] The concepts dearest to the New Frontier—"counterinsurgency," "developing societies," "modernization," "nation-building," and "revolution of rising expectations"—were far removed from traditional diplomacy and State's continuing European orientation.[27]

Yet the department's stodginess was partly intentional. In earlier days, instructions for a board preparing the Foreign Service exam included the following note: "It is possible that FSOs develop a high degree of caution in their statements . . . because of the extent to which their opinions or decisions are subject to review. However if the exam system attempted to select individuals with outstanding initiative and independence of thought and action, these individuals might quickly become unhappy in the Service and might disrupt the service to such an extent as to seriously interfere with its proper functioning."[28]

The best FSOs knew that the higher echelons had much deadwood, promoted upwards by friends, "usually officers with bland records, with no black marks on their efficiency reports, with no history of ever having gotten out of line or rocked the boat or questioned their instructions," in Attwood's words.[29] In contrast, an energetic, conscientious officer may antagonize one superior, even on petty grounds, and be passed over for promotion. Such a pattern has always been common in the department.

Respecting specialization, the New Frontier people were contemptuous of situations like that of the Japanese-language expert never assigned to Japan, or the routine promotion of those blocking or ineptly carrying out decisions. They cited the case of the Rome embassy's number-two man who became an ambassador after obstructing Kennedy's backing of Socialist party entry into the Italian cabinet. A junior officer supporting the administration position was marked insubordinate and only saved from ''selection out'' of the Foreign Service by White House intervention.[30]

So obvious was White House dissatisfaction that by 1963 Kennedy had to reassure FSOs in a White House garden chat, ''In spite of what you read, we love the State Department.'' Nonetheless, White House aides and journalists close to them continued to spread rumors that Rusk would soon be replaced. The secretary was blamed for the department's performance and even for lacking that quality of growing importance in politics—a good image. *The New York Times* summarized the general perception of Rusk as ''a capable but dull executive, ill-suited to the exigencies of high-level negotiation in a period of world tension.''[31] This analysis missed the point: Rusk and State were at their best on traditional types of diplomatic problems, as they showed in the Berlin crisis. Their shortcomings were in coping with the new kind of challenges presented by revolution and upheaval in the Third World. Indeed, Rusk defended State's performance with the maxim that diplomacy is most effective when performed quietly. He told one interviewer, ''The things that are well done are almost by definition not heard of because they don't hit the headlines.''[32]

The department made a number of attempts in the 1960s to cope with these increased demands, including efforts to update management, improve planning, develop a crisis operations center, raise the desk officer's importance, and broaden the Foreign Service corps' character. Many of the reform ventures can be credited to William Crockett, who became deputy undersecretary in 1963, and his aide Richard Barrett.

Money and information were necessary for any assistant secretary to coordinate interagency work on his region. Crockett pressed for a computerized data system providing a complete rundown on all agencies' projects for key countries. But FSOs and other departments, particularly the Budget Bureau, resisted such centralization. Crockett's carefully assembled team was decimated by rotation, resignations, and retirement. After January 1967, when Crockett left State, this program was abandoned. Similar difficulties faced Crockett's attempts to rationalize the personnel system by forecasting and matching future requirements with staff skills. His psychological training and encounter group seminars to broaden the abilities of FSOs were regarded as unsuitable novelties. Without improvements in data collection and assignment procedures, State simply did not have the capability to take the leadership role it had been offered.[33]

More lasting was the new Special Operations Center for handling crises. Teams covering urgent issues staffed it on a 24-hour basis and drew specialists from State and other agencies to gather data on current issues, organize task forces, press for policy recommendations and ensure implementation of decisions. This center continued to exist, though as more of a communications clearinghouse than the nerve center called for in the original ambitious plans.[34]

The civil rights movement, Kennedy and Johnson administration policies, and the emergence of new African states brought challenges to State's employment practices. Rusk called the removal of racial discrimination in the United States "Of fundamental importance to the success of [U.S.] foreign policy. . . . Our own department and our own Foreign Service [must] prove themselves on this point." While 1064 of 4570 civil service workers at State were black, 85 percent of them were in menial positions, compared to only 17 blacks among 3732 FSOs.[35] In earlier decades, many department officials shared their era's prevalent racialist views. Even some of those assigned to Africa in the 1960s evinced such attitudes. FSOs resisted attempts to broaden recruitment, claiming this might lower the staff's quality, but the corps' aristocratic biases had never prevented the influx of many mediocre and incompetent officers. Prejudice against nonwhites, Jews, and women had both excluded talented people and narrowed the range of wisdom and experience that could help in dealing with foreign societies.

The Ford Foundation provided a grant to prepare minority candidates for testing, and the exams themselves were studied for fairness. Eight black ambassadors and a U.S. Information Agency director were appointed between 1961 and 1965 and the first female black FSO was accepted. By November 1967, blacks held 4.3 percent of officer positions and some 246 professional jobs, compared to 45 in 1961.[36] Still, for a variety of reasons, including the continuing country club ambience at State, minorities only gradually came to seek foreign policy careers. Earlier discrimination against Catholics and Jews had disappeared by the 1960s, except that Jews remained informally barred from work relating to Middle East policy and from service in Arab countries, a restriction not ended until the late 1970s.

The situation of women was far worse. Until 1971, women who married had to resign from the Foreign Service. The wives of FSOs were expected to help their husbands' careers—for many years the evaluation of a spouse was put on an officer's record—but only in the 1960s were they encouraged to take relevant jobs in embassy-related activities. A 1967 presidential executive order banned discrimination against women in federal employment, but even in the 1980s, despite a growing number of female FSOs, many sections of the department still did not take women seriously.

While reevaluating internal practices, State was also trying to accommodate itelf to Congress's growing role. Traditionally, aside from budget issues, the legislature played a relatively minor role in State's world, but the need to mobilize congressional support for foreign aid and other policies, and the unhappy experience with McCarthyism, showed the need to improve relations. The Jackson subcommittee hearings and Senator William Fulbright's active leadership of the Foreign Relations Committee made Capitol Hill a new factor in foreign policy.

In a 1963 article, Assistant Secretary of State for Congressional Relations Frederick Dutton described the situation as one of "chronic tension" with "occasional guerrilla warfare." He attributed this conflict to the two bodies' different styles: State "is analytical, tentative, cumbersome as it digests vast detail . . . and cautiously gropes for the real meaning of what is happening in the world." Congress, "regularly faced with re-election, is assertive, often glandular, in its approach to the world" while also showing "a creativeness, vigor and incisiveness often undernourished in the foreign policy apparatus."[37] The Vietnam War would soon bring Congress and State into confrontation.

Long after the end of McCarthy's influence, department security continued to evoke congressional interest. An American Legion report showed, to State's embarrassment, that access to its building was virtually uncontrolled. During the day, visitors were neither stopped nor questioned. This revelation forced improved arrangements to restrict admission to those with proper business in the building. The 1963 battle in which Abba Schwartz, administrator of State's Bureau of Security and Consular Affairs, was unable to fire ultraconservative Frances Knight as passport office director, provided a hotter controversy. Knight had campaigned against granting passports to suspected Communists. Schwartz attempted to reform laws restricting immigration and travel, particularly the ethnically biased national origin quotas. Although Congress did end most quotas and restrictions, Schwartz was forced to resign in 1966.[38]

A very bitter conflict of this period concerned the most tangled firing in State's history. Otto Otepka, chief security risk evaluation officer in 1963, criticized the testimony given before the Senate Internal Security Committee by his superior, Deputy Assistant Secretary for Security John Reilly. In opposing renewal of William Wieland's security clearance during the investigation over State's misjudgment of Castro, Otepka gave classified documents to the committee, defending his actions on the principle that no civil servant can be denied the right "to furnish information to either house of Congress or to any committee or member thereof."

Reilly found evidence of Otepka's responsibility for the leak by having his "burn bag" searched and then ordered installation of a microphone in his office, an action he at first denied when questioned before Congress. Reilly was placed on administrative leave. Otepka refused a transfer to other work, was dismissed, and immediately appealed. The suit, and Otepka's defense by some senators, went on for many years. "Tapping telephones, snooping through trash baskets, locking people out of their offices and searching desks. It must be a charming place to work," said *The New Republic*.[39]

The extent of the uproar produced by the Wieland affair was related to Latin America's high priority for the White House. "I would say we have given more thought . . . to the problem of Latin America than in almost any area involving our foreign policy," said Kennedy. During the administration's first months, a White House Latin America task force under former Assistant Secretary Adolph Berle and the activism of White House aide Richard Goodwin challenged State's role. Berle, who had been harassed as a White House man at State in the Hull era, bitterly wrote that the bureaucracy would attack any presidential influence on policy. But even the Senate Foreign Relations Committee complained about the divided authority. "Too many cooks have been spoiling the hemispheric broth," commented *The New York Times*. Given this record of encroachment, Kennedy had to approach eight candidates before he found one to accept the job of assistant secretary for Latin America.[40]

The administration soon eliminated the duplication by moving Berle into a nominal job as a White House adviser while Goodwin went into State as a deputy assistant secretary. Berle was so suspicious of the Foreign Service that he warned Goodwin that his FSO assistant might be assigned to spy on him and listen in on phone conversations.[41]

The job of coordinating Latin American policy involved much economic work, particularly in regard to the administration's Alliance for Progress program, designed to counter the Cuban revolution's appeal by promoting democracy, development, and land reform. To facilitate operations, the Agency for International Development regional staff and State's regional bureau were put under joint command in 1963, and there was talk of creating an undersecretary of state for Latin American affairs to highlight the region's importance.

The administration's first venture into Latin American policy, however, was Kennedy's approval of a CIA plan for an ill-fated, U.S.-backed Cuban exile invasion to overthrow Castro. During preliminary discussions, Rusk expressed no strong opinions, although Bowles tried to convince him to oppose the project. Few people at State even knew about the operation, nor were experts consulted on the important issue of whether it would trigger a supportive uprising. Goodwin told Rusk, "Maybe we've been oversold on the fact that we can't say no to this."[42]

Ironically, the debacle made Kennedy suspect the military and CIA while depending more on his own staff, which had been just as enthusiastic about the invasion. His administration did far better during the following year's Cuban missile crisis, using a special executive committee as an effective decision-making group. After the CIA's discovery of Soviet offensive missiles in Cuba, the committee met almost continuously for the next thirteen days and almost daily for the following six weeks. Such complete attention by so many high-ranking officials could only be marshaled in the most serious and fast-moving crisis.[43]

When Kennedy decided to stop further shipments of Soviet missiles to Cuba, State Department legal experts suggested the naval operation be called a quarantine instead of a blockade—since no declaration of war was involved—and justified it by hemispheric defense treaties. State's role in the crisis concerned, in Rusk's words, the "extraordinary and complicated . . . diplomatic course: consulting with our allies; preparing the types of communications we would want to make to the neutral and nonaligned countries, getting special emissaries ready to go to see particularly leaders in Europe, getting . . . materials ready for a presentation to the Organization of American States." Rusk himself met with Third World ambassadors to show them aerial photos of the missiles and to outline the president's plans.[44]

Rather than recalling State's successes, however, the White House dwelt on its shortcomings. Attorney-General Robert Kennedy and presidential adviser Theodore Sorensen, disagreeing with State's draft of the key presidential letter to Soviet premier Khrushchev, rewrote it themselves. But when the Soviets began to give way in the face of U.S. determination, John McCloy successfully negotiated a face-saving deal which nonetheless achieved all the American objectives.[45]

The deliberations during the two Cuban crises illustrate an important lesson about a president's relations with his policymaking apparatus. Internal government discussions must be relatively open and freewheeling so that doubts and options are fully voiced. Robert Kennedy noted that he often saw officials change their views to coincide with what they thought John Kennedy, and later Lyndon Johnson, wanted to hear. Subordinate officers at State and Defense, Sorensen recalled, "seemed much more willing to disagree with their bosses when the President was not here."[46]

A staff structure designed to provide the chief executive with accurate information often fails in that task, acting only to reinforce existing views at the top. In that event the entire investment on intelligence and expertise is largely wasted, a tragic flaw consistently manifested during the Vietnam War. While internal conflict can be paralyzing, freer expression of dissent can provide decision makers with a more balanced view of the real situation and lead to productive second thoughts about the path they are following. Of course, leaders must first create a climate in which constructive criticism can be voiced in internal discussions.

After Johnson became president, he chose Ambassador to Mexico Thomas Mann as assistant secretary for Latin America, a position that had already changed hands four times between 1960 and 1963 and would go to three others after Mann during Johnson's administration. Mann, a lawyer who joined the Foreign Service during World War II, had been involved in Eisenhower's Guatemala coup. As assistant secretary, he reflected a traditionalist view of U.S.-Latin American relations, preferring private capital over the faltering government-run Alliance for Progress, and stressing economic growth and military assistance rather than social reforms. State opposed McNamara's attempt to reduce military aid to the region and became involved in the 1964 Brazilian coup against the left-of-center civilian government.[47]

Mann saw the Dominican Republic as a potential new Cuba. When liberal forces in the former nation tried to reverse a military coup in 1965, State wrongly saw this as a Communist-influenced revolution. Its panic was intensified by the earlier humiliating errors in assessing Castro's Communist involvement. In the Dominican case, however, the U.S. embassy's inability to produce serious evidence in response to Rusk's demand for proof of Marxist control did not prevent U.S. military intervention. Once again, department expertise was bypassed rather than used in a crisis. One FSO noted, "On Friday I was Dominican Desk officer; by Friday night Rusk was; and by Sunday noon Lyndon Johnson was."[48]

U.S.-Latin American relations during these years also illustrated two other ongoing problems: the need for research and the tendency toward bureaucratic expansion. The first issue was raised by the controversy around the army's Project Camelot, a study on sources of instability in Latin America. In Chile, the U.S. ambassador only discovered this program when he read accounts in local newspapers. He complained to Rusk, who persuaded President Johnson to order all political research cleared by State. McNamara terminated the project in July 1965.

The Defense Department had stepped in largely due to State's failure to carry out such projects. In fiscal 1966, the department spent only $200,000 on political research, compared to $12.5 million spent by the Defense Department. Rusk, former head of the Rockefeller Foundation, thought such tasks might better be left to private agencies while State doubted the practical value of systematic scholarly study. Academics "could answer every question" except "What shall we do?" concluded one analyst, and this "was the one question for which the men in the State Department had to have an answer."[49] Nevertheless, decisions made without understanding the details of foreign cultural and political systems, areas where top policymakers are often weak, have repeatedly led to disaster. State's technology for organizing and assessing such data was also backward. The department had no computer until late 1962—and then only one for administration. The communications system was little advanced over World War II levels. Information was stored in 17 different depositories.

The experience of the U.S. embassy in Brazil shows the personnel system's uneven performance—too many people in some places, too few in others. Ambassador John Tuthill decided in 1966 that his staff of 920 U.S. citizens and 1000 local employees was excessive. To his credit, one FSO, Frank Carlucci, later ambassador to Portugal and deputy secretary of defense, even recommended his own position be abolished if necessary. Trying to send Americans home, however, was no easy matter. Tuthill only made progress when the ruling junta's new chief expressed irritation at the number of U.S. advisers running around the country. Rusk cabled support for a reduction, promising the backing of the "entire cabinet and one man above the cabinet"; the military, CIA, and AID finally surrendered. By June 1969, personnel was reduced 36 percent, and three years later the number of Americans was 47 percent below what it had been in 1966. Back in Washington, Tuthill was asked, "What are we going to do with them here?"[50]

U.S. embassies in Africa would have been happy to have them. Opening so many missions, said Assistant Secretary Williams, "left the bureau breathless" and often shorthanded, but Kennedy and Bowles prevented State from making Africa a dumping ground for surplus FSOs or those on the verge of retirement. Instead, Africa became an assignment for those who sought adventure, faster promotion, and liked the region despite its chronic unimportance for U.S. policymaking.[51]

The Eisenhower administration had given the region a low priority. In 1960 Washington refused an aid request from Guinea, which had just forced France to grant independence, out of deference to an outraged Paris. The same rejection was repeated with Congo leader Patrice Lumumba after objections from ex-mother country Belgium. In the former case, Communist aid filled the gap; in the latter, ensuing instability led to the U.S.-organized overthrow of Lumumba.

Despite Kennedy's interest in Africa, State's leadership remained far more Europe-oriented, particularly when policy objectives toward the two regions clashed. In the case of Portugal's African colonies, Angola and Mozambique, Washington had to decide whether pressing for independence would endanger the U.S. bases Lisbon hosted in the Azores Islands. Attempts were made to convince Portugal to allow self-determination, since Bowles and others feared an ultimately victorious war of independence would produce conditions for Soviet influence.[52]

Admiral George Anderson, appointed ambassador to Portugal in September 1963, saw the Africa bureau as naive. The bureau thought him overly sympathetic to Lisbon and Ball even asked the deputy chief of mission to keep an eye on him, but Anderson did pursue negotiations over the colonies. After Kennedy's death, however, Africa issues went onto a back burner. While Ball and Harriman supported Anderson's compromise efforts, Portugal rejected them. The United States dropped the matter.

Washington had not fought the issue consistently or at a high enough level and, as often happens, U.S. power and aid could not force even a small ally to do things against its will. A similar frustration resulted over Kennedy's attempts to urge Iran's shah toward reforms. It is not surprising that foreign dictators reject U.S. advocacy of greater democratic practices. Given the unlikelihood of success and the American need to continue to do business with these regimes, argued professional diplomats, such initiatives into other countries' internal affairs were unwarranted.

In contrast, the Kennedy administration did become deeply involved in the Congo, near disintegration after mineral-rich Katanga province seceded in a revolt financed by the Belgian mining company. Liberals in Kennedy's government—particularly Assistant Secretary for Africa Williams, UN Ambassador Stevenson, and the UN affairs bureau—were anti-Katanga, advocated UN action to restore the central government's authority, and insisted that Congolese leader Patrice Lumumba was not pro-Communist. On this last point, Rusk, who had suffered before over such distinctions, was hard to persuade. "I know those agrarian reformers," he said after one briefing. "I dealt with them in China."[53]

Rusk was neutral between the liberals' support and the opposition of the European bureau and congressional conservatives toward UN action against Katanga. As point man on Congo affairs, Undersecretary Ball suggested a U.S.-organized deal for a loose Congolese confederation. Rejecting this, State's liberals and Africanists found U.S. Ambassador to the Congo Edmund Gullion, one of the president's favorite FSOs, a major asset on their side. Ball tried unsuccessfully to neutralize Gullion and Congo desk officer Frank Carlucci. Williams counterattacked with public statements and a torrent of internal department reports.[54]

Ball tried to limit Williams's Capitol Hill and press briefings and, to undermine Gullion, found an FSO in the embassy who doubted the value of UN intervention. Ball brought him to Washington to meet the president, keeping the matter secret to protect the man from charges of having gone over his superior's heads. Kennedy was persuaded to communicate directly with Katanga's leaders; when Gullion tried to block Ball's messenger, the undersecretary trumped the ambassador by citing presidential authority for the mission.

But while Ball had won in the bureaucratic war, the Congolese government continued to fall apart. If U.S. noncooperation were to force the UN out, the regime might seek Soviet help in defeating Katanga. After the murder of Lumumba, with U.S. complicity, Washington had less incentive to undermine the central government. Ball changed sides and Washington helped UN forces defeat Katanga and reunify the country. Joseph Mobutu, favored by the United States, became president. While the Congo (renamed Zaire) hardly lived happily ever after—Mobutu's regime set records for corruption and economic incompetence—the country was held together and the immediate crisis was resolved. The bureaucratic fight over the Congo was the kind of knock-down-and-drag-out brawl that has more than once split the foreign policy apparatus, more likely over marginal issues where leaders take no firm position. The Congo crisis and Portuguese colonial issues reflected State's clientism as well as the administration's liberalism. Those responsible for good relations with a country seek to preserve smooth bilateral links at all costs. Generalists are usually more conscious of competing interests; specialists have the benefit of greater knowledge on a region or country. The former may be ignorant while the latter may be partisan. Consequently, the need to maintain a balance between the two groups is an important task in policymaking.[55]

Gentler conflicts marked debates over responses to European demands for greater power in NATO and its own independent nuclear force. The Defense Department stressed improving existing units while State backed nuclear sharing with a Multilateral Force (MLF). Despite Kennedy's reservations, the political appointees at State, and particularly Ball, pushed aggressively for MLF until the White House finally rejected it. By 1965, Ball and Dean Acheson, brought back to consult as head of a special Europe task force, took strong stands against French President Charles de Gaulle's nationalist, anti-NATO policy. They even quarreled, unsuccessfully of course, with President Johnson, who preferred to settle the differences amicably rather than exacerbate Franco-American relations.[56]

While veteran Cold Warriors emphasized East-West clashes, partly to encourage NATO unity, some Soviet specialists in and out of government, including George Kennan, Marshall Shulman, and Zbigniew Brzezinski, began to speak of possible détente with the Soviet Union. In an October 1963 *Foreign Affairs* article, Walt Rostow proposed the beginning of a major effort "to establish whether or not it is possible for the

Soviet Union and the West to live together on this planet under conditions of tolerable stability and low tension.'' Assistant Secretary of State for Economic Affairs Anthony Solomon, minding commercial possibilities, spoke of the growth potential of East-West trade.[57] But détente would wait for President Nixon. The Johnson administration's energies became more and more absorbed with the Vietnam War.

Kennedy's November 1963 assassination and the subsequent personnel changes ended his experiment to make the NSC staff an alternate State Department. While Kennedy treated Rusk with respect, many White House officials had made little secret of their view that the secretary was ineffective and weak. President Johnson put more faith in his secretary of state. ''Foreign policy had been a passion, perhaps even to a fault, on the part of President Kennedy,'' explained Rusk's special assistant Benjamin Read. ''It was a casual, most casual acquisition of President Johnson.'' One Washington joke ran, ''When did Rusk become secretary of state? On November 22, 1963.''[58]

Meetings of the full NSC, common when Kennedy was president and characterized by lively discussion, became relatively rare under Johnson, reducing internal debate. This narrowing of dialogue became dangerous as Johnson enforced a consensus over Vietnam, closed to uncomfortable realities or ideas contradicting official optimism. In 1966, President Johnson created a Senior Interdepartmental Group (SIG), with Ball as chairman, to monitor developing crises. The SIG's coordinating powers were limited since all decisions could be appealed to the White House, but the committee, which met every Tuesday, took over some of the NSC's original coordinating tasks. Coupled to Johnson's Tuesday lunches with Rusk, McNamara, and his security adviser Rostow, the SIG and subordinate regional interagency committees gave the administration an informal, though somewhat more structured foreign policy system.[59]

Yet the burden was still not easily borne. ''I have not seen the Department so disorganized since the end of the Hull regime,'' Acheson noted privately. Rusk was a good and loyal assistant, Acheson wrote Truman in 1966, but he neither led State well nor prevented Johnson from drifting and from postponing decisions. While Rusk's stewardship was unquestionably competent, State had lost ground in the competition for foreign policy leadership, avoided managerial reform, and continued the lack of planning and direction from the top.[60]

On the most important issue, Vietnam, State held a weak hand as, one observer wrote in 1965, ''the dinghy dragged on behind the Pentagon's yawl.'' Harriman thought that, from the beginning, Dean Rusk gave too much responsibility for Vietnam to McNamara. Rusk's concern for presidential rather than department interests—he cautioned against ''parochial viewpoints or petty bureaucratic 'infighting' "—was a noble attitude, but, ironically, did not always best serve the national interest. Harriman considered it "very fair to say that Dean Rusk's failure to interpose political judgments at the important times was one of the reasons we got so far afield from reality." "I've learned, dealing with the White House," commented Harriman, "that you have a split second [when] the president looks around the room to decide whether you are going to be among those that agree or register your difference." The military's failure to take seriously the war's political, economic, and social aspects—it saw victory only in terms of more manpower and firepower—was a major element in the U.S. disaster.[61] State ran diplomacy, not foreign policy.

The U.S. government consensus always favored helping Saigon fight the Communist-led insurgency. Since 1946, the United States had tried to prevent Communist revolutions or takeovers around the world—this was, after all, the Cold War's essence—but there was much debate about how best to achieve that goal in South Vietnam. This task was made harder by the sad lack of expertise on Vietnam. In the decade after France's 1954 Indochina defeat, State conducted no in-depth training on Southeast Asia and never developed skills in line with new commitments. South Vietnam had three different desk officers and Indochina four office directors between 1957 and 1963. Only three of these seven had served there, and none of these had major influence on policy. The seven assistant secretaries for East Asia between 1957 and 1966 had virtually no Vietnam experience or special training. Many of those in the Saigon embassy were junior officers. As one FSO wrote, "Officers have been rotated in and out, barely able to assimilate the complexities of Vietnam before departing to some other, far-off assignment."[62]

As always, embassy reporting was controlled by the ambassador, which often meant that the weekly (later daily) summaries did not deviate much from the Washington line. Dissenters might get their views across by leaking information during touch football with reporters on Saturdays. The large number of FSOs required for South Vietnam led to pressure on young officers to go, while others were attracted by the excitement and action there. Everyone at State knew the fate of the China hands whose criticism of Chiang Kai-shek and pessimism over his prospects had been deemed pro-Communist sentiments in the McCarthy era. To report negatively on the Saigon regime, hint at the possibility of compromise, or warn about defeat might invite similar reprisals. As one former, strongly dissenting, official put it, "Those who doubted our role in Vietnam were said to shrink from the burdens of power, the obligations of power, the uses of power, the responsibility of power. By implication such men were softheaded and effete."[63] As one FSO who served in Saigon wrote,

"The realization that this pressure was at least in part inspired by perceived domestic political needs adds to the fear that an honest and thorough airing of differing policy alternatives is a fragile possibility under the best of circumstances."[64]

Of course, many of the career people genuinely shared their appointed supervisors' view that Communist forces must and could be prevented from controlling Indochina. By protecting Saigon, they believed, the United States would enhance its international credibility and discourage aggression elsewhere. Even those who might have opposed such commitments in the first place now were faced with the assignment of fulfilling them.

A greater variety of views about Vietnam was allowed in the early days, before decisions foreclosed options. In 1961, when an interagency task force recommended an explicit commitment to defend South Vietnam, Ball softened the proposal to maintain U.S. leverage with the Saigon government. Sterling Cottrell, chairman of the State Department's Vietnam task force, warned in late 1961, "Foreign military forces cannot themselves win the battle at the village level. Therefore, the primary responsibility for saving the country must rest with the government of Vietnam."[65] While Bowles advocated neutralizing Southeast Asia, Undersecretary U. Alex Johnson supported sending several thousand troops, and the Pentagon was already talking about the eventual need for six divisions, some 205,000 men.

Assistant Secretary Averell Harriman, Roger Hilsman, head of the Bureau of Intelligence and Research (INR), and NSC staffer Michael Forrestal argued on lines similar to Cottrell. A December 1962 report from Hilsman's bureau warned that the war would be lost if Saigon did not reform itself and improve the peasants' lot. They felt that the situation was rapidly eroding, largely due to the corruption and mismanagement of Ngo Dinh Diem's regime. All of them were gone by 1965, maneuvered out by Johnson, who had decided to follow the Pentagon's preference for a large-scale infusion of U.S. troops. By challenging the military's assumptions, Hilsman, himself a West Point graduate, incurred their enmity. McNamara and the Joint Chiefs of Staff protested so virulently about an October 1963 pessimistic, though accurate, INR report, that Rusk apologized for circulating it.[66] While the military wanted to focus on supporting Diem and fighting the war, this State Department group militated for political change in Saigon. Walt Rostow and McNamara thought escalation would work; Harriman and his friends at State preferred a counterinsurgency effort.

But lower-ranking men at State could not compete with the Defense Department leadership. Hilsman, like Harriman, concluded, "I can't blame McNamara for pushing his department's view as vigorously as possible. But I certainly can blame Rusk for not pushing his view, or our view. And always over and over again, it ends up with Harriman and Hilsman arguing against McNamara and the [Joint Chiefs of Staff] and [CIA director John] McCone. And that's not quite an equal contest."[67]

By August 1963, with the Saigon government's growing paralysis and its repression of Buddhists, Washington was faced with a major decision. The Vietnamese generals demanded to know what the U.S. attitude would be toward a coup. Harriman, Hilsman, and Forrestal assembled a cable, cleared by General Taylor and by the deputies of McNamara and McCone (who were out of town), hostile to the Diem regime's survival. While McNamara and McCone were reportedly furious about this development, they did not object at an NSC meeting where President Kennedy discussed and approved this policy. Before Diem was overthrown some weeks later, Paul Kattenburg, an FSO working for Hilsman, suggested it was time to consider a withdrawal. But complicity in the anti-Diem coup drew Washington in deeper than ever before.

While some analysts have argued that the policy system worked, allowing decision makers to implement their objectives, this conclusion is vitiated by the slapdash and poorly informed choices being made. "The whole management of the enterprise, the coordination in the field, the team work in Washington, were frankly never very good in the Kennedy Administration," concluded William Bundy.[68]

As the military situation worsened over the next two years, the Johnson administration decided that barring withdrawal meant bombing North Vietnam and deploying more troops. By 1965, as hundreds of thousands of U.S. soldiers were sent to Vietnam, the situation came even more under the Defense Department's control. In the field, AID and USIA men acted as civilian advisers for Saigon's political and propaganda efforts, but these operations took a backseat to conventional military operations. Some lower-ranking American officials in Saigon and Washington challenged the U.S. military's low estimates of the numbers and territory held by enemy forces. Defense Department calculations made victory appear easier and a conventional military strategy seem more attractive, in part because the generals failed to understand the political aspects of the war, both in Vietnam and in domestic U.S. politics. Until his resignation in September 1966, Undersecretary of State George Ball was virtually the sole high-ranking critic of this escalation policy. Ball later wrote, "They regarded me with benign tolerance; to them, my memorandum seemed merely an idiosyncratic diversion from the only relevant problem: how to win the war." To avoid the appearance of internal conflict, Johnson made Ball a semiofficial devil's advocate, and when Ball began to think about resigning, Johnson gave him more responsibility as head of the Senior Inter-Departmental Group.[69]

But Ball's warnings had no effect. In early 1965, the Vietnam interagency working group, under Assistant Secretary of State for Far East Affairs William Bundy—who symbolically came from the Defense Department to fill the job previously held by Harriman and Hilsman—suggested initiating bombing. Increasingly, the main forum for decision making was Johnson's Tuesday lunches with Rusk, McNamara, and NSC adviser Walt Rostow, which, according to a State Department joke, began with a prayer and ended with a selection of bombing targets in North Vietnam. The optimists included high-ranking U.S. officials on the scene, senior military officers in the Pentagon, and most of those working on the issue at State. Only gradually did those who backed the war realize that no amount of military effort would shake enemy resolve or shore up Saigon.[70]

Once policy had been set at the top, career people were not prepared, by either training or habit, to maintain criticism of the war. For its part, the White House believed that a change in policy would damage U.S. foreign and domestic credibility. This proved Harriman's wisdom in an earlier crisis: "We must never face the President with the choice of abandoning Laos or sending in troops. This is our job, to keep him from having to make that choice."[71] Given political views at the time, the latter option would inevitably be selected, and once made, that decision would be virtually irreversible.

The highly secret nature of decision making on Vietnam meant that relatively few people at State were involved. The Far East bureau generally had some influence. Assistant Secretary William Bundy had to go to the head of Rusk's staff, Ben Read, to keep posted, and his deputy, Phillip Habib, was also quite influential.[72] When top career people were consulted, they sought to stay on the team and accept administration premises. In June 1965, William Bundy wrote five U.S. ambassadors in Asia—Edwin Reischauer in Japan, Winthrop Brown in Korea, Edward Rice in Hong Kong ("our best China man"), Graham Martin in Bangkok, and William Sullivan in Laos. Bundy suggested, "I think the situation is going to hell in a hand-basket, and we face the choice of a lot more bombing [and] forces . . . or letting it go and seeing whether there's a fall-back line in Southeast Asia." The response was, Bundy recalled, that there was no other line in Southeast Asia. Martin and Sullivan stated flatly, "There is no holding in Thailand."[73]

Rusk and the department leadership remained loyal and supportive of Johnson's Vietnam policy even when Defense Department officials developed doubts. Before leaving in 1967, McNamara was disgusted and anxious about the policy he had helped create. At a State Department luncheon, he condemned the bombing: "It's not just that it isn't preventing the supplies from getting down the trail," he complained. "It's destroying the countryside in the South. It's making lasting enemies. And still the damned Air Force wants more." Rusk silently stared down at his drink. By early 1968, Deputy Secretary of Defense Paul Nitze and McNamara's successor, Clark Clifford, also concluded the war effort was futile. No one really backed a proposal to give Saigon an ultimatum to reform or face a U.S. policy reevaluation.[74]

Loyalty to superiors, bureaucratic inertia, civil service routine, careerism, narrow vision, and sincere belief all fed an inability to reconsider the Vietnam War. Once again, the task of honestly analyzing a policy's effects had become hopelessly entangled with the job of implementing it. State gave the president the support he wanted—as was its duty—but, in doing so, did not provide what he needed. Johnson, faced with advisers who argued that the war was winnable and that the United States could not afford to withdraw, carried on the battle until he gave up the presidency in 1968. Rusk was the president's star defender on Vietnam, willing to be the target of public, press, congressional, and international criticism. By the end of the Johnson years, Rusk's men ruled State, a sharp contrast to the early Kennedy era, working well with the Defense Department and White House. Perhaps a little more conflict would have been better for the country.

Both a grasp of reality and open internal debate fell victim to a loyal drive to administer a flawed policy. "They talk all the time," said one of Rusk's colleagues about the department's top levels, "but only on the pragmatic business of the hour. I doubt if they've ever had a serious general discussion, an exchange of views, about anything, not even Vietnam." Another wished Rusk "would use the same force in asserting his views to the President within the administration that he does in protecting the President against critics outside the administration."[75]

Certainly, Rusk's prophecy to Ball when they first entered office was amply fulfilled: "You and I are going to have a miserable life. Almost everything that happens in the world touches U.S. interests, and, since at any moment of the day or night two-thirds of the world is awake, someone somewhere will be committing some outrage to cause you and me trouble." Ball later agreed, "Hardly a week went by in the next six years, that one of the four telephones by my bed did not ring sometime in the night, requiring an instant response and often prompt and frequently complicated action."[76]

In the Kennedy-Johnson years, and in memoirs written later, State was often made the scapegoat for foreign failures. The career staff's rejoinder was best given by John Campbell, a brilliant FSO as well as a cogent critic of the institution's faults. "It is easy for Kennedy and Johnson White House staffers to blame the shortcomings of their bosses' policies on bureaucratic incompetence and opposition but . . . it can at least be argued that the main errors of judgment in foreign policy in the past decade emanated from the White House itself."[77]

Both parts of Campbell's analysis are undoubtedly correct. Most of the blame for policy failures must go to the White House; but that very same White House, as judge and jury, would also find State Department performance inadequate. Ironically, the department's fault in the case of the Bay of Pigs and Vietnam—hardly one of opposition—was in failing to question a policy approved by the president and implemented by the Defense Department and the CIA.

6

The Contemporary State Department

Up to the 1940s, foreign policy had always been a low priority for the U.S. government, but the country's emergence as the world's strongest power required a more sophisticated and effective policy process. As a result, from 1945 to 1969 Washington experimented with a number of forms and innovations, and at the end of that period, State had attained the structure it retains down to the present.

In 1969, the newly elected Nixon administration began a third era by using the NSC and White House staffs to give the president an alternative analysis and decision-making center to the State Department. Nixon's successors tried, and failed, to revive the past when, at least in theory, the secretary of state and his agency held primacy. This outcome was, in part, the costly result of State's earlier inability to adjust successfully to changing requirements. Thus, at this point it is useful to digress from the policy process's historical development to explain State's structure in this third era of U.S. foreign policy organization, the 1970s and 1980s.

The State Department's hierarchy consists of three groups. At the top are the seven *principals:* the secretary of state, deputy secretary, four undersecretaries (political, economic, management, and security assistance), and a counselor. The middle echelon is formed by about 20 appointed assistant secretaries, or their equivalents, who run the bureaus and the approximately 110 deputy assistant secretaries who help them. Finally, comes the *working level:* the hundreds of office directors, desk officers, and other FSOs or civil servants who work in Washington. State itself has 24,000 employees, of whom over 10,000 are local citizens working in overseas missions. The Americans are split evenly between Washington and posts abroad.[1]

Each job's relative importance is largely determined by the ability of the person who holds it as well as his or her relationship to superiors. The deputy secretary may be his chief's partner or a suspiciously received president's man. The undersecretary for political affairs, always an FSO, may be a valued voice of experience or an unwelcome emissary of the career staff. Access to the top and to key meetings determines an assistant secretary's influence. In short, a successful official at any level must not only formulate proposals, but must also become a policy entrepreneur to "sell" them. Personal networks and bureaucratic maneuvering are all-important in this process.

What kind of people aspire to the taxing but fascinating positions available? Members of the foreign policy community—internationally oriented lawyers, professors, career officials, researchers, consultants, corporate executives, and journalists—meet in politics, government, conferences, and read each other's books and articles. They constantly change and exchange posts. Along the way, through all this interaction, they supposedly learn who works and thinks most effectively while they themselves gain bureaucratic skill and substantive knowledge.[2]

Political appointees have used an "operator's" personality to gain the confidence of the governing politicians. Being a successful operator is not necessarily consistent—it may even be contradictory—with understanding international problems. Operators tend to be aggressive and opinionated, spending their time building good relations with other powerful figures rather than studying the subject.

Unfortunately, the study of foreign affairs is more burdened with charlatanism than most professions. "All you need to pass yourself off as an expert," says one cynical State Department employee, "is a world atlas and a copy of *The New York Times.*" Many of them cultivate an air of dignified experience and discretion that hides a profound ignorance. Obfuscatory language covers over a lack of ideas; tight-lipped sobriety conceals an absence of inside information. Sadly, journalists and the general public often cannot tell the difference. Such phonies are evenly distributed among FSOs and appointees, as well as outside critics and consultants waiting not so patiently for their turn in the arena.

The most successful FSOs, particularly those who reach the higher career posts, combine operators' characteristics with career staff virtues. More often, however, friction occurs when the political operators' personality meets the permanent staff's bureaucratic mentality. The latter group is far more cautious, wedded to channels that the operator likes to short-circuit or manipulate, oriented toward neutrality and deference since it must coexist with different administrations, and slower to take responsibility. Many bureaucrats, of course, are quite skilled at getting what they want and some are would-be operators in their own right, but their methods are still quite different from appointees whose past environments stressed independent action and the open exercise of power.

In addition to operators and bureaucrats, a third group—scholars and area experts—evinces what might be called an outsider personality. These people, secure in their own field of knowledge, are often less interested in building personal alliances, gaining power, and making the kinds of moral and intellectual compromises that politics require. They are more willing to give their views frankly, with accuracy a higher priority on their list than action or conformity. Such individuals are among the most valuable and well-informed staffers, but they rarely rise to the top. As appointees, their effectiveness rests on whether they can apply their knowledge practically and work as part of a team.

There are no conspiracies embodied in the Council on Foreign Relations, Trilateral Commission, or other groups to propagate a single political line among the elite. Personality conflicts, competition for honors and posts, and ideological differences among the politically active create a whole range of viewpoints and counterposed alliance systems inside and outside of government. Still, the need for a basis of shared perceptions also carries the danger of excessive conformity, making dissent or even skeptical inquiry seem almost subversive. One official critical of the Vietnam War in the 1960s wrote, "No one was prepared to discuss why we persisted in a war . . . in pursuit of an objective that seemed every day to have less reality. Men with minds trained to be critical within the four walls of their own disciplines—to accept no proposition without adequate proof—shed their critical habits and abjured the hard question 'why' once they caught hold of the levers of power in Washington."[3]

Behaving responsibly does imply, especially for the career staff, working to implement even policies of doubtful value. Otherwise, U.S. diplomacy would hopelessly fragment.[4] But refusal to question policy may have more to do with career ambitions than with dedication to the national interest. The same is true with another diplomatic attribute: emotional detachment. It is shockingly easy to forget that foreign policy deals with real people and life-or-death situations. Peace of mind and objectivity sometimes require such an attitude, but this amnesia also endangers the humanitarian content of policy.

If policymakers were too sympathetic with the needs of other countries and peoples, they might not serve their own nation well. Yet foreigners' lives and desires are too often an abstraction or of low priority to policymakers, making U.S. actions brutal, ignorant, and ineffective. Too often, the process becomes a game conducted for the benefit of policymakers' careers. Under such circumstances, a high official's "realism" can be quite irrelevant to reality.

When process makes bureaucratic and personality considerations dominant, this may produce the populist nightmare vision of icy, isolated men flinging around continents or playing global chess. More often, reality coincides with the insider view of overworked, harassed people trying to cope with an endless Chinese-firecracker series of problems. "This was the almost invariable pattern: a crisis occurs, and everybody stays up all night and fires off cables around the world and worries like hell," a former undersecretary of state recalled. "This goes on for two or three hectic days. You come up with some answers as best you can . . . and the crisis passes. Another crisis occurs two months later, then the third and fourth, and after you have about five crises, you have a policy."[5]

Historically, political appointees have generally been corporate lawyers or investment bankers, since it was thought that the best foreign policymakers were cultured, cosmopolitan generalists experienced in dealing with a wide range of people and crises. Many of them were energetic, adept at analyzing information and producing decisions, willing to take initiative and responsibility. Yet they often had little specific knowledge about other peoples and world views, particularly outside of Western Europe. Self-confidence could easily become arrogance and blindness. Lawyers often saw international affairs as a courtroom competition, bound by rules in which victory was rewarded for the best brief, rather than a rough-and-tumble, life-and-death political contest. FSOs like to think that, once in office, the political-type lawyers and the outsider-type academics learn that the careerists know best about navigating the complex maze of foreign policy.

In the last 20 years, growing numbers of university-trained specialists have filled top policy jobs. They are more knowledgeable than the lawyer/businessman type and are often more politically astute, innovative, and supportive of administration ideology than the career person. Such types, including Henry Kissinger, Zbigniew Brzezinski, Walt and Eugene Rostow, and McGeorge Bundy, among others, are bureaucratically skilled, despite their outsider characteristics. But reserved, self-consciously dignified FSOs think this type possesses, as one staffer claimed, "distressing personality failings, of which absurd presumptions and self-importance, together with a gauche abrasiveness in dealing with people, are the most common." Worst of all, he tends to "treat the constructs and abstractions of his own creation as reality—harmless and indeed desirable in an academic context'' but disastrous ''on the prudent, pragmatic evolution of policy.''[6]

Diplomats have their own strengths and weaknesses as policymakers—experience with issues and the training to make balanced assessments—though career reflexes may undermine their effectiveness in a highly political game. The question of whether diplomats make good policymakers has been hotly debated. Chester Bowles felt strongly that the top people at State should not be FSOs, but rather those ''who were prepared to take risks and to think imaginatively and freshly.''[7] Veteran FSO Charles Bohlen disagreed: ''My experience had been that Foreign Service Officers were just as willing, in fact more so, to stick their neck out than were political appointees.''[8]

Roger Hilsman, a good example of the outsider/academic type, served at two assistant secretary positions in the Kennedy administration. He doubted that an FSO ''can adequately do a 'hot-seat' Washington job'' because he would have to act in such an aggressive and partisan manner that his career could not continue under another president. ''If he behaves in a way that permits him to be used in the next administration as an ambassador, he will not have done that job adequately.'' Hilsman thought policymakers must battle other agencies: ''You've got to tread on their toes; you've got to make them toe the mark. You've got to fight with the Joint Chiefs of Staff. You have to do hard, political things. You have to get out in Congress, and you've got to make some enemies.'' If an FSO were to behave this way, he would destroy his career. Bohlen and Kennan were at times exiled because of their activist role in top posts.[9]

In short, there is no ideal background or personality type for policymakers; a mixture is necessary. It is equally impossible to generalize about the quality of those who staff the policymaking apparatus. The most dedicated, competent, and energetic people can be found at all levels alongside pompous, ignorant deadwood.

The contrasts among subordinates pose another problem for the secretary of state, whose power has already been diminished by the diffusion of authority among other institutions and policymakers, nevertheless, his task remains one of great importance and complexity. He must advise the president, coordinate and compete with other agencies, and represent administration policy to foreign countries, Congress, the media, and the public. He is responsible for managing the department and determines the relative influence of its various bureaus and officials. There are endless facts to digest, conferences to attend, people to see, and papers to read. If this job is not beyond an individual human being's physical and mental capacity, it must come very close to the limit.

The 133 U.S. embassies around the world included in the secretary's domain have three main jobs: As representatives, they maintain American presence in each country, facilitate communication between the two governments, explain Washington's policies, protect U.S. interests, and seek local support for American initiatives and goals; as negotiators, they carry out bilateral exchanges over treaties and other agreements; as reporters, they observe and analyze events and viewpoints within the host country, sending home information and advice.

Today, the embassy's reportorial duty is often ranked foremost among its responsibilities. A failure to explain developments to the home office, or Washington's refusal to listen to warnings, is the prelude to disaster. A competent envoy will assemble a good team and pay attention to its conclusions. Mediocre ambassadors, perhaps an arrogant, ignorant political appointee or a worn-out career man, are tempted to tell superiors what they want to hear while stifling differing views within the embassy.

On one hand, FSOs consider many noncareer ambassadors as transient and poorly informed amateurs. FSOs passionately resent outside appointments that inevitably block their own promotion to ambassadorships. On the other hand, defenders of political appointees believe FSOs are too eager to maintain cordial relations with the country in which they serve, becoming more concerned with that nation's requirements than with U.S. interests. They also claim that appointees have better access to the White House and, consequently, more credibility with the host country. But the practice of using such posts to pay off campaign contributors and political supporters has produced some ambassadors of extremely low quality.

The embassy staff is divided into four parts. The political section deals with the local foreign ministry and studies the country's government, parties, and policies. The economic section monitors domestic and bilateral business, trade, and financial matters. Consuls protect U.S. citizens abroad and issue visas to foreigners. Administrators keep the embassy running, ministering to the needs of all agencies represented, obtaining required supplies, managing personnel, and hiring local employees.

From all sections of the embassies, Washington receives a tidal wave of cables and airgrams containing reports, requests, comments, memos, or diplomatic notes from other governments. Ambassadors and lesser officials from foreign embassies in Washington send their proposals, inquiries, offers, and warnings directly to the State Department. Meanwhile, the president and his chief lieutenants give orders that must be implemented and informed by the middle and working levels.

In this era of rapid communication and travel, ambassadors and embassies have lost some importance as officials, including the secretary of state, seem to set out from Washington daily to handle every detail of foreign affairs. Certainly, U.S. embassies abroad are routinely passed over as Washington or special envoys conduct negotiations, gather information, and maintain contacts. FSO Thomas Hutson resigned his post in the Moscow embassy in 1980, commenting, "We don't need an ambassador in Moscow . . . because he has nothing to do." The peripatetic Secretary of State Henry Kissinger allegedly claimed, "Ambassadors don't count anymore."[10]

In contrast is the view of retired Ambassador Ellis Briggs: "The person who is content to carry messages is in fact a messenger boy, but he has no business being an ambassador. How an envoy delivers a message can be as important as the communication itself." For two states to understand each other's position and objectives, particularly where suspicion or cultural differences intervene, is a surprisingly difficult task. As embassy director and international intermediary, the ambassador plays an essential part in the success or failure of U.S. efforts in the country.[11] At State, when an FSO begins looking for his next post, the quality of individual ambassadors and embassies, as well as the living conditions in different countries, are well-known and frequently discussed matters.

Assistant secretaries and deputy assistant secretaries who run the bureaus in Washington are often chosen from outside State. They may be "professional amateurs" who usually follow nongovernment careers but are named to posts when their party is in power, or FSOs who can later become ambassadors if the next administration is not too hostile about their service to a predecessor's policies. An assistant secretary must build alliances with other agencies, and even in Congress, to dominate the policy process in his jurisdiction. When his issue area does not draw the attention of higher-ups, he will have extensive control over U.S. policy in that sphere; when his beat becomes a crisis area, he may brief and influence superiors but will lose direct authority.

Political appointees often consider FSOs, by training or concern for their future career, too timid or too uncommitted to administration ideology to function effectively in these jobs. As one former official put it, "An assistant secretary must be tough, expendable, willing to take risks and force the implementation of decisions down through the department desks."[12] On the other hand, FSOs frequently consider political appointees as too opinionated or ignorant. Nevertheless, the two groups usually try hard to work together amicably.

On the working level, the office directors and deputy office directors are the link between the line officer and the middle level. This involves managerial skill—a quality different from the FSO's usual earlier experience at lower-level jobs—overseeing the office's work and deciding what problems and data to take to superiors. Such slots may be held by an able officer on his way up the ladder or it can be a plum given to a senior, but relatively ineffectual, one.

For FSOs, a country desk job in one of the regional bureaus is a preferred position. Most of them are in their middle to late thirties, rotated in from the field, though they may not have served in the nation they cover. A desk officer reads the cable traffic from the U.S. embassy, drafts instructions for it, writes background papers, and answers questions for the higher levels. In addition, he talks with other bureaus, agencies, and outside interests concerned with his country: Congress and Congressional Relations, journalists and Public Affairs, AID and USIA, the Defense Department and Politico-Military Affairs, businessmen and the Department of Commerce or Agriculture, the CIA and the Bureau of Intelligence and Research, as well as an endless variety of other people.

His main focus is not as an expert on the country he covers, but as a data bank on bilateral relations between the United States and that government. Haphazard personnel shifts mean the desk officer may not know very much about his subject when he starts out; rapid rotation can deny him the time to learn enough or shuttle him off to an unrelated post as soon as he becomes knowledgeable. At the same time, U.S. embassies sometimes feel themselves forgotten castaways, cut off from department activity. The desk officer can be a lifeline communicating their viewpoints and information to decision makers and explaining policy developments to them. A good desk officer can even supply bootleg copies of Washington memoranda not intended for dispatch to the embassy.

All these officers are literally drowning in paperwork—about 3350 incoming and outgoing cables daily plus the pouches delivered by 63 diplomatic couriers. Reports pour in from the 133 U.S. embassies and 101 consular posts, plus the liaison office with Taiwan and interest sections in countries—Cuba and South Yemen, for example—with which the United States has no diplomatic relations.[13]

The Executive Secretariat attached to the secretary of state's office decides which papers to send the overworked principals, produces summaries for them, and acts as a catalyst for moving policy and implementation forward, helping to coordinate 19 bureaus and other offices including protocol, press relations, the legal staff, and the U.S. mission to the United Nations. Since jurisdictions overlap, persuading the relevant offices to compromise and "clear" proposed decisions is no easy task.

State's functional bureaus deal with administrative duties, liaison tasks, and technical issues. The ever-unpopular Bureau of Personnel is charged with handling assignments and evaluations. The Bureau of Administration copes with budget and management problems as well as with communications, supplies, housing, and other services for overseas posts. Its "housekeeping" responsibilities are expansive since they include maintaining the large non-State contingent in overseas embassies: employees from the U.S. Information Agency (USIA) and the Agency for International Development (AID) as well as military attaché and advisory groups, CIA officials, and representatives of many other agencies. In 1971, less than 15 percent of all U.S. government personnel stationed abroad (excluding military forces) were State Department employees, and a decade later the figure was still only 23 percent.[14] In some places, the very size and visibility of the U.S. presence can be provocative. Iran was one case in point. The construction of an enormous high-rise embassy building in overstaffed Cairo—despite warnings that such a visible profile would stir local anger—shows a refusal to learn vital lessons.

The average American has the most contact with the Bureau of Consular Affairs, which distributes millions of passports, protects Americans living or traveling abroad, screens visas for foreign visitors or immigrants, and was once responsible for State's internal security. Given the many foreigners wishing to come to the United States and the complex restrictions involved, this last task involves a huge amount of labor. Many Americans have complained about consuls' failures to provide more aid, sometimes due to the diplomats' desire to appease the local government—although, in fairness, tourists often have unrealistic notions about Washington's ability to intervene with foreign courts and procedures.

Other bureaus maintain liaison with important institutions. The Office of Congressional Relations copes with the legislative branch, a priority symbolized by the location of its assistant secretary's office just down the hall from the secretary of state. High department officials spend a remarkable amount of time testifying before Congress. The office tries to keep senators and representatives happy by organizing briefings, providing responses to their constituents' letters, arranging overseas trips, and helping them prepare speeches and legislation.

The Bureau of Intelligence and Research (INR) is State's link to the intelligence community—the CIA, the Defense Intelligence Agency, and the National Security Agency—and provides a group of people divorced from decision making who have in-depth expertise. Many of its analysts are civil servants who spend their careers following a particular country. Generally, the INR country officer is best informed on events and leaders in his jurisdiction, while the regional bureau's desk officer is most knowledgeable about U.S. policy toward that country. Because INR is seen as a specialized and research-oriented bureau outside the policymaking chain of command, many FSOs consider an INR assignment detrimental to their careers.

The ''Secretary's Morning Summary,'' produced in conjunction with the Secretariat, is INR's most important product. It gives the secretary of state a survey of the latest world developments, plus key incoming cables chosen by the regional bureaus, and brief analyses or longer background pieces by INR analysts. William Rogers preferred oral briefings; Henry Kissinger demanded full documentation, so his version was much larger. Cyrus Vance wanted his summary ready by 5 A.M. with two four-page sections: the first with six to eight items drawn from incoming messages, the second with two or three longer items. Alexander Haig had his summary cut shorter.[15]

Charged with liaison to the Defense Department, the small Bureau of Politico-Military Affairs (PM) has been gaining increasing importance. It monitors foreign wars and U.S. military actions abroad, military aid, base negotiations, and arms control issues. It advises the secretary of state on his relations with the secretary of defense—a key link in the policy process—and works closely with the semiautonomous Arms Control and Disarmament Agency (ACDA) and the U.S. team negotiating with Moscow over nuclear weapons. During the Carter and Reagan administrations, Politico-Military developed a particularly close relationship with the secretary.

The Bureau of Public Affairs (PA) has a high profile since it is responsible for dealing with the U.S. citizenry and media. It publishes documents, provides speakers, and answers public correspondence. To promote its policies, the government provides much information to journalists. On the other hand, reporters are searching out stories about political events and internal debates that the administration would prefer not to divulge. Officials know that one misstep with the media can destroy their careers, but they also have an interest in using the press to express their personal viewpoints or to popularize a factional position. As a result, attempts by presidents to restrict media access to policymakers are as unsuccessful as they are common. Leaks are an integral part of the system.

Secretary of State Dean Rusk once told CBS anchorman Walter Cronkite that the gulf between State and the media was inevitable since the business of diplomacy was to prevent crisis headlines, not to make them. To reporters, ''no blood'' was ''no news.''[16] But Henry Kissinger took a different approach. By systematically cultivating reporters and feeding them what he wanted broadcast, Kissinger used the media as an important part of his diplomacy and as a way to build his power and prestige.

PA's most important activity is the daily press briefing. By disseminating the department's point of view, PA sends signals to other countries and seeks public support for State's efforts; but inquiries often receive answers so carefully worded that numerous probing questions are required to ascertain the most basic facts. PA's job is to say only what it is told to say. Consequently it is often caught between reporters, who demand to know more and are adept at finding it out, and officials at other bureaus, who limit spokesmen's information or circumvent them with direct media contacts.

In recent years, high officials have increasingly resorted to background briefings where they cannot be quoted. Reporters are often contemptuous and readers are puzzled by this approach, but if a secretary of state makes a statement under even the thinnest of covers, other governments will not have to react to it as they would if he were speaking for the record. In fact, *The New York Times* and *The Washington Post* can be almost semi-official bulletin boards where policymakers can see what colleagues are saying and doing. Foreign leaders read these newspapers carefully, allowing the State Department to use them for launching trial balloons and sending messages.[17]

Perhaps the part of State with the most innately difficult jurisdiction is the Bureau of International Organization Affairs (IO), which handles U.S. involvement in the United Nations, other international organizations, and about 800 conferences a year. Its central problem is with the post of U.S. ambassador to the UN, whose rank and political assets are superior to those of the IO assistant secretary. This dichotomy, and the problem of coordination between Washington and the U.S. Mission to the UN in New York, has led to some embarrassing errors and clashes in recent years.[18]

Ambassadors Andrew Young and Jeane Kirkpatrick, nationally known figures with special access to presidents Carter and Reagan, respectively, are symbols of the ambiguous chain of command. Following this pattern of conflict, Kirkpatrick collided with Secretary of State Alexander Haig and IO. Kirkpatrick complained of the effort "to dictate tactics from Washington, by people who have no personal knowledge or 'feel' for the politics of the issue." She called for the Mission to take the lead, since IO's supremacy "practically ensures that mistakes in strategy and tactics will occur."[19]

Other parts of State deal with special topics: the Bureau of Educational and Cultural Affairs (to 1978), the Bureau of Oceans and International Environmental and Scientific Affairs, the Bureau of International Narcotics Matters, and the Bureau of Human Rights and Humanitarian Affairs.

The Bureau of Economic and Business Affairs (EB) is involved in issues concerning international trade, finance, resources, transport, and telecommunications. Of all the department's bureaus, EB has the most contact with domestic interests, which are closely touched by international economic developments. It has been the most hard-hit by other agencies' turf aggression: the U.S. trade representative (a cabinet-ranked presidential adviser who took away many of State's powers in negotiating international trade agreements), Treasury, Commerce (which has its own overseas representatives, the Foreign Commercial Service), Labor, and Agriculture (which also has its own attaché system). President Johnson's EB Assistant Secretary Anthony Solomon was the last official in that position to dominate international economic policy.[20]

The Policy Planning Staff (SP) is another part of State that has found it hard to live up to its potential. SP, intended as a planning, research, and idea-generating group serving the secretary of state and coordinating State's input into the NSC, went into decline after a brief period of success under Truman. SP's small group of FSOs and outside appointees are expected to do longer-range thinking, an art held in low regard at an institution where day-to-day considerations predominate. A lack of direct authority also limits its influence. Therefore, SP is often used to reinforce other bureaus and write speeches. But if the director of Intelligence and Research, Politico-Military, or Policy Planning develops a close relationship with the secretary, he may be called on more often to conduct special studies and propose policies, winning his bureau much greater influence.

The heart of the policymaking apparatus consists of the five regional bureaus: Europe, the Near East and South Asia, East Asia and the Pacific, Africa, and Inter-American Affairs.[21] They are directly responsible for analyzing events, dealing with foreign states, and developing options. Their assistant secretaries chair interagency regional groups that feed information to the NSC and implement White House decisions.

While the department often discourages FSOs from specialization, a number of them serve much of their career under the jurisdiction of one of the regional bureaus. The selection will be affected by the officer's own interests, estimate of career possibilities, language skills, and State's needs, as well as by the Personnel Bureau's whims. There is no shortage of stories about people with advanced degrees in Soviet studies and fluency in Russian being sent through a long round of Latin American posts, or Japanese-language speakers and specialists being dispatched to Africa. Breadth of experience is certainly necessary, but this procedure may make it hard to find someone with real knowledge of a country's politics actually serving in the relevant embassy or desk.

The Bureau of European Affairs (EUR) is historically the largest and most prestigious of the regional bureaus and retains the highest esprit de corps. Europe—both Western Europe as allies and the USSR and Eastern

Europe as a foe—is always seen as a priority area for U.S. policy. This is the central irony of EUR's position: Because of Europe's importance, management of relations is rarely left in EUR's hands. There have been few memorable EUR assistant secretaries. Furthermore, the desirability of ambassadorships to European countries draws a far higher percentage of political appointees to those capitals than anywhere else, making it more difficult for EUR FSOs to rise to the top.

Despite these problems, EUR still considers itself an elite within an elite and as State's foremost practitioner of traditional diplomacy. If the old stereotype of Foreign Service anglophilia—though West German tours of duty seem more important nowadays for rising Europeanists—and snobbishness is still alive, it is in this bureau. EUR has few internally divisive issues but many outside competitors.[22] Great bureaucratic struggles from the 1930s through the 1960s pitted EUR, favoring alignment with its colonialist clients, against the Third World bureaus. This problem has faded in recent years but periodically reappears, as when EUR and the Inter-American Affairs bureau clashed over U.S. policy for the 1982 Falklands War between Britain and Argentina.

Since the late 1940s, U.S.-European relations have focused on the North Atlantic Treaty Organization (NATO) alliance, nuclear and conventional military strategy, the balance between U.S. leadership and European independence, and on relations with the Soviet bloc. Economic and trade questions have been particularly important as have "third country" issues involving coordination of U.S. and European policies toward the rest of the world. A good EUR assistant secretary or deputy assistant secretary must be adept at consoling European complaints over a lack of U.S. consultation with the allies.

While relations with Western Europe take up most of EUR's time, it is also responsible for the USSR and Eastern Europe. Surprisingly, given the importance of those countries, they are dealt with by a single office under a deputy assistant secretary. Beginning in the 1930s, the department produced an impressive corps of Soviet experts including Charles Bohlen, Loy Henderson, George Kennan, Foy Kohler, and Llewellyn Thompson. Although State still has many officers trained in Soviet studies and the Russian language, the rise of the NSC staff and the CIA provided alternative sources of expertise. Given Soviet secrecy and disinformation, analytical and intelligence methods usually tell policymakers more than ordinary diplomatic reporting and conversation. Bilateral negotiations are of such importance as to be closely controlled by the White House.

This pattern means that the U.S. embassy in Moscow and State's Soviet section are frequently bypassed, so that those directing U.S. policy understand less about their adversaries than do Soviet leaders. To cite only one example, Ambassador Malcolm Toon recalls that when Secretary of State Cyrus Vance visited the Kremlin in March 1977, Soviet Foreign Minister Andrei Gromyko was flanked by experts who spoke fluent English and had over 20 years' experience dealing with the United States, while Toon was the only one in the U.S. delegation who even spoke Russian.[23]

Observers and FSOs currently rank the quality of the Bureau of Near Eastern and South Asian Affairs (NEA) on a par with that of EUR. The bureau oversees a great swath of territory, stretching from Morocco to India, which, given its particular features, requires special knowledge of convoluted problems and a proficiency in difficult languages. Consequently, it is the most ingrown regional bureau, with fewer outside appointees and more continuity through changing administrations.

NEA's principal preoccupation is with the Arab countries, through which a group of specialized FSOs (often called "Arabists," although this term, strictly speaking, applies only to the minority fluent in Arabic) circulates without having to make great cultural or linguistic adjustments from one to the other. As a result, NEA's non-Arab states—India, Pakistan, Iran, and Israel—are often treated as secondary interests. It is simply not to an FSO's advantage to specialize in one unique state when he can be seen as an expert on over a dozen.

The Middle East has been an area of great White House, NSC, and State Department activity in recent years. The NEA assistant and deputy assistant secretaries work closely with the highest decision makers on the hottest issues and crises. Such proximity to power and the limelight means NEA officers work hard and have opportunities to build personal relations with senior officials. These factors have built up NEA's reputation, attracting some of State's best young recruits, who know the bureau is where action and promotions can be found.

Still, tendencies toward provincialism and clientism remain, meaning that while NEA's technical and reporting work is much in demand, its advice is frequently ignored. The emotional Arab-Israeli conflict, increasingly the bureau's main business since the late 1940s, has made NEA itself very controversial. The staff's view that NEA's main task is maintaining good relations with the Arab world makes them nervous about policies that might antagonize those countries. Thus, something of a bureaucratic bias works against U.S. support for Israel, an alliance that tends to make their jobs more difficult and even dangerous. The bureau's sensitive dealings with domestic politics tends to make its officials feel beleaguered and somewhat bitter, and they frequently complain that the

White House, Congress, and the pro-Israel lobby constantly overcome their preferences.

Because of opposition to a number of U.S. policies, NEA's territory has been the site of most mob and terrorist attacks on embassy personnel—in Lebanon, Iran, Afghanistan, and Pakistan, for example—and FSOs daily hear local notables criticize U.S. actions. Since personal relationships are important for an effective work and social life, the diplomat learns to express sympathy and is under pressure to apologize for his own government's decisions, and with NEA FSOs spending their lives studying and living in the region, they often develop emotional commitments.[24] Therefore, Arabists have historically refused assignment to the U.S. embassy in Israel because such a posting might damage their careers. Some also have gone on to lucrative jobs or consultancies with Arab companies or U.S. corporations doing business in the region.

Given these problems alongside the region's complexity and mutability, State's reading of economic, cultural, religious, and ideological factors has frequently been inadequate. Middle East reporting, said Raymond Hare, an FSO who served as NEA assistant secretary in the 1960s, is like being hypnotized by a monotonously turning wheel, making it easy to miss an important change in the pattern of events.[25] There is, however, much variation among individuals. Today, the most interesting work is that of dealing with both sides in the Arab-Israel conflict and playing a role in the difficult, tense negotiations that have preoccupied NEA since 1967.

For decades, while NEA was still a backwater, the Bureau of East Asian and Pacific Affairs (EA) ranked beside EUR as the most prestigious regional bureau. U.S. interests in Asia consistently made it Washington's number-two area of concentration, but the McCarthy-era attack on EA's China experts was a traumatic experience, and many fine FSOs were purged or transferred. Further, those interested in China could not even visit the country for a quarter-century. Just as an earlier generation of Soviet experts, before the opening of U.S.-Moscow relations, avidly sought assignment to the Riga, Lithuania, listening post, would-be China hands served in Hong Kong. The corps of specialists was only gradually rebuilt.

When Indochina became the main focus of U.S. policy during the Vietnam War, EA had few area specialists. Hundreds of FSOs served in Indochina; EA assistant secretaries in the Kennedy and Johnson administrations devoted overwhelming attention to the issue, though the Defense Department and NSC held the major brief for directing U.S. activities there. During the early Nixon years, EA Assistant Secretary

Marshall Green put his senior deputy assistant secretary, William Sullivan, in charge of all Vietnam aspects of the bureau's work. By the time EA gained expertise on Vietnam, however, the U.S. withdrawal and the war's end once again returned this area to relative obscurity.

Meanwhile, EA's other main issue was preempted by National Security Adviser Henry Kissinger, who did not tell it much about his negotiations to reopen U.S.-China relations. During the Carter and Reagan years, East Asia was a medium priority for the more chronic, rather than dramatic, problems concerning China and U.S.-Japan economic competition. The bureau was now dominated by China specialists who lobbied for strong relations with their new client, Beijing (Peking). At the same time, EA was starved for Japanese expertise and State had little influence on commercial issues. The Department's Japan experts complained that their skills were unappreciated.[26]

The Bureau of African Affairs (AF) has almost always been the least important of the regional bureaus, a distinction due mainly to the nature of the region's problems and the low priority accorded it by top policymakers. The White House under presidents Kennedy and Carter tried to make Africa a special area of attention, enhancing AF's energy and morale, but the relative lack of U.S. strategic and economic interests in the continent has resulted in AF's difficulty in gaining a hearing among senior officials. A crisis involving global factors like Soviet involvement will be taken out of AF's hands, as happened with the Congo in the 1960s, the Angola civil war of the mid-1970s, and the Ethiopia-Somalia conflict later in that decade. AF retained a large measure of autonomy over other questions and southern Africa issues—which attracted little sustained White House interest—and generally lobbied for a tilt toward the positions of its main clients, the black-ruled African states.

Otherwise, AF's human rights and development matters, dubbed "humanitarian" issues, are short on funding or the high-level backing needed to galvanize the rest of the bureaucracy, unless a powerful patron brings them to White House attention. UN Ambassador Andrew Young, through his access to President Carter, successfully lobbied on African issues. Chester Crocker, an area specialist who became the Reagan administration's assistant secretary, gained relative independence on African issues, given both his own skill and his superiors' disinterest.

Almost all African capitals are hardship posts, meaning dangers to health and problems in finding schools for FSO children. Still, AF holds some advantages, since small, growing posts can mean faster promotion and better chances for an ambassadorship. At the same time, AF's unspoken problem is the policymaker's patronizing attitude toward the continent. One Africa expert recalls that colleagues would make drumming noises on the table when he entered a meeting. A middle-level Europeanist, reading cables during a recent crisis, suddenly blurted out, "Where is Chad anyway?"

Everyone knows where Latin America is, but the Bureau of Inter-American Affairs (ARA) is not well regarded at State. Despite disclaimers, Washington takes the Western Hemisphere for granted unless, as in the case of Africa, crises develop a U.S.-Soviet dimension. Historically, ARA has presided over a relatively slow-paced area of foreign policy, but its posts offered good living conditions, an easily learned language, and an active social life among the local elite. With the onset of terrorism and economic shifts, luxurious living standards have declined, and the White House has alternated between neglecting and purging the bureau. While some FSOs like the region, many shun ARA or leave it as soon as possible for more exciting or rewarding fields.

To make matters worse, ARA's assistant secretary position has been a revolving door. Liberal adminstrations have seen ARA as soft on the reactionary status quo and incumbent dictators, while conservative presidents suspect the bureau of sympathy with revolutionary forces. In reality, ARA has remained fairly constant while U.S. regional policy shifted more radically through successive administrations then it did for any other part of the world. All these factors wear out and demoralize ARA personnel.

From the American revolution down to 1940, the simplicity and sparseness of U.S. interests around the world made for an uncomplicated policy process. Routine business was left to career officers; rare important matters were handled on an improvisational basis by the president or the secretary of state. America's rise to a central role in world affairs necessitated a more sophisticated policy system involving new channels and agencies, but with the secretary of state in at least nominal control.

The Nixon administration marked the onset of a third era. The White House believed that the NSC staff furnished an institutional alternative displacing State; other agencies also demanded an equal or greater voice than the department. While Henry Kissinger was strong enough to maintain primacy as both national security adviser and secretary of state, succeeding presidents faced multisided, confused battles for control that threatened to make policymaking unmanageable.

7

State of Decline: The Nixon-Ford Administrations

1969–1976

The Nixon administration forged a new policy system based on longstanding criticisms of State from the White House, Congress, academics, and the media. The Jackson subcommittee hearings and a score of other panels and reports had failed to effect reform and the NSC staff, Defense Department, CIA, and other agencies had grown bolder, challenging State's tottering preeminence.

Now State was shoved into a clearly subordinate status. National Security Adviser Henry Kissinger dominated the system through a close partnership with the president, a relationship previously the secretary of state's exclusive domain. From 1969 to 1973, Kissinger made his NSC staff the power center, excluding Secretary of State William Rogers and the department from important decisions and information. In the fall of 1973, Kissinger was given Rogers's job and simply transferred his team and method to State's office building.

In subsequent administrations, when national security advisers found Kissinger's monopoly on authority hard to match, State made a partial comeback; but the department never regained even its pre-1969 position. At best, it could only win a costarring role in the bureaucratic battles around the president. The resulting power vacuum sharpened the conflict, creating a level of confusion and competition that has characterized the policy system down to the present day.

The reduction of State's influence is not implicitly bad for effective policymaking. Indeed, White House mistrust of State and its search for alternatives have been familiar features of the Washington scene for so long as to seem the norm rather than some heretical deviation. Furthermore, Kissinger mostly used State's personnel on his NSC staff and promoted a group of able officials who would play important roles in both the Carter and Reagan administrations.

Circumventing the bureaucratic apparatus of the department gave the White House far greater maneuverability and allowed a consistency of strategy and a much finer tuning of tactics, perhaps best seen in the opening of relations with China. But this approach also destroyed State's function of anchoring policy—forcing more careful consideration of risks and pressing an array of issues on a leadership that preferred to concentrate on pet goals. Many talented people were excluded from making any contribution to foreign policy. Kissinger created more bitterness, even hatred, toward himself at State than any other secretary in U.S. history.

Nixon and Kissinger achieved their goal of mastering the bureaucracy and tested the theory that foreign policy would function better if State were pushed out of the way. Kissinger's philosophy, best articulated in a 1968 lecture, warned that "the actual decisionmaking process leads to a fragmentation of the decisions. . . . A series of moves that have produced a certain result may not have been planned to produce that result." Bureaucrats, according to the Kissinger theory, are set in their ways, narrow in their focus, and incapable of thinking strategically. "The day to day operation of the machine absorbs most of the energy" of State's leaders "and the decisions that are made depend very much on internal pressures of the bureaucracy."[1]

The president, Kissinger continued, was besieged by people arguing different positions and it was impossible for him to know whom he should heed. Even if Kissinger persuaded the chief executive that "his whole bureaucracy was wrong and I was right," the president would still not be able to implement those suggestions. "Unless you can get the willing support of your subordinates, simply giving an order does not get very far." So, since "management of the bureaucracy takes so much energy and . . . changing course is so difficult," important decisions could best be made in "a very small circle while the bureaucracy happily continues working away in ignorance. . . ." An unpopular decision may be fought by "brutal" means, such as leaks to the press or to congressional committees. Thus, the only way secrecy can be kept is to exclude from the making of the decision "all those who are theoretically charged with carrying it out."

Under Kennedy and Johnson, Kissinger argued, channels had been vague and ineffective; dissenting officials had competed in pushing their favored policies. What was required was a way to have "the procedural regularity of Eisenhower with the intellectual excitement of Kennedy." This is precisely what Kissinger tried to do. He also borrowed a page from McNamara, who, he claimed, "got control of the Defense Department by flooding the various agencies with questions which they could not answer and which gave him good information. It took several years to bring home to them that the usual bureaucratic double-talk wouldn't go."

Finally, Kissinger saw State as wanting "what is negotiable" over "what is desirable," preferring short-run success over more serious objectives. "State Department training is in the direction of reporting and negotiation, not of thinking in terms of national policy. They are trained to give a very good account of what somebody said to them. They can give a much less good account of what this means."[2]

So Kissinger, at Nixon's direction, set out to build a small, efficient NSC staff, outside State's complex structure and burdened routine. State would be kept busy producing information papers, but isolated from important matters in which it might seek compromise and good relations with other countries as ends in themselves. Americans now needed to play a tougher, more strategy-minded game by rules long used by the Western Europeans and Soviets. "For the first time in American experience," Kissinger wrote in a Bicentennial article, "we can neither escape from the world nor dominate it. Rather we—like all other nations in history—must now conduct diplomacy with subtlety, flexibility, persistence and imagination."[3]

Kissinger's sense of realpolitik stood on its head the traditional American concept that the best and most successful diplomacy operated by minimizing conflict. As Kissinger explained his differences with Secretary of State William Rogers, "Rogers believed it desirable to reassure nervous adversaries that we intended them no harm. My view was the opposite, that once we were embarked on confrontation, implacability was the best as well as the safest course. Rogers thought calming the atmosphere would contribute to its resolution; I believed that it was the danger that the situation might get out of hand which provided the incentive for rapid settlement."[4] Repeated, similar debates took place in the post-Kissinger era between NSC Adviser Zbigniew Brzezinski and Secretary of State Cyrus Vance on the Iran hostage crisis, and between Secretary of State Alexander Haig and NSC Adviser William Clark over Israel's 1982 siege of Beirut.

Nixon, thinking himself a great realpolitik statesman, held views similar to Kissinger's. He had become famous as the nemesis of Alger Hiss, a symbol to many of State's effeteness and disloyalty. Nixon told Eisenhower that while he had met some fine FSOs on trips abroad, an "astonishing number of them have no obvious dedication to America and to its service—in fact, in some instances they are far more vocal in their criticism of our country than were many of the foreigners.'' Nixon thought FSOs were loyal to past Democratic administrations and evinced ''an expatriate attitude,'' saying he often heard them proclaim, ''I hope I never have to go back to the United States.''[5]

In phrases obviously descended from his earlier thoughts, President-elect Nixon promised, in 1968, to ''clean house'' at State, removing ''the routine men who have been the architects of the policies of the past'' and creating ''a Nixon-oriented State Department.'' Returning home from his first presidential trip abroad, Nixon told his staff, ''The trouble with too many FSOs is that their first loyalty is to the Foreign Service. Always playing it safe. Incredible.''[6]

This attitude was very much in evidence during his first talk to the newly formed NSC staff. ''We were in the Cabinet room, sitting there smoking our pipes,'' Roger Morris later recalled. ''And Nixon said, 'Look, I know what you have to deal with.' He pointed at the State Department. 'That place is impossible. . . . They were always screwing up when I was Vice-President.''' Nixon apparently did not stop to think that about 70 percent of the audience were FSOs on loan from State. Nixon concluded, ''Henry and I will end the war. I want you guys to run the rest of the world.''[7]

Nixon had decided that he would keep control of foreign policy in the White House through National Security Adviser Kissinger, draining power from the cabinet departments. Of course, the president always had the ultimate decision-making power, but the growth of his own staff for the first time made possible the exercise of this authority on a daily basis. From 1968 to 1972, the number of employees in the president's Executive Office increased by 20 percent to 5000. Only about 10 percent of them worked as White House staff and only about 10 percent of those were at the NSC, but the centralization of authority on foreign policy was even more marked than on domestic issues.[8]

Early in the administration, Senator Stuart Symington complained that Rogers was a laughingstock in Washington: ''They say he's only the Secretary of State in name.'' Kissinger was the ''actual architect'' of policy. Nixon lamely denied what came to be the capital's worst-kept secret.[9] When Rogers resigned in 1973, Nixon, himself jealous of Kissinger's power and prestige, intended to maintain his original structure by appointing Deputy Secretary of State Kenneth Rush to replace Rogers. In the midst of Watergate, however, Kissinger was virtually the administration's last remaining asset. So, in September 1973, Kissinger became secretary of state while also remaining national security adviser until well into 1975. The locus of power moved with Kissinger to State, but the authority remained personal rather than institutional.

Kissinger's and Nixon's mutual suspicions never prevented them from maintaining a close working relationship. In later years, Nixon described his diplomatic aide as an intellectual giant who was also moody, secretive, capable of outrageous private remarks, intensely protective of official prerogatives, and greedy in claiming credit for achievements.[10] All these negative characteristics applied equally to Nixon himself and to many others in an administration riddled with bitter enmities. Never in modern American politics has there been a government in which so much energy was focused on petty conspiracy and gratuitous character assassination. Kissinger's designation as chief of foreign policy, however, kept down the level of bureaucratic warfare, since he would win every fight with the president's backing.

Dissatisfaction with administration policy moved Congress to intervene more actively in foreign policy matters. From the late 1950s on, Rep. John Rooney's appropriations subcommittee annually tormented State in its budget hearings, and visiting congressional delegations sometimes harassed FSOs overseas. The new activism was more substantive and, ironically, Congress now became defender of State's prerogatives against Kissinger, since department officials, unlike the NSC staff, were answerable to Capitol Hill through testimony at hearings and Senate confirmation of presidential appointments. The Johnson administration's "credibility gap" on Vietnam had made Congress distrust the executive branch and begin building its own expertise through the Congressional Research Service and committee staffs. During the Nixon-Ford era, laws were passed requiring State to report on human rights in countries receiving aid, limiting presidential authority to engage U.S. troops, allowing Congress to block military sales, and cutting off bombing in Indochina.

The NSC staff's primacy also produced increased friction with State. Staffers can easily become officious, claiming to speak for the president. "Frequently," recalled Benjamin Read, who handled State's liaison with NSC for Rusk, "it's the most junior, new member of the White House staff who says that, and he doesn't have any more idea of what the President wants than you or I do." The recipient of such claims at State wonders how much urgency and authority lies behind it, while an NSC counterpart, dissatisfied with the response, complains, "Look what the State Department sent over this time."[11]

But relations were usually better on the staff level than on higher planes. Much of Kissinger's staff, after all, was from the State Department. Viron Vaky, his Latin America expert, had previously been acting assistant secretary for ARA and still maintained close contact with his old bureau. In contrast, NSC Africa specialist Roger Morris had strong policy differences with State's Africa bureau. Still, since Kissinger and Rogers were not on speaking terms—the latter even suggested reprisals against FSOs returning from NSC assignments—the working levels were often left to bridge the differences.

From his first meetings with Kissinger, Nixon expressed his preference for White House primacy. Weeks before the inaugural, Kissinger, Defense Department official Morton Halperin, and retired General Andrew Goodpastor, who had helped organize Eisenhower's system, devised the new framework. President Johnson's State-dominated Senior Interdepartmental Group was replaced as the main formulator of options and reports by the Kissinger-chaired Senior Review Group (SRG). The SRG requested studies and supervised the implementation of policy. In theory, this was at the behest of the full NSC, but in practice, all decisions were made privately by Nixon and Kissinger. The SRG was assisted by regional interdepartmental groups chaired by assistant secretaries of state and attended by representatives from the NSC staff, Defense, CIA, the U.S. Information Agency, and other relevant agencies.[12]

The SRG was the key body, and through it Kissinger judged the work of the interdepartmental groups, molded options for Nixon, presented issues to NSC meetings, and oversaw the carrying out of orders. He also accumulated control of other important committees: the Washington Special Action Group, for handling crises; the Forty Committee, for approving covert operations; the Verification Panel on Soviet adherence to arms control agreements; the Vietnam Special Studies Group; the NSC Intelligence Committee; and the Defense Program Review Committee. Formal organization was, as always, secondary: Major decisions were predetermined by Nixon and Kissinger, not by the debate and membership of these bodies.

One high-level official recalls, "It was hard to tell where Henry began and Nixon left off because Henry would use his name so freely." In contrast, the State Department was totally demoralized. An FSO complains, "It was demeaning to spend four years working for a secretary of state who always caved in rather than fighting and a president who missed no opportunity to criticize the Foreign Service."[13]

At meetings of the Washington Special Action Group in the White House basement, Kissinger's technique could be seen to full advantage. The members, including U. Alexis Johnson, representing State, arrived promptly at 2 P.M., but Kissinger would come late, apologizing that he had been with Nixon. The first subject was the latest leak, creating an atmosphere of mutual distrust among participants. Kissinger would then say he'd had a good discussion with the president, who wanted various things done. The discussion, then, was not to be about the substance of policy but only about implementation. When someone brought up the former, Kissinger replied that their job was to carry out presidential decisions and not to quarrel with them. An aide who continued to raise issues was not invited back.

The NSC staffers, as well as Kissinger himself, were quite different from the Eastern Establishment lawyers, bankers, and businessmen from wealthy families and prep schools who had traditionally ruled State. This new group was composed of foreign policy professionals with scholarly training and expertise, from poorer, often immigrant, backgrounds. Even many of the career State Department men on the NSC staff differed from the old pattern—like Viron Vaky, born in Texas of Greek immigrant parents, or Helmut Sonnenfeldt, a German-Jewish refugee, the staff's Europeanist.[14]

Particularly remarkable is the extent to which this team dominated policymaking positions not only during the Nixon and Ford years, but in the succeeding Carter and Reagan administrations as well. Consequently, their experiences with the Nixon-Kissinger policy framework affect U.S. diplomacy down to the present day. The staff's bipartisan background was demonstrated by the appointments given so many of them by the Carter administration. Under Carter, Vaky would be assistant secretary for ARA; Middle East staffer Harold Saunders, an earnest and soft-spoken NSC career official promoted to director of INR under Nixon, became assistant secretary for NEA. Kissinger's aides Fred Bergsten and Richard Moose became, respectively, undersecretary of state for economic affairs and assistant secretary for African Affairs under Carter. Anthony Lake became Policy Planning staff director.

Kissinger protégés attained their highest offices during the Reagan administration. Laurence Eagleburger, an FSO who was Kissinger's executive assistant at NSC and his deputy undersecretary for management at State, became Reagan's assistant secretary for EUR and later undersecretary for political affairs. Richard Allen, the right wing's representative at NSC until ousted by Kissinger, became Reagan's NSC adviser for a short time. Alexander Haig, Kissinger's deputy at NSC and later Nixon's chief of staff, was Reagan's first secretary of state. One Kissinger assistant, Peter Rodman, was made head of the Policy Planning Staff under Reagan, and another, John Lehman, became navy secretary.[15]

Most of them paid a high personal price. The work itself was high-pressured and exhausting, and Kissinger enjoyed humiliating subordinates. His devious behavior as well as policy disputes could offer staff members a choice of maintaining their powerful positions or their self-respect. Halperin, Laurence Lynn, and Daniel Davidson, who worked on

Vietnam, left the government in protest over Kissinger's methods, including the tapping of their telephones in his intense hunt for the source of leaks to the press. Kissinger's obsession on this subject apparently sprang from a conviction that there was an inevitable temptation for bureaucrats to use the media to sabotage policies they disliked. Other subordinates quit because they were cut out of policy. Eagleburger's health broke down under the strain. Lake, his successor, as well as William Watts and Morris resigned over the Cambodia invasion.

If competition within the NSC was fierce, so was the battle among the agencies. Kissinger's insecurity about his acceptance within the administration added to his aggressiveness. Despite his close working ties to Nixon, Kissinger knew that he was an outsider in the White House, detested by John Haldeman and John Ehrlichman, the president's chief staffers. Aware of the unprecedented nature and vulnerability of his position, Kissinger felt that the loss of even a single battle would undo him.[16]

Given Rogers's passivity, Undersecretary for Political Affairs U. Alexix Johnson played an important part in providing leadership at State. Johnson had joined the Foreign Service in 1935, subsequently serving as ambassador to several countries, including Japan, where he had first been stationed three decades earlier. According to one cabinet member, Johnson was "the only man at State who could frustrate Kissinger," daring to draft cables instructing embassies without going through the pyramidal committee system. But he was not able to convince Rogers to challenge Kissinger's system at the beginning, and State soon fell to a position where it was largely restricted to fulfilling NSC requests and to handling lower-priority areas.

The first thing Kissinger wanted was information. His National Security Study Memoranda (NSSM) ordered State and other agencies to produce studies on major issues. "It is the dream of the bureaucracy to give [the president] yes-and-no answers," Kissinger said. Unless prevented from doing so, he argued, the bureaucracy would get its way by only offering unattractive alternatives to its preferences. Therefore, he refused to accept papers that did not meet his standards. Many State Department officials viewed the rain of NSSMs as designed to keep them too busy to "interfere" in policymaking. Though this may have been part of his rationale, Kissinger also used the NSSMs as a crash course for himself on existing problems and possible remedies.

State's performance on these early studies disillusioned anyone in the NSC or White House not already predisposed to doubt the department's abilities. Before Nixon's first trip abroad, to Europe, State produced a tardy and overlong briefing book—a typical product of the clearance system—dealing with every conceivable issue on a lowest-common-denominator basis. A review of U.S.-USSR relations was sarcastically summarized by the NSC staff as presenting three options: (1) Ignore Moscow; (2) carefully negotiate agreements based on U.S. interests; or (3) seek all possible agreements. Obviously, the intention was to have the middle choice accepted, but it was of little use in making operational decisions. All such reports were returned for rewriting.[17]

Basic department philosophy dictated this coy approach. Career diplomats tend to see the world in terms of day-to-day problems to be coped with by clever mediation. Longer-term strategy, much less solutions, are impossible to formulate because of the large number of factors that are virtually impossible to predict and harder to control. This view, quite different from Kissinger's confident global focus, was why State had never taken planning very seriously. Administration appointees who failed to shake State's empiricism, Kissinger and his associates believed, would become its victims. This same factor of world view made individually able FSOs collectively incompetent.

Kissinger employed an integrated strategy to counter Moscow's actions, enhance U.S. credibility, and maintain a balanced détente designed to limit friction. The administration sought to manage policy so that even the smallest detail would be orchestrated toward the overall goal. Bombing strikes in Vietnam were to be carefully coordinated with diplomatic moves at the Paris peace negotiations, for example, or trade and passport regulations on China were to be adjusted in light of progress on relations with Peking.

This approach required secrecy, superb timing, and efficiency. Nixon and Kissinger believed the State Department was incapable of the required performance and circumvented it in a hundred different ways. For instance, Kissinger told U.S. ambassadors visiting Nixon not to take notes lest they detail the meeting for the secretary; and State was assigned studies without knowing the research's purpose. The White House also developed its own secret back channel to Moscow using Soviet Ambassador Anatoly Dobrynin. Said one Kissinger aide, "We were always afraid that through a slip of the tongue Dobrynin might tell Rogers about secret meetings with Henry."[18]

But the victories of Nixon and Kissinger also produced problems. Secrecy sometimes sabotaged coordination among agencies. Increasingly, options sent up reflected decisions already made at the top, meaning that analysis and intelligence were being shaped to meet the preconceptions of powerful but ill-informed policymakers—perhaps the greatest single internal cause of recent U.S. foreign policy disasters.[19]

Political Scientist I. M. Destler suggests that the Kissinger system was most effective with countries having powerful leaders able to make deals, on simple matters involving bilateral relations, and on issues where U.S. leaders had a large degree of control over events. This explains its relative success in governing Washington's direct dealings with the USSR and China.[20] A fourth qualification would be that the structure worked best on issues to which Kissinger devoted personal attention, while missed opportunities and future crises arose from neglected matters. Kissinger and Nixon were capable of courageous positions on some issues, but their belief that Third World upheavals were merely offshoots of a global East-West conflict often meshed poorly with real issues. Roger Morris called the result a contradictory mélange "of enlightened initiative and sophistication, of ignorance and impulsive savagery."[21]

The centralized White House authority so basic to this program meant that State Department morale, which had long seemed to be moving in the downward direction, sank even deeper. An FSO returning from abroad found that "the never excessively nimble pace of work in Foggy Bottom is in fact even more lethargic than formerly, and that the officers one encounters are nowadays even more discouraged."[22] One high official later said State's function was limited to "carrying out directives received from Kissinger." Some consoled themselves with jokes at Kissinger's expense. One asked, "What does Henry say in his private meetings with the president?" The answer: "Yes sir, yes sir, yes sir. . . ." But no one at State could doubt that Kissinger's working relations with Nixon were far superior to those of Rogers. In the midst of the one-sided Kissinger-Rogers struggle, Dean Acheson, asked how he would cope with such humiliation, responded, "I would have ceased to be secretary of state."

Rogers tried a number of stratagems to preserve at least his honor. He would only take White House orders with which he disagreed if they were transmitted by the president, not Kissinger.[23] When Egon Bahr, representing West Germany's Social Democratic government, visited Washington, Kissinger was forced to agree that he would let Rogers and EUR Assistant Secretary Martin Hillenbrand handle the formal talks. Bahr, however, was sneaked into the White House basement for more important meetings with Kissinger.[24]

State's European bureau proved particularly resistant to Kissinger, who claims it once took him a full year to get an option paper from them. Both sides viewed themselves as area experts with special proprietary rights to the continent, and EUR regarded itself as the pick of the Foreign Service elite. The cornerstone of EUR's policy had long been support for European economic and political cooperation. By 1970, however, the Agriculture, Commerce, and Treasury departments, with congressional support, complained that European commercial competition was damaging U.S. interests. Kissinger, for once in accord with EUR, gave the neglected, State Department-chaired Undersecretaries Committee responsibility for the issue in order to prevent protectionist forces from creating diplomatic conflicts.[25]

Europe was never a major problem in the Nixon years, but State's desire to maintain good links with European allies was made subsidiary to administration strategy. For example, State accepted West Germany's détente policy vis-à-vis the Soviet Union, while the White House urged restraint, not wanting Bonn to get too far ahead of Washington. Kissinger's effort to focus on U.S.-Europe relations faltered partly because he failed to use department resources. When other problems distracted his attention, the responsibility for shoring up alliance relationships was simply forgotten. As for economic issues, State's neglect and Kissinger's disinterest usually produced a vacuum. The Council on International Economic Policy was founded in 1971 to fill the gap, but it fell short of expectations. Under aggressive Secretary of the Treasury John Connally, that department eroded State's jurisdiction and produced about 70 percent of the economic studies for the NSC.[26]

The White House snubbing of State took many forms. Instead of the department, Nixon used Kissinger's staff to prepare him for press conferences. Rogers was told little about developments in the SALT talks, Vietnam strategy, and the opening to China until the last minute. One of the more absurd forms of friction was over an annual foreign policy report to highlight the integrated nature of administration diplomacy. A tug-of-war between State and NSC led to the preparation of two reports; much time was wasted in this exercise, but Kissinger seems to have correctly judged State's effort as excessively abstract and lacking a strategic sense of priorities.

Many career officers within State were also critical of their department's performance. For the first time, a reform effort developed from the bottom up, reflecting the decade's rebellious spirit. The activists were tired of the endless frustrations faced by, as one of them put it, "First-rate people having to operate in a third-rate system."[27]

In 1965, a group of FSOs had formed the Junior Foreign Service Club to discuss department problems. A year later, it submitted a memorandum explaining that "a feeling of professional uneasiness and uncertainty now appears prevalent among Junior FSO's which, justified or not, tends to lower morale and create a climate for resignation." In January 1967, it called a general meeting which drew an overflow crowd. Out of the efforts came a task force to study internal problems and State's declining influence over foreign policy.[28]

Later that year, leaders of the movement, by now dubbed the "Young Turks," decided to seek election to the board of the American Foreign

Service Association, the hitherto low-key FSO guild. They sent mailings, recruited supporters to bring out the vote in overseas posts, and swept into office with large majorities. The new board chairman was Lannon Walker, 31 years old, from the Executive Secretariat; Philip Habib, who worked on the Vietnam negotiations, was president. They lobbied the incoming Nixon administration, which, Walker told colleagues, "We are convinced . . . is serious about implementing reform." But Kissinger's ideas for improving the policy process were different from what FSOs wanted.[29]

William Macomber, the new deputy undersecretary for management, appointed task forces to recommend administrative improvements. "Despite many brilliant performances along the way," he admitted, "we have not met the challenge of foreign affairs leadership as successfully as we might have. Our failure to do so has caused frustration. And it has raised a clear prospect: either we will do this or it will be done for us." Much work went into these studies, with Macomber's able staff providing the energy, but results were disappointing. Their report dealt with persistent problems such as improving creativity and access to top policymakers, assessing U.S. interests and setting priorities, and evaluating the implementation and correctness of decisions. It recommended more rational assignments policy, greater incentives for specialization, mid-career tenure, performance ratings based on achievement rather than personality, and better pay and allowances.[30]

The departure of pro-reform Undersecretary Elliot Richardson from State in 1970 and Macomber's later appointment as ambassador to Turkey undercut these efforts. Lack of continuity has been one of the greatest enemies of State Department reform; the difficulty of gaining needed funds and congressional legislation to launch new approaches was another roadblock. At best, the Macomber-era proposals could only improve the department's ability to implement, rather than to participate in, policymaking.[31] More significant for this last objective had been Richardson's attempt to create a small planning and coordination staff capable of strategic thinking, which would also supervise State's response to Kissinger's queries. State's group made little headway against Kissinger's more prestigious and powerful staff and was disbanded after Richardson left.[32]

Within the department, personnel matters created tremendous controversy. The Foreign Service was top-heavy after the Wristonization transfers: There were more FSOs over 45 years old than under 35 by 1960, and twice as many officers in the top four career ranks as in the bottom four. Promotions were slow, and many FSOs believed that those that were given were unfairly dealt out. One of them wrote, "Foreign Service personnel operations have deteriorated into arbitrary and capricious rewards and punishments, lacking essential elements of due process in grievance procedures. Gossip circulating in secret channels among management officials determines careers.''[33]

A tragic demonstration of this was the case of Charles Thomas, an FSO with 19 years' experience, who, failing to win promotion, was ''selected out'' at the age of 45. Arguing that his records had been misread and misplaced, Thomas spent two years trying to win a review. Unable to collect a pension or earn a living, he committed suicide in April 1971, making it possible for his wife and children to collect a government annuity. The embarrassed State Department gave his widow a job and promised that in the future no one with long service would be terminated before being eligible for a pension. Congress passed a bill restoring Thomas posthumously to active service.[34]

FSOs' bitterness in dealing with a bureaucracy so entangled that it could drive a man to suicide produced heated antagonism toward Macomber and Personnel Director Howard Mace. When Mace was proposed in 1971 as ambassador to Sierra Leone, the Foreign Service seethed. One ''selected-out'' officer, John Hemenway, said, ''Mr. Mace has destroyed the careers of hundreds of men far better qualified to be an ambassador than he.'' Thomas's widow also testified against the appointment. Others complained that Mace had denied them grievance hearings warranted under departmental regulations. Macomber and several retired ambassadors defended Mace, and the department hierarchy pressed an internal petition drive to support him. So great was the outcry, however, that the nomination was finally withdrawn and Mace was given a post that did not require congressional confirmation.[35]

''People are afraid to argue with their bosses,'' said one FSO of the prevailing atmosphere, ''because, if they do, it will be reflected in their next efficiency report.'' Publicity and the possibility that Congress might legislate new regulations forced some improvement in grievance procedures, including the right of FSOs to seek correction of inaccuracies or prejudicial statements in their files.[36]

There were also changes in State's policy toward women. While the Mace controversy raged, FSO Alison Palmer was filing the department's first sex discrimination case after three ambassadors to African states refused to accept her as a labor officer. One of them wrote, ''Believe me, the savages in the labor movements would not be receptive to Miss Palmer, except perhaps her natural endowments.'' When Palmer complained, she received a letter from an Equal Opportunity Employment officer warning that Mace expressed ''apprehension'' that the protest might hurt her chances for promotion. Macomber finally decided in Palmer's favor. Other rulings ordered that wives of FSOs no longer be assessed in their husband's evaluation reports and that State would try to assign married FSO couples to the same post or give one of them leave without loss of benefits during the tour of duty. By 1975, some 100 such teams were stationed all over world.[37] But even here improvements were limited. A decade after Palmer's victory, women were only beginning to be fully accepted at State and the grievance system was still rigged in favor of "management."

If State could not move decisively on even these questions, it certainly could not compete with Kissinger on handling major foreign policy issues. Thus, the department was given only the leftovers: trade and aid (though not security assistance to major countries); cultural, scientific, and information programs; UN and other international organization matters; and low-priority regions, which included Africa, Latin America, and the Middle East. State often spoke with conflicting voices, except when Undersecretary Johnson coordinated its positions; while Kissinger knew exactly what he wanted and frequently invoked the president's name. The department could not easily circumvent Kissinger to reach the Oval Office. As time went on, the number of high-level committee meetings were reduced, and those held became increasingly dominated by Kissinger monologues, while State was kept busy responding to 164 NSSMs between 1969 and 1972. When it came to the important issues—including Indochina and relations with China and the Soviet Union—State had a clearly secondary role.[38]

The State Department had always played a subordinate role on Vietnam policymaking. President Johnson preferred to turn to elder statesmen, the NSC, and the Defense Department for information and advice. Rusk belonged to President Johnson's inner circle, but he accepted the Pentagon's claims of progress in winning the war. State's task was to await the appropriate signals indicating North Vietnam's willingness to negotiate.[39] President Johnson and Secretary of State Rusk had become obsessed with Vietnam. By 1968, the war had become virtually the only foreign policy issue for them. As George Ball later recorded, top officials "progressively constricted their vision like a camera focused sharply on a small object in the immediate foreground but with no depth of field, so that all other objects were fuzzy and obscure."[40]

Nixon and Kissinger viewed the war as unwinnable and, while withdrawing U.S. troops from South Vietnam, sought to extract the maximum political price from Hanoi. But they also believed that an excessively rapid pullout or collapse of Saigon would damage American credibility throughout the world. So they continued the war to keep up the pressure for enemy concessions, a strategy that seemed to damage U.S. credibility just as seriously while producing periodic military escalation and the war's extension into Cambodia.

Ironically, State's involvement diminished further at precisely the moment when the services of career diplomats would have been most useful in the negotiations process. Yet Nixon's and Kissinger's distrust of State as inept, prone to leak, and dovish meant its exclusion from these efforts. "Many in the State Department," Kissinger wrote, "shared the outlook advocated by the leading newspapers or the more dovish figures in the Congress partly out of conviction, partly out of fear." Rogers's attempt to make rapid progress in 1969 by offering concessions to North Vietnam convinced Kissinger he was giving away major points without reciprocity.[41]

The fact that State was responsible for winning passage of aid and appropriations bills made it more sensitive to Congress's thinking, but expertise also prompted State's pessimism about administration strategy. Congress might also punish individual FSOs identified with Indochina policy, as when the Senate Foreign Relations Committee rejected Ambassador to Laos Godley's appointment as EA assistant secretary.[42]

State was only allowed to handle the nominal Paris peace talks while Kissinger carried out the real negotiations with Hanoi via his own back channels and trips to Paris. The department's Vietnam task force did not become involved in the latter operation until near its end in the fall of 1972. Kissinger finally obtained his diplomatic goal—if not his preferred outcome—in peace accords, which Rogers, who had little to do with their formulation, signed in January 1973.

During the war, about 600 FSOs, nearly 20 percent of the entire Foreign Service, served in Vietnam, about half with the pacification program, nominally under AID but in practice under military command. Such assignments to another agency provided a way around the reductions in State's overseas personnel, ordered in the late 1960s to save foreign exchange. Some new officers resigned rather than go to Vietnam. While FSOs were told that a Vietnam tour would help their promotion prospects, one top official who later studied this process concludes that it actually slowed down advancement.[43]

Some of the FSOs sent to Vietnam were relatively untouched by the experience; others faced personal dilemmas over whether to report atrocities, dissent on policy, or take the safer career path of going along. They split sharply on their attitude toward the war, but everyone saw the distortion and suppression of information and statistics, hardly the best experience for teaching them to be honest reporters. The U.S. embassy in Saigon, which grew to about 2000 people by 1971, was particularly responsible; a Senate Foreign Relations Committee staff report charged the embassy with altering field reports and withholding information.[44]

The department warned embassies to ensure that dissenting reports were not leaked to Congress or the press. Even within approved channels, would-be critics feared they would be branded as troublemakers. Those who disagreed were simply not consulted again. When the CIA or State were skeptical about the general picture or specific operations, the administration preferred to believe the military and the embassy in Saigon. Such experiences spread cynicism in FSO ranks.

White House mistrust of the State Department was most clearly shown in the events leading up to the U.S. invasion of Cambodia in May 1970. State was unenthusiastic about the decision to support a junta that had overthrown Prince Sihanouk, Cambodia's neutralist ruler. Although the eccentric prince had never been loved by the U.S. embassy, State recognized his popularity in Cambodia and the widening of the Vietnam War that would follow his deposal. Rogers's reservations were overcome by his loyalty to Nixon, but, to Kissinger's chagrin, State dragged its feet on allowing preparations for delivery of U.S. aid to the new regime.[45]

EA Assistant Secretary Marshall Green was one critic who did not give up easily. His first assignment had been as Ambassador Joseph Grew's private secretary in Tokyo in 1939. After the war he held a series of high-level posts in South Korea, Hong Kong, and as ambassador to Indonesia. After Green had tried to block General Park Chung Hee's takeover in Seoul, the new dictator thanked him, "You have made it so difficult for me to pull a coup d'etat that I don't think anybody will try it again."[46]

Now Green sent a memo to Kissinger and Rogers arguing that the United States should seek a diplomatic solution to maintain Cambodian neutrality and predicting that U.S. involvement would bring escalating Communist attacks. He also warned that Congress might react to an intensification of the war by restricting aid to Vietnam, which is precisely what happened. The White House resented Green's criticisms and later gave him an ambassadorship to Australia instead of the EA prize embassy in Tokyo.

The U.S. offensive into Cambodia in May 1970 triggered a huge wave of demonstrations across the nation and provoked the greatest internal dissent in State Department history. Some 250 FSOs sent Rogers a petition protesting the invasion; he refused a White House demand for the names in order to protect their careers.

One FSO's career, however, greatly benefited from these events. Thomas Enders, a tall and intellectually imposing officer, possessed the traditional diplomat's blue-blood background. Coolness and arrogance repeatedly got Enders into trouble; ability helped him to escape it. In 1970, at age thirty-nine, Enders was already deputy chief of mission in Yugoslavia, but a feud with Ambassador William Leonhart produced so much friction that, according to department legend, Leonhart shortened the legs of his office guest chair to prevent the towering Enders from using it. Finally, Enders was recalled to Washington with a black mark on his record.

Within a year, however, Enders persuaded U.S. Ambassador Coby Swank to take him as deputy chief of mission to Cambodia. Swank, like Green, was skeptical about administration policy and asked too many questions for Kissinger's taste. General Brent Scowcroft, Kissinger's deputy, commented, "We felt Swank's attitudes were not healthy. He was pessimistic and therefore a bad influence on the [Cambodian] government. He had a negative attitude towards what we were doing; didn't put his heart into it." Kissinger's aide, Alexander Haig, visited Cambodia and found Enders more eager to execute White House wishes and willing to bypass State's chain of command.[47]

Enders pursued the war, controlled the Cambodian government, and told Washington what it wanted to hear. He ignored embassy political officers who thought that only major political reform could save the regime. In April 1974, Enders returned to Washington and Kissinger rewarded him with a promotion as assistant secretary for economic affairs. Some members of the Senate Foreign Relations Committee, angry at Enders's attempts to frustrate their investigation of U.S. bombing in Cambodia, delayed approving his nomination for six weeks.[48] But Enders's downfall was a quarrel with the undersecretary for economic affairs, who insisted that one of them must leave. So, in 1976, Enders was made ambassador to Canada. Haig, on becoming Reagan's secretary of state in 1981, recalled Enders's previous services and made him assistant secretary for Latin America.

State was even more excluded from the successful administration effort to open relations with the People's Republic of China. Again, Kissinger felt his strategic view conflicted with the department's shortsightedness. He claimed that State feared rapprochement with China because it made Moscow nervous, which was, after all, one of Kissinger's main objectives. Nixon and Kissinger laughed at State's Soviet specialists for suggesting the Russians be kept informed about U.S.-China contacts.[49] Nevertheless, Kissinger made better use of State's technical skills on China than he had on the Indochina war. One of the NSSM-mandated studies had called attention to possible U.S. benefits stemming from the Sino-Soviet split. Previously, the press of immediate business had prevented any serious consideration of this vital issue. Kissinger's skillful maneuvering made possible a carefully orchestrated change in China policy and a major U.S. victory in its rivalry with the USSR.

Kissinger secretly and slowly signaled his intentions to the Chinese by easing trade and passport restrictions. Neither Rogers nor Undersecretary Johnson were told about Kissinger's plans and his use of Pakistan as a channel to Peking. Kissinger successfully handled the secret talks, overshadowed Rogers on Nixon's triumphant first public visit to Peking, and played the principal role in composing the historic joint comminqué. Johnson, unable to use his contacts to inform Japanese leaders of the pending dramatic breakthrough, was left to console an upset Japanese ambassador when the surprise public announcement was made. As Washington accepted the Chinese Communist regime, State made peace with those who had accurately predicted its coming to power. The department held a luncheon in January 1973 to honor the China hands who had suffered so severely in the 1950s; Davies, Service, Emmerson, and Vincent's widow were welcomed back to Foggy Bottom at an emotional ceremony.

Around the world, FSOs and U.S. embassies were under a more deadly assault than they faced during the McCarthy era: terrorism and political violence against Americans and U.S. facilities. Jordanian students and PLO gunmen attacked the embassy in Amman in April 1970; FSOs Morris Draper and Robert Pelletreau were briefly kidnapped and a U.S. military attaché was killed. Eggs were thrown at the ambassador to Sweden. Two embassy aides narrowly escaped kidnappers in Uruguay, while a consul in Brazil evaded an ambush by driving over one of his assailants. A bomb exploded in the parking lot of the Athens embassy; the home of an attaché was burned in Buenos Aires; the Marine Corps guards' house in Bolivia was sacked; and the consulate in Toronto and the embassy in Cambodia were firebombed.

In 1971 alone, the embassy in Ceylon was attacked by anti-Vietnam War demonstrators, the ambassador's residence in Paris was bombed, and Turkish terrorists attacked the Istanbul consulate. A Marine guard in Sudan was wounded during a coup, guerrillas attempted to assassinate the ambassador to Cambodia, and Mexican terrorists were caught while planning to kidnap the ambassador. Between 1968 and 1979, U.S. ambassadors were murdered in Guatemala, Sudan, Cyprus, Lebanon, and Afghanistan.

The U.S. embassy in Moscow faced a particularly peculiar threat. In the mid-1970s, it was discovered that its offices were being bombarded with intense radiation, probably a Soviet attempt to monitor conversations. Although repeated protests failed to shake Soviet government denials, the problem gradually disappeared after U.S. experts discovered that ordinary window screens provided effective shielding. Despite State Department assurances, the staff worried about possible health hazards, but none accepted reassignment elsewhere.[50]

Spy technology had reached frightening levels, making embassy life in Moscow a constant round of precautions. In addition to traditional electronic bugging devices and wiretaps, the impulses of word-processing equipment might be read from the power lines, and even the very vibration of window glass from human speech could be interpreted through special equipment. Furthermore, travel restrictions and surveillance were always in force. The embassy staff found it impossible to gain access to more than a handful of Soviet officials; many of the private citizens they met reported to the KGB. All in all, it was a stressful assignment, but one eagerly sought by Soviet specialists.

Certainly, if the White House did not trust State over the Vietnam and China issues, it was even less inclined to do so on the single most important relationship in U.S. foreign policy. Early in the administration, Nixon told Soviet Ambassador Anatoly Dobrynin to deal directly with Kissinger on important matters, rather than with State. Dobrynin's privileged position was constant irritation to U.S. ambassadors in Moscow who could not even get appointments with Soviet leaders. Kissinger's strategy on U.S.-Soviet relations was to promote détente, exchanging trade and other benefits for Soviet restraint elsewhere in the world. But Kissinger complained that U.S. policymakers spent as much time negotiating among themselves as with the Russians.[51]

Kissinger briefed Rogers on the SALT-1 treaty only when details were completed, though Arms Control and Disarmament Agency (ACDA) Director Gerard Smith played a key negotiating role. Paul Nitze resigned from the SALT-2 delegation, citing administration refusal to trust its own negotiators. One reason for Kissinger's secrecy was policy disagreements among government agencies. Rogers, eager for diplomatic progress, angered Kissinger by telling Dobrynin prematurely that Washington was ready to open arms control talks.

Kissinger's greatest problems, however, came from the Defense Department. Defense Secretary Melvin Laird wanted a tougher stance than Kissinger's SALT deal accepting superpower equivalence. Indeed, many observers, including Kissinger himself, later concluded that the SALT-1 treaty put the United States in a weaker position. Perhaps more consideration of the Defense Department's position would have produced better results.[52]

Although Kissinger had his way on the SALT talks, Laird's successor, James Schlesinger, and the Pentagon attacked him for alleged softness on the Soviets. This campaign made Kissinger unpopular with the Reagan camp in the Republican party. So great was the institutional mistrust that the Joint Chiefs of Staff even had a Navy enlisted man secretly passing them NSC documents during the early Nixon years.

Handling all foreign policy matters simultaneously placed an enormous work burden on Kissinger. He made 13 secret trips to Paris as part of Vietnam negotiations, 6 trips to China, and 5 to Moscow between 1971 and 1973. State came to expect his domination over the most important issues like Vietnam, the USSR, and China. The greatest clashes between State and NSC occurred over crises in other parts of the world, where the department still hoped to play the leading role. Yet Kissinger outmaneuvered everyone in almost all corners of policy. He even took over the Latin America interagency group on the Chile issue, pushing aside the ARA assistant secretary.[53]

Perhaps the most passionate NSC-State clash took place during the 1971 Bangladesh crisis. When Bengali nationalists in East Pakistan won national elections, the Pakistan analyst at State's INR, Joel Woldman, predicted that their demands for self-determination would lead to civil war and partition. Woldman's paper slowly drifted through the bureaucracy, but without effect.[54]

As the central government instituted bloody repression in East Pakistan, FSOs at the U.S. consulate in Dacca reported that they were "mute and horrified witnesses to a reign of terror" and "selective genocide" by the Pakistani military. Their telegrams urged an immediate high-level protest from Washington. In contrast, the U.S. embassy in Pakistan's capital found these events "regrettable," but recommended against any "premature" judgment about this "internal affair" of a "staunch ally." In a cable sent through the "dissent channel"—perhaps the most courageous demurral since the China embassy disagreed with Ambassador Hurley in 1945—nineteen Foreign Service, AID, and USIA officers, supported by Consul-General Archer Blood, criticized the passive U.S. policy as one which "serves neither our moral interests broadly defined, nor our national interests narrowly defined. . . . Our government has evidenced what many will call moral bankruptcy." These remarkably strong words, for diplomats, were supported by junior officers on the Pakistan desk, who wrote a memorandum to Rogers suggesting an aid embargo on the Pakistani military regime.[55]

Kissinger supported the Pakistani government largely because of its intermediary role in secret ongoing U.S.-China contacts. State, knowing nothing of this, could not understand the basis of his policy. When Kissinger asserted that he would not change course, the Interdepartmental Group did nothing, awaiting signals from above. Dissenting officers could only leak news of the pro-Pakistan tilt to the media; Blood was brought home to a job in the Personnel Bureau.

As India increased support for separatist Bengali guerrillas, the United States condemned the action and ordered a naval task force to the area to prevent, Kissinger said, the destruction of Pakistan by New Delhi. NEA Assistant Secretary Joseph Sisco thought Indian objectives far more limited and played down Soviet involvement on India's side, although he loyally followed administration policy in blaming India for the crisis. But Kissinger complained that State sabotaged him by freezing arms supplies and aid to Pakistan without White House clearance.[56]

Even if action had been needed to prevent Pakistan's disintegration, Washington's refusal to intervene earlier to soften Pakistani policy toward the Bengalis hardly helped this goal. Rather, the U.S. stance encouraged the central government to fight what turned out to be a losing battle.[57] The affair ended on a somewhat craven note not untypical of the Nixon administration. The White House produced its own press leak claiming that State was at fault for ignoring the crisis and letting it get out of hand. The president supposedly only stepped in after it was already "too late in the day."[58] Bangladesh won independence anyway, but the Soviets gained nothing from the affair.

While Kissinger sarcastically remarked that "the President is under the 'illusion' that he is giving instructions," State was under the illusion that the White House was paying some attention to its information and judgments.[59] Although bureaucratic politics played a large role in mishandling the issue, it was in a rather different manner from that indicated by Kissinger. Actually, the administration's secretiveness, its overemphasis on the East-West factor, and its facile belief in the concept of power politics made matters worse. As noted in one government study, this was "an extreme instance of the lack of engagement between global and regionally oriented policymakers, between generalists and professionals." Those at the top were given—and ignored—alternative evaluations and options by the working level.

Similar problems marked the policymaking process over Chile. Ironically, ARA had more influence in its client countries—where the tradition of U.S. ambassadors playing a direct and detailed role in local politics was still alive—and less in Washington than the other regional bureaus. Western Hemisphere governments were frustrated and humiliated with being shuttled from agency to agency as major decisions were delayed for months or even years.[60]

In most cases, White House disinterest in the region contributed to these problems. While Kissinger was dealing with Cambodia, a senior official tried to focus his attention on the El Salvador-Honduras war.

"Kissinger asked how it started," the man recalled, "and when told it began in a riot over a soccer game, well, you can imagine what he said." More important, a 1972 Salvadoran military coup against the elected Christian Democratic government was also ignored, although it began a spiral of violence leading to the Salvadoran civil war that later became the Reagan administration's main foreign policy problem and, ironically, the subject for study of a Kissinger-led commission.

During the Nixon administration's early months, there was not even an ARA assistant secretary. Charles Meyer, who held the post by the time of the Chile crisis, had no previous government experience. Consequently, he let Deputy Assistant Secretary John Crimmins run the bureau. The NSC's able Latin America specalist, Viron Vaky, kept contracts with ARA "over Secretary Rogers's dead body," as one FSO put it, but the NSC had little time for the region and was dissatisfied with ARA's responses to its inquiries.

Chile was the great exception. Salvador Allende, Socialist party leader, was a strong candidate for president in the September 1970 elections. U.S. Ambassador Edward Korry suggested plans to defeat him either in the three-way election itself or by bribing Chilean parliamentarians, who choose the winner when no candidate has received a majority of the popular vote. State, believing Allende would lose, advised only limited actions. Temporarily, this advice prevailed over those favoring more active intervention, including the Defense Department, CIA, and Korry himself. ARA did not understand, complained CIA Director Richard Helms, that Allende was a "real Marxist" and the United States could not "throw in the sponge." In contrast, one department official said that promoting subversive efforts as Korry or the CIA recommended "assumed too much reliability from people over whom we had no control. We were doing something culpable and immoral. Why take these risks?"[61]

White House anger that Allende won despite State's prediction is reflected in Kissinger's memoirs: "No agency called our attention to the gravity of the situation. . . . Chile indeed is a classic example of how major events can unfold without the White House's knowing because the line agencies cannot agree on their significance." He blamed ARA for favoring the moderate Christian Democrat over the conservative candidate, confusing "social reform with geopolitics," splitting the anti-Allende vote among two opponents. Abandoning "covert support for foreign democratic parties, which had for so long been a central feature of our Chile effort," demoralized anti-Allende forces who thought passivity toward Allende's election signaled Washington's indifference, and ran "the kind of unacceptable risk that policymakers are hired to avoid." For

Kissinger, the risk was the regional repercussions of a successful Allende government in damaging U.S. influence and boosting the prestige of the USSR and Marxism.[62]

Therefore, Korry's delicate electoral maneuvers were too tame; the White House decided to support efforts to keep Allende from ever taking office. The local CIA station chief lied to Korry about these plans and the skeptical State Department delegate was no longer invited to the Forty Committee meetings discussing covert operations. The early, ham-handed political maneuvers backfired, however, and Chile's parliament elected Allende president in October 1970. The NSC produced three U.S. options: "Make a Conscious and Active Effort to Reach a Modus Vivendi," "Adopt a Restrained, Deliberate Posture," or "Seek to Isolate and Hamper Allende's Chile."[63] The administration chose the third approach. U.S. economic and political pressure—and covert activities—helped lead to Allende's overthrow in a September 1973 military coup.

The Foreign Service obeyed White House decisions and took the blame. Harry Shlaudeman's proposed appointment as ARA assistant secretary—he was deputy chief of mission in the Chile embassy during the crisis—ran into a controversy similar to that of the earlier Godley and Enders nominations. Shlaudeman had once falsely told the House Foreign Affairs Committee that "the U.S. government adhered to a policy of non-intervention in Chile's internal affairs during the Allende period," but he was finally approved. *The Washington Post,* a strong critic of the Chile policy, editorialized, "We do not think it can rightly be held against a career civil servant that he followed the policy of the administration he served."[64] (There is always a possibility, however, for department officials to interpret policy by their own behavior. Bob Steven, an FSO serving in Chile during and after the coup, took a personal interest in the junta's political prisoners and invited their relatives to his home to see a CBS television film on the subject. It was an emotional moment for those who had not seen their relations in months.)[65]

In the case of Chile, Kissinger's definition of *acceptable risk* was strangely abstracted from the actual situation. Was it a greater risk to create genuine crises to prevent possible ones, to exacerbate U.S.-Chile tensions in the name of avoiding other political friction? Did the treatment of Allende push other dissidents in the region from peaceful to violent tactics? Social reform is often an important element in geopolitics, after all, and it is dangerous simplistically to classify liberal or left-of-center forces of change as irredeemable enemies. These were reasonable considerations formulated in the State Department and elsewhere.

Lack of knowledge about a region or country increases the policy-maker's sense of risk, since he is unsure about the factors involved and the probable results. The desire to avoid possible dangers sometimes leads to more dangerous risk taking. Intervention may wrongly seem attractive because the policymaker trusts his own efforts rather than awaiting the "unpredictable" actions of "incomprehensible" regional or local forces. He may be so eager to confront a perceived threat that he unnecessarily creates or reinforces one.

In contrast to Indochina and Chile, the Middle East provides an example of how the NSC staff and State moved from conflict into a period of relatively fruitful cooperation. State's NEA was dominated into the late 1960s by Arabists whose views seemed to mirror those of their client governments.[66] But just as the 1967 war brought the Middle East to center stage, it also ushered in the dominance of the Arab-Israeli specialists. The key figure was Joseph Sisco, an FSO whose career was spent working on international organizations rather than in the Arab world. There he came into contact with Middle East issues and met Rogers, who made him NEA assistant secretary in 1969. Sisco was the only one of State's assistant secretaries who could deal with Kissinger as an equal. He was energetic, arrogant, decisive, and full of ideas, behavior contrasting sharply with cautious colleagues who clung to the established wisdom like a life raft. Sisco transferred many of the Arabists into the field as ambassadors and changed NEA from a reporting bureau into an active diplomatic enterprise. His right-hand man was Roy Atherton; their NSC staff counterpart was Harold Saunders. There was a great deal of continuity: Atherton succeeded Sisco at NEA, managing the bureau while Kissinger and Sisco shuttled around the region, and Saunders held the post under President Carter.[67]

NEA professionals after 1967 were gloomy about U.S. prospects in the region. They believed that continuation of the Arab-Israeli conflict played into Soviet hands, while U.S. support for Israel isolated Washington and radicalized the Arabs. There was an urgent need for a negotiated solution that would satisfy Arab needs, they argued, and pressure on Israel was the key to this goal. "As to the quality of the [projected] peace agreement," writes NSC veteran William Quandt, "the standards to be applied to Arab commitments were not overly rigorous. From Israel's perspective, an 'evenhanded' American policy was tantamount to being pro-Arab." This school of thought produced the 1969 Rogers plan that both sides rejected. When Nixon saw it, he asked, "Do you fellows ever talk to the Israelis?"[68]

The White House view was different. As in other issues, Kissinger thought conflict could be turned to advantage. He thought it both dangerous and impractical to make concessions to adversaries at the expense of friends, since this removed the other side's incentive to make a deal. In

Kissinger's words, the "Arabs had to be taught that using the Russians against us would go nowhere. Once the Arabs learned the Soviets couldn't deliver they would turn to us." The Arab states could expect U.S. help only when they were ready to move away from Moscow and to negotiate seriously.[69]

Thus, while State frantically sought negotiations, trying every available combination and route, Kissinger remained skeptical that they would get anywhere. State was allowed to run Middle East policy for the administration's first three years but was entirely on its own. When Sisco obtained an Egyptian-Israeli cease-fire along the Suez front in 1970 Kissinger told him, according to one witness, "Joe, you got remarkably far given the fact you didn't have White House backing." Even here, Nixon and Kissinger felt State did not perform well. Failure to take proper photographs of Egyptian positions, for example, led Washington mistakenly to deny that Cairo was violating the agreement by moving up more equipment.[70]

In terms of broader Arab-Israeli issues, Kissinger counseled U.S. patience, branding the State Department effort "Activity for its own sake amid self-generated deadlines that could be met only by papering over irreconcilable differences that, in turn, made a blowup all the more inevitable." In June 1971, Donald Bergus, head of the U.S. interests section in Cairo, gave the Egyptian government detailed notes as advice on a negotiating approach toward Israel. The Egyptians used these to formulate their stand, presuming they represented the official U.S. position. State had to disclaim the proposals. To Kissinger this was, at best, another example of department ineptitude. "A negotiation can succeed only if the minimum terms of each side can be made to coincide," Kissinger later wrote. "During Nixon's first term, neither side would state anything other than its maximum program—Israel unwilling to forego wholesale alterations of frontiers, the Arabs demanding total withdrawal and reluctant to undertake significant commitments for peace."[71]

The situation changed with the appearance of several new factors. Nasser's death in late 1970 and the succession of Anwar al-Sadat as Egypt's president introduced a creative and determined actor able to absorb the lesson Kissinger was trying to teach. Five years' cumulative deadlock without progress also played a role. The winding up of Vietnam negotiations and the successful breakthroughs on China and SALT freed Kissinger's attention. Finally, a good team was assembled on the U.S. side. In addition to Sisco, Atherton, and Saunders, there were a number of rising junior officers. Michael Sterner became office director for Egyptian affairs in 1970. Stationed in Cairo in the early 1960s, Sterner convinced the then obscure Sadat to take his first trip to the United States, and escorted him around the country. The personal link gave an extra edge to the bilateral relationship.

Still, as on other issues, Kissinger began by opening his own secret channel to Sadat, which State only discovered from Egyptian and Saudi sources. But when the long-awaited signal from Egypt came, Kissinger was unprepared. Sadat had been told by Washington that he might show his readiness for progress by expelling Soviet advisers from Egypt, a step he took in July 1972. When Washington did not act, Sadat set a course leading to the 1973 Arab-Israeli war. It was, one FSO bitterly remarked, a "typical crummy performance by the U.S. government." Kissinger simply missed the signal, just as he did not comprehend the looming revolution of nationalization and price rises of Middle East petroleum that would lead to the dramatic events of 1973.[72]

By the time these two crises blew up, though, Kissinger was dealing with the world from a somewhat different perspective. In August 1973, Rogers finally resigned and Kissinger became secretary of state as well as national security adviser. With the Nixon administration enmeshed in the far-reaching Watergate affair (of which one element was the sale of ambassadorships to campaign contributors), the usual musical chairs' rotation of high positions accelerated.

Kissinger was involved in the scandal through his approval of telephone taps, aimed ostensibly against press leaks. His enemies suggested that they were also intended to provide intelligence on rivals at State, Defense, and even on some of his own more independent-minded staffers. Among those tapped was Richard Pederson, State's counselor and a Rogers confidante. The long-suffering Rogers was not unhappy to see Kissinger's discomfiture. "It is very important," said Rogers, "for the United States not to become so obsessed with security matters that laws are freely violated."[73]

Nixon had chosen Kenneth Rush to succeed Rogers. Rush, a successful corporate executive, had once taken a year off to teach at Duke Law School, where Nixon had been his student. The teacher must have made a good impression for, though they rarely saw each other in the intervening years, Nixon named Rush ambassador to West Germany, deputy secretary of defense, and to State's second-ranking post. At the last minute, ironically, the Watergate crisis necessitated making Kissinger secretary of state, both to improve the administration's image—he was its only asset at that point—and to protect its foreign policy.

The October 1973 Arab-Israeli war began almost immediately after he moved into his new office.[74] Kissinger had been awaiting a new development that would convince the local parties to negotiate and would allow U.S. leverage to be used. "A reputation for success tends to be self-fulfilling,'' he later wrote. ''Equally, failure feeds on itself. . . . By early January 1974, the positions of the two sides were approaching each other; both feared the penalties of failure.'' When Kissinger first arrived at Cairo in January 1974, a U.S. diplomat assured his Egyptian counterpart of Washington's commitment to the peace process. ''Yes,'' said the Egyptian, ''but it took a war to get you here.'' Still, Kissinger succeeded, with his proper sense of timing and able assistance from Sisco, Saunders, and others, in obtaining Egypt-Israel and Syria-Israel disengagement agreements.[75]

Kissinger obviously appreciated the strange circumstances that made him leader of an institution he had so frequently outmaneuvered and humiliated. The fact that he continued to hold his old post as well as the newer one prevented anyone else from following the same pattern. The loyal and capable Gen. Brent Scowcroft ran the remaining NSC staff as his deputy, and when Kissinger was forced to give up the security adviser job two years later, succeeded him. With the president preoccupied by Watergate, Kissinger's flanks were protected. Nixon ''would sign memoranda or accept my recommendations almost absent-mindedly now,'' recorded Kissinger, who had become a virtual ''Presidential surrogate.''[76]

As secretary of state, Kissinger continued to operate using his tight inner circle; to others—rarely admitted into any real policymaking role—he was tough, dictatorial, and tempermental. When he so desired, Kissinger could make people feel they were in his confidence, that they were the only ones who really understood what was happening. But his propensity for expressing far lower opinions of these same people behind their backs was extreme even by Washington standards.

Kissinger was skilled, however, at surrounding himself with able lieutenants: Helmut Sonnenfeldt, a veteran INR and NSC official, became counselor; the popular Robert McCloskey became ambassador-at-large, supervising congressional, public, and media relations; Eagleburger was Kissinger's chief of staff in managing the department; Winston Lord, a Kissinger favorite who specialized on Asia, became Policy Planning director; and William Hyland, a brilliant and personable Soviet specialist, ran INR. Kissinger was thus assured of good advice, efficient administration, and a group used to his methods. Sisco became undersecretary for political affairs, and other FSOs who had earned Kissinger's good opinion, often through cooperation with him during Rogers's era, were raised to assistant secretary posts. He pushed everyone to improve the department's ''product''—the reports, studies, and options' lists called colloquially the ''pieces of paper''—arguing that better performance was the only way to prevent the very institutional conflicts on which he had risen. ''The work done in the Department of State,'' Kissinger told employees, "has to be so outstanding that the issue of who is the principal adviser to the President does not arise." On his frequent trips abroad, he kept tight rein on the department, using the upgraded communications system to obtain cables almost instantaneously.[77]

Despite feelings of exclusion—"You might as well turn the bottom six floors into a warehouse," said one disgruntled FSO—morale rose, since at least the department's building was once again at the center of foreign policy. Kissinger could be as eager in spotting younger talent as he was energetic in punishing dissenters. When Ronald Spiers was given his first ambassadorship in the quiet, sun-soaked Bahamas, he asked for transfer to a more active post even at a lower rank. Kissinger was pleased: "I'm more impressed by people who want to do something rather than be something." Spiers landed the prestigious number-two position at the U.S. embassy in London and went on to major assignments. Kissinger also tried to attack complacency by ordering each geographic bureau to transfer 20 percent of its personnel to other regions, the so-called "Glop," Global Outlook Program.

Under Kissinger, many of the turf battles that play such an important role in U.S. foreign policy went on as usual. An interesting example was the struggle over the new position of undersecretary of state for security assistance. The 1971 Foreign Assistance Act mandated that military assistance be separated from development aid and put into the State Department. State's Office of Politico-Military Affairs, backed by Undersecretary Johnson, who had helped create the office, worried the new job would duplicate its jurisdiction. He preferred that an FSO fill the slot; the Office of Management and Budget wanted a skilled administrator. The White House chose a friend of the president who possessed neither experience. The new office was limited by a tiny staff, no authority, and little access to information. Kissinger replaced Nixon's original choice with an equally inexperienced investment banker, who was similarly bypassed and soon resigned. Kissinger then appointed his personal lawyer. Down to the present, State has never figured out what to do with the prestigious but somewhat irrelevant position.

While Kissinger made good use of a Middle East war to spur negotiations, U.S. inattentiveness to a nearby problem encouraged a destructive and avoidable crisis. The United States had long enjoyed close political ties to Greece. When conservative sectors threatened a violent response to the victory of George Papandreou's center-left party in the 1967 elections, State ruled out endorsement of "extraconstitutional tactics" but decided Washington's response to a coup "would depend on circumstances." Some FSOs in Athens worried that the United States would one day pay dearly for this ambiguity, which opened the door to military dictatorship. In fact, U.S. military attachés, without the embassy's knowledge, were already supporting the Greek colonels' successful takeover plans.[78]

Andreas Papandreou, George's son and a future prime minister, wrote about U.S. diplomats in Athens in terms applicable to those in other places: "On the whole they were intelligent men. They took their work seriously; indeed they were literally immersed in it, so much so that in most cases they had lost their detachment. Socially, the American contingent belonged to the Greek establishment circle. They had not necessarily sought this but they had been sought after. Yachts, island vacations, sumptuous dinners, apartments in the countryside were placed at their disposal by the Athenian elite. The Americans could hardly refuse such courtesies but in this fashion they had been literally assimilated by the ruling class. Their contacts with the politicians who belonged to the Right were frequent and intimate. In contrast, contacts with the Center deputies were infrequent and strained."[79]

Nixon's first ambassador to Greece, Henry Tasca, went out of his way to befriend those in the new junta, sent favorable reports on it, and avoided meeting opposition leaders. Tasca told one FSO, "We have two policies toward Greece. One is Nixon's and the other is Secretary of State Rogers's. I work for Nixon. I am his personal representative and I am going to carry out his policies." Tasca later commented, "What is one to say to the opposition when it is the policy of the United States, for overriding reasons of the national interest, to support a government which that opposition is fighting?" This is indeed a difficult and delicate problem, but "Nobody expected [Tasca] to sabotage the policy," comments *Washington Post* correspondent Dusko Doder. Rather, the problem was that "Mr. Tasca became divorced from both the U.S. and Greek political worlds which are far broader than a narrow group at the top who held power at the moment." The embassy became identified with a pro-junta stance, unnecessarily alienating and snubbing a popular and democratic opposition whose leaders were friendly to the United States.[80]

Local public opinion, sensitive to every nuance of U.S. policy, saw the dictatorship as Washington's creation and puppet. Meanwhile, Tasca reduced the embassy's sources of information to those most favorable to the junta. As a result, Washington wrongly estimated, and was unprepared for, coming crises. In the summer of 1974, against deluded reporting from the U.S. embassy, Cyprus desk officer Thomas Boyatt told superiors of an impending Athens-sponsored coup against Archbishop Makarios' Cyprus regime. Boyatt later won an award for "intellectual courage, activity, and disciplined dissent" on the issue, but his warnings, finally taken up by the CIA as well, produced only limited efforts.

Makarios had cooled the chronic tension among Greek and Turkish Cypriots that produced friction between Athens and Ankara but was on bad terms with the Greek junta and Washington.

Although Sisco ordered Tasca to warn the junta's leader against subversion on Cyprus, the ambassador only gave the message to lower officials and refused to see the general, whom he personally disliked. Washington repeatedly pressed Tasca for action but took no other steps, such as calling in the Greek ambassador, to stem the danger. When Makarios was overthrown in August, State reacted mildly, somewhat relieved at his fall and determined to preserve good relations with Athens.[81] Events moved quickly: Sisco's shuttle mission was too late to ease the crisis. The Turkish army occupied part of Cyprus in retaliation; the usurping regime there and the Athens junta collapsed, and Tasca was removed from his post. Cyprus has been divided ever since.

FSOs did not want to become subject to more congressional oversight or retaliation—their memories of the 1950s were still fresh—but many were equally disturbed about the department's performance on the Cambodia, Chile, and Cyprus issues. "What a contribution," one retiring officer bitterly remarked on his three decades of service since World War II. The United States had gone from being "the arsenal of democracy to the arsenal of dictatorship in one career."[82]

In light of the Watergate scandal, the mishandling of Cyprus, revelations on intelligence shortcomings, and administration duplicity in Chile and elsewhere, Congress launched investigations on covert operations and foreign policy in the mid-1970s. Kissinger resisted its attempts to obtain information from State, telling one foreign visitor, "There's no government in Washington right now. [Senator] Jackson has more secret documents than I do!" In response to congressional requests for Boyatt's memos, Kissinger answered, "Recommendations by junior officials to their seniors should not be submitted to congressional committees, because it would lead to a situation in which every official would be afraid to make his recommendations. . . ." Rep. William Lehman, a Florida Democrat, commented, "Kissinger has the edge on us in public relations. . . . His image at this point will be the knight in shining armor protecting the middle-level people in his department against the attacks."[83] The executive and legislative branches of government had reached a degree of conflict over foreign policy rarely equaled in U.S. history.

Africa policy produced another major conflict with Congress and a further example of poor U.S. comprehension of events. Policymakers had long debated whether African issues should be viewed primarily in a regional or East-West context and whether relations with black African states or strategic interests should dictate the U.S. stance toward South

Africa. State's Africa and International Organization bureaus took the former positions while the CIA, Commerce Department, National Aeronautics and Space Administration, and Navy, all with their special interests in Pretoria, tended toward the latter.

Throughout the late 1960s, State's Africanists opposed an NSC-Defense Department alliance favoring a noninterventionist stand on South Africa, Rhodesia, and Portugal's colonial empire in Angola and Mozambique. The NSC's Roger Morris argued that more communication and less public criticism of the white-ruled states, plus economic development, would contribute toward progress on racial and political issues. State Department representatives replied that the chance of fostering reform was too slight to warrant endangering U.S. relations with black Africa.

After months of debate in 1969, the working group of NSC, State, CIA, and Defense representatives agreed to present five options: normalization of relations with the white regimes; relaxation of arms' sale sanctions against them plus increased aid for black Africa; continuation of current policy; decreased contacts with the white regimes; or severing U.S. ties to the area. The first and last were obviously to be rejected; the second and third were the real alternative positions. The Africanists at State, fearing the second would prove inescapably sticky, dubbed it the "tar baby" option.[84]

"Battled out over verbs and commas, vague intelligence estimates and nuances of diplomacy that would never be practiced," wrote NSC staffer Morris, "the conduct of the review was alternately childish, venomous, dull, colossally wasteful of official time, and very much the daily stuff of government in foreign affairs." At a full NSC meeting, officials from State who had earlier argued against "tar baby" rushed to assure Nixon, who favored the idea, that it could work. Participants displayed appalling ignorance about the region. One aide remarked to Haig that the meeting had been "unbelievable." "Not only is it unbelievable," Haig replied. "It sounds like one of the best they've had." Nixon and Kissinger approved Morris's second, "tar baby" option.[85]

Despite the fuss, the new stance simply allowed the United States to avoid dealing with the issue until 1974, when Portuguese officers rebelled against the colonial war and overthrew their own government, an eventuality never considered in the U.S. policy debate. They handed power to the Movement for the Liberation of Angola (MPLA), the group with the closest relations to Moscow, setting off a civil war between MPLA and its two rivals.

In the summer of 1975, Kissinger chose Nathaniel Davis as AF assistant secretary. Davis's service in Chile during the anti-Allende coup wrongly caused liberals and Africans to consider him a hardliner. Actually, he sought a diplomatic solution in Angola while Kissinger, perceiving a U.S.-USSR test of wills, was determined to prevent MPLA's triumph, particularly after Cuban forces intervened in the power struggle. Kissinger complained that his order for CIA covert assistance to the two other Angolan groups—one of them a corruption racket, the other politically damaged by a U.S.-encouraged South African intervention on its side—was blocked by Davis.[86] Davis was soon sent off as ambassador to Switzerland, but Congress reacted to Kissinger's high-handed style by passing the Clark amendment, barring U.S. covert intervention in Angola. South Africa's subsequent withdrawal and the MPLA's Cuban help did the rest. As in the cases of Bangladesh and Cyprus, ignorance of regional politics and a late start produced a policy that actually exacerbated an international problem and damaged U.S. interests.

The Nixon administration's policy process freed the White House of constraints, allowing outstanding successes and costly failures. Like the nursery rhyme, when it was good—in the Middle East, the USSR, and China—it was very, very good, and when it was bad, it was horrid. Sometimes Kissinger was right, sometimes the State Department was right, and sometimes the whole spectrum of opinion within the U.S. government did not match stubborn realities. No one agency had a monopoly on truth or effectiveness. However, Kissinger's policymaking revolution demonstrated the potential effectiveness of the national security adviser and the NSC staff as an alternative to the primacy of the secretary of state and the State Department. The Carter and Reagan administrations would have to deal with this development. They now had a way of streamlining the policy process, but the method could also produce a vacuum of power and out-of-control government infighting.

8

Divided Counsels: The Carter Years

1977–1981

Jimmy Carter and Ronald Reagan, elected in the aftermath of Kissinger's eight-year experiment in realpolitik, differed strongly with that orientation; both wanted to break with what each saw as the discredited strategies and methods of their predecessors. Despite dramatic ideological differences, the two administrations faced similar problems organizing the policy process and coping with the heritage of Kissinger's revolution.

The secretaries of state and national security advisers under Carter and Reagan found the responsibilities and relationships of their jobs increasingly open to dispute and redefinition. National security advisers were tempted to model themselves on Kissinger by challenging the secretary of state's power. The resulting bureaucratic warfare produced unprecedentedly vicious infighting. Institutional frictions were enhanced by the fact that, despite apparent overall ideological harmony, leading figures within each administration held sharply differing political philosophies.

Dramatic political changes also unsettled the policy system. Alongside some veterans of Kissinger's staff, the Carter and Reagan teams included a large number of first-time appointees determined to blaze new trails. Global economic and political developments and the heated domestic debates of the Vietnam years had broken the historic foreign policy consensus. Congress and the media were now playing a greater role than ever before in defining the issues. The ensuing uncertainties and poor performance of policymakers and their system gave both administrations a reputation for ineptness. Frequent changes in the power balance between competing government agencies and constant personnel shifts at the top—plus the contrast between Nixon's cynical realpolitik, Carter's idealistic liberalism, and Reagan's combative conservatism—further undermined continuity.

When he took office, Jimmy Carter lacked experience in national government and foreign policy. His attitudes typified some classical American beliefs on diplomacy: The United States would do better to be liked rather than feared; and promotion of democracy and reform would win friends. As a self-proclaimed outsider and critic of realpolitik, Carter was suspicious of both the State Department and Kissinger's strategy. At the beginning of his term, Carter told State's staff he favored "open, frank discussions" and even "tough, sharp debate in the Cabinet meetings. . . . I don't want to ever see a concentration of complete authority within one person, because when that is done, there is a great neglect of that reservoir of talent and ability that exists among all of you. . . ." But, Carter concluded, "I think, to be perfectly frank, that the State Department is probably the Department that needs progress more than any other."[1]

Despite such self-conscious efforts to combine pluralism with discipline, the Carter administration gained a reputation for amateurism, internal schism, and disorganization. The way decisions were implemented often sabotaged their value. By the end of Carter's term, typical newspaper headlines read: "U.S. FOREIGN POLICY IN DEEP DISARRAY," "ERRORS AND CRISES: A FOREIGN POLICY AGAINST THE ROPES," and "EUROPEAN ALLIES VIEW CARTER WHITE HOUSE AS UNPREDICTABLE AND INSENSITIVE." Carter's bitter response was to blame State, which, he said, had not produced a "new idea in 20 years." He criticized Secretary of State Cyrus Vance as being too passive, while praising National Security Adviser Zbigniew Brzezinski and the NSC staff as innovative. When Vance resigned in April 1980, one aide said the event "hit people hard because it somehow symbolizes the fact that we're not listened to. And if we're not listened to, then what's the point of staying on?"[2]

How did this sad situation come about? Kissinger had alternately built up his power as security adviser and then his own position as secretary of state, never allowing the existence of any rival. In this very effort to unify the process, Kissinger created the basis for deep division. It was as graphic an example as possible that organizational schemes or general prescriptions for effective policy are unimportant compared to the strengths and interactions of the people who occupy each post.

Carter wanted no omnipotent Kissinger on his team, partly because he wanted to be an activist president himself. His administration was the first in which two equal power centers—the State Department and NSC staff—constantly contested the direction of foreign policy. Carter's selections for the two chief foreign policy posts, Vance and Brzezinski, were incompatible, despite the fact that, on paper, they seemed to provide a good balance. In practice, the president would face the difficulty of choosing between them on every disputed issue and the government would seem vacillating and slow to respond to crises.

Vance's popularity at State was partly due to memories of his predecessor—tales of Kissinger mistreating FSOs and ashtray-throwing tantrums are legion. Although officials might miss the excitement and theatricality of Kissinger's performances, they welcomed a duller but more staid chief. Yet Vance also had something of an FSO approach, recalling Dean Rusk's virtues and faults. Kissinger perceived this in his description of Vance as "the epitome of the New York corporation lawyer, meticulously executing his assignments, wisely advising his clients."[3]

Vance's uncombative nature cost him dearly in his bureaucratic rivalry with Brzezinski. His legal career, based on calming and resolving disputes, conditioned him to considerable patience and restraint, while Brzezinski's instinct for action was often exercised without careful consideration of the consequences. On policy questions, Vance put greater emphasis on détente with the USSR and good relations with the Third World. Brzezinski called the appointees at State "ideologically one-sided" and too soft on these issues. He took the realpolitik view that threats, force, and the threat of force were highly effective ways to gain successes for the United States.

Brzezinski thought Vance a victim of the "Vietnam syndrome," fearful of needed interventions. But it was Vance's experience as President Johnson's army secretary that made him suspicious about simplistic notions on the effectiveness of military power. The United States, Vance commented, had "felt that by the gradual application of force the North Vietnamese . . . would be forced to seek a political settlement of the problem . . . that rational people on the other side would respond to increasing military pressure and would therefore try and seek a political solution. We did not sufficiently understand the North Vietnamese, nor what would motivate them."[4] In dealing with the USSR, Iran, and other issues, Vance remembered that power might only bring heightened conflict rather than the other side's capitulation.

The NSC staff was reduced and Brzezinski kept a low profile at first, promising Vance he would not meet secretely with foreign ambassadors or use covert channels to other governments, although he later did both. But Brzezinski presumed his institutional role as the White House's man would soon expand his power. "Carter's desire to be an active, dominant President," he later wrote, "automatically enhanced my role as his stand-in on national security matters." Since Vance was not a "good communicator,'' Brzezinski claimed, the president urged his national security adviser to speak up publicly.[5]

During the first two years, the State and NSC staffs cooperated effectively on the Panama Canal treaties and the Camp David talks. Carter usually settled conflicts in Vance's favor; the policy process was much less rigged in the national security adviser's favor than in Kissinger's day. Brzezinski was dependent on a more uncertain relationship with Carter than Kissinger's virtual blank check with Nixon, but soon after Camp David, events moved from diplomacy to confrontation. The Iran hostage crisis and Soviet invasion of Afghanistan seemed to fit Brzezinski's world view better. The power balance shifted sharply in his favor in Carter's last two years in office.

Brzezinski's bureaucratic skill also helped him come out on top, leading one State official to charge the national security adviser ''never won on substance.'' On decisions he did not like, Brzezinski persuaded Carter to await consultations with Capitol Hill or to take a gradual approach, with Brzezinski blocking implementation at some point in the process.

One of Brzezinski's favorite tactics was to leak stories favorable to himself, often to *New York Times* correspondent Richard Burt, while complaining to Carter that State was criticizing him in the media. Brzezinski told the department spokesman, Assistant Secretary Hodding Carter, ''I don't think reporters are getting their stories from the janitors there.''[6]

In February 1979, Carter called in about 25 State Department officials to warn, according to Leslie Gelb, director of politico-military affairs there, ''If there are any more leaks in . . . your areas, I will fire you, whether or not you are innocent.'' Gelb thought the blame for leaks lay elsewhere: ''While Mr. Vance played by Marquis of Queensbury rules . . . Mr. Brzezinski was more of a street fighter.'' While Vance discouraged leaks about the NSC staff, journalists and members of Congress were told by the NSC adviser and his staff that State's leaders had been mesmerized by Vietnam, were afraid to use force, did not understand power, and wanted to make far-reaching concessions to Moscow.[7]

Media leaks could not be stopped because they were an expression of the internal policy struggle.[8] For example, President Carter's March 1978 address on U.S.-Soviet relations, written by a Brzezinski aide, signaled Moscow that its continued military buildup could jeopardize future relations. Vance allowed Marshall Shulman, his Soviet affairs adviser, to urge the Soviet embassy to note also the speech's conciliatory passages. Burt quoted a White House source as saying, ''I'm sure [Moscow] took [Shulman's message] to mean that they didn't have to take the President's statement seriously.''[9]

Such public squabbles might seem petty, but they damaged administration credibility with Soviet leaders, European allies, and the U.S. public. Perhaps the Soviets actually took these signals to mean that they need not take the president's foreign policy team too seriously. No wonder Vance later asserted that the only way to avoid confusion was for the national security adviser to "act as a coordinator of the various views. But he should not be the one who makes foreign policy or who expresses [it] to the public. That is the task of the president and the secretary of state."[10]

In fulfilling as much of this task as could be wrested from the NSC staff, Vance chose loyal assistants. He worked as closely with his number-two man, Deputy Secretary of State Warren Christopher, as any secretary before or since. Christopher, a fellow lawyer, "makes even Vance look like a flamboyant personality," one official observed.[11] Christopher monitored policy implementation at the working levels, especially on human rights, and lobbied on Capitol Hill for the Panama Canal treaties. Later, he played a major role in handling the Iran hostage crisis. To those FSOs critical of Carter's policies, Christopher became a disliked symbol of the administration line.

The senior career job, undersecretary of state for political affairs, went first to Philip Habib, an old Asia hand, and later to David Newsom, a former AF assistant secretary. There were many new faces on the assistant secretary level with a larger-than-usual number of outside appointees, giving the administration closer control over State's operations and ideology. Some of these selections, like Policy Planning Director Anthony Lake, EA Assistant Secretary Richard Holbrooke, and AF Assistant Secretary Richard Moose, had earlier left government to join the expanding nongovernment foreign policy establishment.

Lake had resigned over the Cambodia invasion and Kissinger's tap on his phone, after a rapid rise through the Foreign Service ranks, to run International Voluntary Services. He had served as staff assistant for ambassadors Maxwell Taylor and Henry Cabot Lodge in Saigon. In Carter's administration, Lake's ties with Vice-President Walter Mondale and Vance made Policy Planning an important bureau. During the late 1960s, Holbrooke also worked on Vietnam. He left the Foreign Service to become Peace Corps director in Morocco and then managing editor of *Foreign Policy* magazine, a new liberal quarterly. The knowledgeable Richard Moose, another former Kissinger staffer, had resigned from the Foreign Service to work on the Senate Foreign Relations Committee staff.

In short, the foreign policy field outside government—think tanks, university centers, and membership groups—had grown large and influential enough for FSOs to quit the service, find employment, and later return to government at higher posts than colleagues who had stayed in the bureaucracy. This development marked another downward step for the Foreign Service: Now, continued membership could even be detrimental to an international relations career. FSOs had to contend not only with appointed ambassadors, but also with new groups of competitors for mid-level and NSC staff positions.

The Carter administration chose no more career ambassadors than did its predecessor. It tried to raise standards by establishing a board to make recommendations on nominations. Although fewer big campaign contributors were named, there were still unqualified choices from among Carter supporters and out-of-office Democratic politicians. Many senior FSOs were left walking the halls without assignments commensurate with their skills.[12]

When Vance pressed to have either a woman or a minority group member as an assistant or deputy assistant secretary in each regional bureau, FSOs faced still another reservoir of candidates competing for policymaking spots. While some such appointees, like ARA Assistant Secretary Terence Todman, a black, came from FSO ranks, even he was opposed by Hispanic groups, which wanted one of their own for the post.

Two other controversial appointees were Assistant Secretary of State for Human Rights and Humanitarian Affairs Patricia Derian and Ambassador to the United Nations Andrew Young. Derian, a civil rights activist in the South, lobbied to gain influence for human rights, which she called "the basic tenet of foreign policy under this administration," as a policy consideration. Many FSOs, like Todman, found the whole subject so alien to traditional diplomacy and protection of their client states as to resist its applications.[13] One FSO complained that activists "seem to regard criticism of the human-rights policy or caveats about its implementation as bordering on immorality or disloyalty to the Administration." But another career man working in the Human Rights bureau commented, "We're not in the business of trying to overthrow governments. We are trying to get governments—within their own limits—to treat their own people better."[14]

Both the positive and negative sides of Carter's "outsider" appointments can be seen in UN Ambassador Young's performance. As a political ally, Young had excellent access to the chief executive and became a virtual ambassador-at-large to the Third World. Unquestionably, Young did a good job improving U.S. relations with many of those countries, particularly in Africa.

The UN ambassador can treat his post as an ordinary embassy—taking a low-key diplomatic approach—or use it as a platform for expressing policy views. Most recent UN envoys have taken the latter course. Young argued that the way to deal effectively with other nations was to be willing to listen to their viewpoints, accept them as equals, allow others to take the lead on issues of mutual agreement, and not to worry about anti-American rhetoric. He represented an important strain in administration thinking, wanting to deal with regional conflicts and problems on their own merits, not just as part of the global power game with Moscow. This was a clear rejection of Kissinger's and Brzezinski's "zero sum" view that any apparent loss for one superpower meant a gain for the other. The more confrontational school, represented by Nixon's UN Ambassador Daniel Moynihan and later by Reagan's Ambassador Jeane Kirkpatrick, feels that power must be exercised to be respected and that anti-American verbal attacks must be countered and punished.

"The UN job is like the vice-presidency," says a former assistant secretary of state. "Many who claim they don't want it are eager for the job." But operating under State's discipline can prove frustrating for someone who regards himself as a national political leader. Washington closely controls its UN delegation, demanding to approve every word in U.S.-supported resolutions and every theme in the ambassador's speeches.

Young, who had rarely used a prepared text, agreed to do so in speeches to the UN Security Council. Still, his more informal statements caused considerable trouble because he did not follow the public caution that usually characterizes diplomacy. Young's comments that the Cubans brought "a certain stability and order" to Angola, that Swedes were "terrible racists who treated blacks as badly as they are treated in Queens," and that Britain still possessed an "old colonial mentality," neither helped U.S. policy nor contributed to solving problems. In July 1978, he told a French newspaper that there were "hundreds of political prisoners in the United States" just as Vance was holding a major meeting with Soviet leaders. When Young conferred with a PLO representative at the UN, contrary to U.S. policy, and then gave State a misleading version of the affair—because the less they know, he explained naively, the less they could be held responsible—it was the last straw. He resigned in August 1979, to be replaced by his deputy Donald McHenry, a black FSO.[15]

While permitting such freewheeling activities, Carter's policy structure was designed to split authority between the secretary of state and the national security adviser. Two main committees oversaw decision making: A Policy Review Committee developed positions on issues, while a Special Coordinating Committee orchestrated the different parts of government to provide options for the president, implement his decisions, and manage the response to crises. The former committee was chaired by the head of the agency most concerned with the issue being discussed—usually the secretary of state. Brzezinski led the latter. Defense Secretary Harold Brown was often the swing vote between the two.[16]

The lifeblood of these groups were presidential review memoranda, requesting studies of issues, and presidential directives, ordering action. One detail especially strengthened Brzezinski. When either committee failed to agree, the national security adviser produced a summary report for Carter; if there was consensus, he drew up a draft presidential directive. Brzezinski thus wrote the reports for both committees as well as the president's instructions. In order to avoid leaks, other cabinet officials were not allowed to review these before they went to Carter. Vance thought the reports were sometimes inaccurate; his staff believed they were slanted to favor Brzezinski's arguments. When Edmund Muskie became secretary of state during the administration's final months, Vance advised him to demand review rights.[17] Other channels included a daily report from Vance and a weekly report from Brzezinski that went directly to the president.[18]

Alongside the formal setup, of course, arose an informal system for influencing the president: weekly foreign affairs breakfasts involving Carter, Mondale, Vance, and Brzezinski, as well as another weekly meeting between Vance, Brzezinski, and Brown. Philip Odeen, a management expert conducting a study for the Carter White House, felt such ad hoc decision making led agency heads to "come away with differing perceptions of just what the agreement was." Odeen concluded that the lack of discipline among the NSC staff and constant debate produced internal conflict and confusion.[19]

Within State, Carter and Vance preferred to allow mid-level officials greater participation in shaping decisions. "In the old days there were three options," said an FSO. "Now there are seven options." Documents and discussion were spread among a host of interdepartmental committees and working groups, as well as informal discussion circles that formulated useful ideas on Europe and the Middle East.

This was another paradox of policymaking. More debate enhanced creativity and could improve the quality of suggested options, while also threatening coherence and consistency. Formal decision making reduced the possibility of misinterpretation, but also could be slower and more rigid. As a comparison of the Kissinger and Carter systems shows, there were no solutions—only choices, each involving strengths and weaknesses.

Relationships between individual NSC and State staffers varied according to their personalities and viewpoints. The NSC's William Quandt and NEA Assistant Secretary Harold Saunders cooperated closely on the Middle East (the two had briefly worked together on Kissinger's NSC staff), as did counterparts Richard Funk and Assistant Secretary Richard Moose on Africa. The Latin America situation was more variable—Carter had three ARA assistant secretaries in four years—and the NSC's Robert Pastor and Assistant Secretary Viron Vaky held strongly different views on the revolutionary upheaval in Nicaragua. Assistant Secretary Richard Holbrooke at the East Asia bureau, considered by critics as abrasive and ambitious even within the Washington context, did not blend well with Brzezinski's deputy David Aaron, who had a special interest in the region. Those with a good personal relationship usually worked better together, quietly resolving problems that produced clashes at higher levels.

NSC staffers charged that many assistant secretaries still thought that decision making ended when they sent a memorandum to the secretary of state; in fact, this was now only the beginning of the process. One NSC man, describing the evolution of relations between himself and his counterpart at State, said that at first the assistant secretary would suggest frequent meetings—dinner twice a week, breakfasts, and more formal conferences; but, gradually, the department official became less eager for contacts, realizing that interaction meant power sharing.

Carter sometimes used his prerogative to overrule all his advisers, although usually not for the better. Both State and NSC opposed his plan to withdraw U.S. troops from South Korea, favored increasing defense spending by 3 percent in real terms, and worried about his promise to demilitarize the Indian Ocean. Public reaction forced the president to change course on all three issues. At other times, Carter had great difficulty in deciding among their conflicting positions or blending them into a coherent policy, particularly during revolutionary crises in Iran and Nicaragua. As early as the end of his first year in office, however, there was a growing feeling in and out of government that President Carter was not fully in control.[20]

Following the fierce executive-legislative power struggle of the Nixon years, Vance needed to be attentive to relations with an activist Congress. About 25 percent of his and Deputy Secretary Christopher's time was spent in preparing testimony, testifying, or briefing congressional committees. Preserving a healthy relationship between Congress and State was particularly difficult given the traditionalist attitude at the working level. Asked how laws on international economic relations might be improved, one official commented candidly, "We never think about how to improve the laws, only how to implement them or get around them." This attitude reflects both the patronizing and deferential side of State's attitude toward Congress, stemming from the career staff's reluctance to meddle in lawmaking or encourage congressional activism. As one observer has put it, "'Congress won't buy it' usually sounds the death knell for the option since it is assumed that no one will sell it."[21]

The human rights issue arose as a legislative reaction to Nixon's policies and methods. Congress mandated regular reporting on nations receiving U.S. assistance, aid restrictions on those violating human rights, and appointment of an assistant secretary of state to deal with these questions. While the Carter administration was strongly committed to this approach, much of the State and Defense departments were uncomfortable with both the theory and practice of such measures.

State had long followed the classic diplomatic approach that a government's treatment of its own citizens was an internal affair of no concern to other states. The realpolitik view is also that a nation's foreign policy should be set exclusively by strategic, political, and economic interests. Washington often has limited capacity to affect such actions, even allies' human rights performance. In contrast stands the particularly American belief that morality should play a central role in foreign policy, and thus contribute to the extension of democracy. This idea has been reinforced by a pragmatic consideration: Allies that are unpopular dictatorships may be prime candidates for revolution. Mitigating repressive policies can prevent anti-American upheavals and thus further U.S. interests. By the same token, an active human rights policy can expose the totalitarian nature of Communist states and reduce their international influence.[22]

Human rights, coupling as it did conservative anti-Communism with liberal criticism of authoritarian U.S. allies, was a perfect consensus-building issue for the Carter administration. Assistant Secretary of State Derian and Deputy Assistant Secretary Marc Schneider lobbied the rest of the department to consider the human rights factor in formulating aid and other policies, as mandated by law. Since the United States could do more to pressure allies than enemies, the administration's stance tended to affect friends more than rivals. FSOs disliked both the techniques and objectives of the internal human rights lobby, feeling that such considerations could compromise vital interests. Restricting U.S. assistance to clients lessened their willingness to give the United States things it wanted. "More than a human rights bureau," said one FSO, "the administration needed a bureau of realpolitik."[23]

The Human Rights bureau was equally dissatisfied with the career staff's performance, particularly apologetic reports on countries' human rights performance that understated abuses and overstated positive trends. Further, the officer in each embassy assigned to monitor human rights was often obstructed by superiors. For example, despite a conscientious FSO's efforts, the Philippines mission sent cables ignoring corruption and overstating improvements in living standards under dictatorial President

Ferdinand Marcos. Glowing reports were sent from Zaire in the midst of a repression campaign. Human Rights bureau officials had to battle the Iran desk, just before the revolution, on whether internal documents should say that there was "discontent" rather than "dissatisfaction" with the Shah's rule. At times, the bureau had to obtain information unavailable from its own embassies through independent organizations like Amnesty International.

Assistant Secretary Derian's bureau did win some victories. The U.S. ambassador to Liberia helped free a badly beaten political prisoner who later became that country's foreign minister and thanked the United States for saving his life. In Bolivia, the Dominican Republic, and elsewhere, Washington made a significant contribution to reinstituting electoral systems. Much of the bureau's work was through behind-the-scenes diplomacy, instructing ambassadors to approach foreign leaders confidentially with requests. The effectiveness of these efforts might depend on whether a U.S. envoy signaled the local goverenment not to take the message seriously or strongly endorsed the position.

Much of the effort was aimed at Latin America, where the United States had great influence. Since most dictatorships there seemed to face no serious internal or foreign threat, security could not be used as a rationale to justify U.S. inaction. Military sales or aid were cut to Argentina, Chile, El Salvador, Haiti, Nicaragua, and Paraguay. "ARA hated us," said one Human Rights bureau official. Assistant Secretary Todman finally lost his job after publicly disagreeing with administration policy on human rights in a February 1978 speech. In Argentina, the U.S. embassy played down the disappearances of dissidents; the FSO responsible for reporting human rights, F. Allen "Tex" Harris, was harassed by his own superiors. Yet, after the return of democracy, President Raul Alfonsin praised the Carter policy, which had saved the lives of some of his colleagues.

Given the mid-level disputes, as many as 50 cases involving shipments of military equipment, riot-control gear, or other assistance were held up awaiting arbitration by State's leadership. Other issues frequently took precedence over human rights, but Deputy Secretary Christopher closely watched these priorities and often sided with Derian's bureau. "I'm here to make sure we don't violate the law," he would explain.

The laws governing the Foreign Service were also changing. The 1980 Foreign Service Act provided a new pay schedule, mandatory retirement at age 65, and a tenured Senior Foreign Service. It also revised promotion and retention standards based on performance, provided bonus pay for outstanding service, and made grievance procedures slightly more favorable to the career staff.[24] State Department recruitment was drastically broadened. Blacks and other minority groups were given five-point advantages on the Foreign Service entry exam. Of 200 new FSOs accepted in 1979, 39 were minority group members and a sizable portion were women. There were also more ambassadors from these groups.[25]

The real test of the Foreign Service, of course, was the quality of its work. In June 1977, Vance cabled U.S. embassies, "In fast-breaking situations, we need authoritative, objective reports . . . promptly. We also need, however, your analysis of the implications of these situations for U.S. interests, your predictions of the possible course of events, and your suggestions as to steps we might take." A year later, after Iran's revolution clearly revealed shortcomings in the reporting process, Vance admitted that about 25 percent of the data sent to Washington was of little practical use and that channels for alternative views were not fully used. State often did not communicate much better with its embassies. Malcolm Toon recounts learning "mostly from the press" of an impending U.S. arms control proposal to the Soviets. When Vance finally asked Toon what he thought of the new U.S. proposals, he replied, "I would be in a better position to reply if I knew what they were." After a briefing, Toon predicted the Soviets would reject them, as indeed they did. "Anyone with limited prescience and some knowledge of Soviet attitudes could have made the same prognosis," Toon concluded.[26]

Toon's observation on the lack of expertise was backed up by the facts. Due to economy measures, positions for political officers abroad had been reduced by 21 percent between 1971 and 1976; regional bureaus lost 8.2 percent of their slots worldwide in the same period. In addition, lower promotion rates had a crippling effect on morale and persuaded good officers to resign. Other trends posed additional problems for FSOs. The dollar's decline meant poorer living conditions overseas. There were more hardship posts, greater dangers, and a lower pay scale than comparable private employment. Wives of FSOs, many with their own careers, were becoming reluctant to leave Washington. "While life in the Foreign Service is stimulating and has undeniable rewards of personal growth, travel and international friendship," said Leslie Dorman, president of the Association of American Foreign Service Women, "we experience the alienation of culture shock, the isolation of language inadequacy, the hazard [of] climate and endemic disease, the trials of evacuation and the pervasive fear of terrorism."[27]

Between 1968 and 1978, there had been 252 attacks on U.S. diplomats or property. As one spokesman noted, "This country is not in a position to provide fortresses overseas in which to operate: we have to depend on the host government." But "those assurances may not be worth as much as they once were."[28] During Carter's term, U.S. diplomats were violently attacked in Iran, Afghanistan, Pakistan and Colombia. Many Americans interpreted these events as humiliating assaults on national honor; President Carter would pay dearly for such perceived insults in the 1980 election.

During Carter's first two years in office, however, there were more successes than failures. The new administration wanted to "hit the ground running" with lots of initiatives: Panama Canal treaties, a new U.S.-USSR arms control accord (SALT-2), Israel-Egypt talks at Camp David, normalization of relations with China, and the human rights policy. The small group of decision makers had to deal with most or all these issues at the same time.[29]

The first Brzezinski-Vance confrontation occurred in early 1978 when the Soviets and Cubans were providing aid and troops to Ethiopia's military junta after the country had been invaded by Somalia. Somalia sought U.S. assistance, but received little sympathy from the rest of Africa since it was the aggressor and could end the fighting by withdrawing from the disputed territory. Brzezinski, however, saw the issue as an East-West confrontation, worrying that Moscow's support would seem more valuable than Washington's friendship and consequently enhance Soviet credibility.

To pressure Moscow to desist, he wanted to postpone SALT negotiations and to dispatch U.S. naval forces to the area. Vance, for whom progress on arms talks was the highest priority, tried to avoid a U.S.-Soviet confrontation. In opposing Brzezinski, the secretary was supported by the Africa bureau. Secretary of Defense Brown also failed to see the utility of sending a U.S. aircraft carrier. Carter ruled in Vance's favor and Somalia began to withdraw at U.S. urging.[30]

Two old friends, Assistant Secretary for Africa Moose and Policy Planning Director Lake, author of a book criticizing Kissinger's southern Africa policy, worked closely together. Using Lake's channel to Vance—and with Young's clearance on UN-related matters—Moose gained top-level attention for African problems. With the NSC staff and State's Politico-Military Affairs bureau monitoring the Ethiopia-Somalia conflict, Moose spent most of his time on white-ruled Rhodesia's transition to a black majority government. Washington complied with UN economic sanctions except on nickel, where Congress legislated an exception, and refused to recognize the settler regime's unilateral declaration of independence from Britain. Moose had to fend off congressional conservatives who sought to end adherence to the trade embargo. Britain, with U.S. help, finally negotiated a settlement ending the long war and gaining independence for the country, renamed Zimbabwe.

The Africa bureau also dealt with an invasion of Zaire by dissidents in

Angola, a coup in Liberia, and the regional intrigues of Libyan leader Muammar el-Qadafi. Negotiations over independence for South Africa's colony of Southwest Africa (Namibia) through the UN were handled mostly by State's International Organization bureau.

Brzezinski convinced Carter not to recognize Angola's government, provoking a typical tactical disagreement between the NSC staff and State. The national security adviser wanted to punish Angola for its continued hosting of Cuban troops, while State thought normalization of relations would encourage the regime to send the Cubans home.

As a sign of U.S. disapproval for South Africa's system, black American diplomats were assigned there, producing tension in the U.S. embassy in Pretoria. Some FSOs felt the black officers sought to provoke local police by deliberately breaking racialist laws; black officers thought that in doing so they were raising the morale of the black Africans. Black FSOs serving in South Africa frequently moved on to plum jobs in Washington as rewards.

U.S.-Soviet differences on Africa, the discovery of a Soviet military brigade in Cuba, and other issues made it more difficult to pursue détente and the SALT talks in 1978 and 1979. To Brzezinski, these events represented a "worldwide thrust of Soviet assertiveness" and "misuse of détente" to improve their geopolitical and strategic position. Vance believed that better bilateral relations and negotiations could solve these problems; Brzezinski felt the USSR would be deterred by greater use of gunboat diplomacy, closer U.S.-China relations, and a linkage of Soviet behavior to progress on SALT. Vance wanted to move slower on normalization with China to avoid increasing friction with Moscow, but Carter sided with Brzezinski, although not to the extent Brzezinski desired, on laying the basis for a strategic relationship with Peking. Both Vance and Brzezinski sought allies, including Vice-President Walter Mondale, Secretary of Defense Harold Brown, and White House staffers, to win Carter's support.[31]

The president's energy in the first half of 1978, however, was focused on the Arab-Israeli conflict. Egypt's President Anwar al-Sadat, who had earlier caught Kissinger off guard by expelling Soviet advisers, surprised Carter in November 1977 with a major peace initiative: He visited Israel. Carter was annoyed at not having been consulted, but his ensuing decision to bring the three countries' leaders together for a summit produced the Camp David accords.[32]

If much diplomatic work is tedious, repetitive, and buried in detail, this is especially true of complex negotiations like Camp David. Endless hours must be spent analyzing the changing positions of all sides. Proposals must be drafted, with each word carefully weighed. A strategy that took weeks to develop may have to be discarded in a few minutes. To be able to master detail and still produce a creative synthesis able to satisfy all parties is an important skill.

Harold Saunders, then head of INR at State (and later NEA assistant secretary), developed many of the ideas that led to breakthroughs during the negotiations. Saunders, said one colleague, had "tremendous powers of concentration" and could work up to 16 hours a day without becoming mentally stale. He knew when to move from conversational accord to written formulation and how to persuade people gently toward a consensus. After he supervised production of the first drafts, they were inspected by attorneys from State's legal division, compared to existing Israeli and Egyptian positions, and related to the U.S. actions and aid necessary for implementation. Briefings also had to be prepared at each stage for the Defense Department, Congress, and the media. When President Carter actually met with President Sadat and Israeli Prime Minister Menachem Begin at his Camp David retreat, this was only the midpoint of long and complex staff work.

At the top level, Carter, Vance and Brzezinski, with advice from Mondale and Brown, headed the U.S. team. Their staff, including NEA assistant Secretary Roy Atherton, the U.S. ambassadors to Egypt, Hermann Eilts, and Israel, Sam Lewis, and the NSC's William Quandt, worked well together without disruptive turf battles, partly due to good personal relationships. The NSC and State staffs were fused into a single team. While Saunders headed the drafting group, Atherton tried out U.S. ideas and Quandt elicited reactions from both sides. Eilts and Lewis analyzed Egyptian and Israeli positions.

The main diplomatic error was the U.S. embassy in Saudi Arabia's mistaken report that the Saudis would support the Camp David agreements. Eilts, Saunders, and Quandt were skeptical, but Carter preferred to believe the ill-informed U.S. ambassador, John West. Another problem with so many senior officials working on Camp David was that no one at the top had the time to comprehend Iran's developing revolution.

The Iranian revolution began in January 1978 and ended with the Shah's replacement by an Islamic fundamentalist government in February 1979. On November 4, 1979, the U.S. embassy in Tehran was seized by student militants and much of its staff was held hostage until January 20, 1981. The Iran crisis, then, presented U.S. policy with two separate problems: How should Washington react to a revolution against an important allied leader and, later, how might the hostages be freed?[33]

The alliance between the United States and the Shah began in 1946 when Washington decided to help protect Iran and its oil from the Soviet neighbor to the north. In 1972, the Nixon administration broadened this relationship by aiding the Shah's ambition to be the Persian Gulf's policeman, protecting that region from radical Arab forces allied to Moscow. Toward this end, the Shah was permitted to buy as many U.S. weapons as he wanted. When oil prices increased dramatically in 1973, the Shah could afford to purchase vast quantities of military hardware and to implement a misconceived, excessively rapid modernization program.

By the time of Carter's election, Congress was worried about Iran's military buildup, the presence of tens of thousands of U.S. advisers there, and reports about human rights abuses. But defenders and critics of U.S. policy were debating Iran's disruptively expansionist potential in the Gulf and the Shah's authoritarianism at home, not the possibility that his rule might collapse entirely. These preconceptions of the Shah's invulnerability reinforced complacency about Iran. Once the Nixon-Kissinger policy of building Iran as a regional power was set, the bureaucracy was directed into the familiar pattern of neglecting information that conflicted with U.S. policies. Consequently, as riots mounted from January 1978 on, the U.S. embassy and State were slow to understand or transmit the repercussions.

In addition, the reduction of contacts with the Iranian opposition, in response to the Shah's objections, meant the U.S. embassy was increasingly dependent on information from his secret police. This factor also distorted U.S. perceptions on the strength and nature of dissidents. At the same time, as retired Ambassador Martin Herz noted, contact with anti-regime forces can also present problems, including keeping "a sense of proportion under the avalanche of derogatory information about the regime." Even anti-American groups energetically try to gain hearings at U.S. embassies, but contact does not necessarily signify persuasion. As Herz put it, "The fact that a group . . . that appears dangerous to our interests comes to power after insufficient contacts with the United States government, does not prove that if such contacts had been cultivated more assiduously, the policies of that new government would have been more to our liking."[34]

A balanced assessment is also needed on the repeated claims of some academics to have out-thought government analysts on any given issue. "When it comes to crystal balls," said Gary Sick, the NSC staffer then responsible for the Persian Gulf, "the academic community doesn't have to commit itself to action, it can think what it wants to and then denounce that later on. . . . People don't go back and hold them to account."[35]

Although the failure of top policymakers to recognize the seriousness of the revolutionary challenge to the Shah or to understand opposition forces were major problems, equally culpable was a U.S. government so bogged down in wishful thinking and bureaucratic considerations that truth became lost in the shuffle. Officials lacking knowledge about Iran, imagination, or courage slowed the flow of accurate information. Policymakers smugly set in their prejudices were uninterested in hearing facts that contradicted these views.

Recognizing some of these shortcomings, State told Brzezinski, "Over the last few years there has been a steady decline in the number of political reporting officers in the Foreign Service, the number of analysts in the Bureau of Intelligence and Research, and the funds available for local travel by political officers and analysts abroad. At the same time, the requirements we have placed on our missions for non-political reporting and analysis tasks have mounted steadily." State had stressed daily events over research and analysis. A decade earlier, Kissinger himself had chosen Iran as an example of the personnel system's damage to expertise: "If an officer works on the Iran desk, he can tell you every detail about Iran. Then he gets transferred to another job and you can never get him to talk about Iran again, but he will know everything about Austria or whatever."[36] Ironically, Kissinger later accelerated rotation.

The main fault lay with the White House and appointees rather than State's career staff. "You couldn't give away intelligence on Iran," recalled one analyst. A congressional investigation concluded, "Policymakers were not asking *whether* the Shah's autocracy would survive indefinitely; policy was premised on that assumption." The Nixon, Ford, and Carter administrations did not demand studies on the Shah's stability or the effect of U.S. arms sales on Iran. Leaders' attitudes inhibited intelligence collection, reduced their appetite to receive such materials, and deafened them to available warnings like those seeping into State's morning summary for the president.[37]

After all, some of the U.S. embassy reporting was quite accurate. As early as February 1978, it described the main opposition leaders and forces, noting, "If additional incidents involving the religious community, such as firing upon marchers, either occur or can be generated, religious fervor could be activated to provide the mob manpower for demonstrations." But there was also an air of unreality in the dispatches from Tehran even as late as October 1978. They stressed the importance of the Shah's liberalization program, though by then it had no effect on the situation, ignored the fact that time was on the opposition's side, and understated the radicalization of antigovernment forces.[38]

Henry Precht, the office director for Iranian affairs in NEA, was the highest Washington official handling the crisis for much of 1978. By July 1978, Precht was warning that the Shah could not survive politically, a judgment the administration was not yet prepared to accept. Ironically, after the Shah's fall, Precht became a scapegoat for hard-liners who thought his view was a self-fulfilling prophecy. This turn in events, a not atypical fate for FSOs analyzing controversial issues, led one top State Department official to comment, "The only thing wrong with Precht was that he was right."[39]

Nevertheless, word of the true state of the Shah's troubles spread only slowly through the government apparatus. Some of the FSOs in Iran most aware of the danger, particularly the perceptive and conscientious U.S. consul in Tabriz, Michael Metrinko, were discouraged from filing such reports or found them criticized as "anti-Shah." If Precht was convinced of the regime's instability in July, NEA was not persuaded until August or September, while U.S. Ambassador Sullivan did not begin to broadcast the danger signal until November. As late as October 19, the embassy reported, "There are encouraging indications that the Iranian crisis may have passed a fever point and opened some prospects for its constructive resolution. . . . The Khomeini star seems to be waning."[40]

In November, President Carter selected former Undersecretary of State George Ball as an outside consultant to recommend responses. Carter, unable to choose among his own advisers, obviously doubted the bureaucracy's ability to cope with the crisis. By accepting the views of Saunders and Precht, Ball provided these mid-level officials with a channel into the White House. By the time of Ball's report in mid-December, however, Washington still did not have any coherent Iran policy or adequate coordination among different policymaking groups. Carter neither settled their intensifying debates nor formulated one position to rally around. Consequently, each U.S. move was dictated partly by chance, partly by the relative strength of various personalities, and very much by immediate reactions to events in Iran. By the time Carter and Vance decided, in late December, to support a transfer of power from the Shah to a moderate civilian government—the idea advocated by Ball, Saunders, and Precht—it was too late for Washington to affect developments very much.

A central problem in these delays was Brzezinski's recklessly independent activities. The national security adviser never understood the breadth of the revolution in Iran and the exigencies of local politics there. Long after the Shah's authority had crumbled and the Iranian military was cutting its own deals with Khomeini, Brzezinski was still advocating a military coup to save the monarchy. He chose to accept Iranian Ambassador Ardeshir Zahedi's self-serving accounts rather than the U.S. embassy's more accurate ones.

In December 1978, Precht cabled Ambassador Sullivan: "There is real concern in this building about back-channel communications from the White House directly to the Iranians, notably the Brzezinski-Zahedi channel. . . . I met with Brzezinski myself 2 or 3 weeks ago in a private session in which he queried me about Iran in general and my pessimistic views of the future. I did not tell him what I have since tried to convey through [his staff]: That is that I consider Zahedi to be a disastrous counterpart in dealing with the Iranian crisis. . . . I regret that I believe his counsel has been one of the strongest factors working on opinion in the White House.'' Zahedi was arguing that a strong U.S. line and personal encouragement could still save the monarch, but Brzezinski's attempt to contact the Shah directly had no positive effect.

Precht and the State Department had concluded, in contrast, that it was necessary to find ''a graceful exit for the Shah while gaining a fair amount of credit in doing so for the U.S.,'' perhaps by supporting a committee of notables, including representatives of Khomeini, to work out some new government. When the White House sent a questionnaire to the Shah requesting his views on solving the crisis, the document was not even shown to Precht. Precht attributed this to the ''level of distrust that exists in the White House towards the State Department (and . . . myself). I am afraid that we are losing valuable time and that events may sweep us by, depriving the United States of the opportunity to recoup its position in Iran.''[41]

Precht summed up the emerging department view in a message to Saunders, his superior: ''I believe the Shah's position has eroded more rapidly than our perception of it.'' Analysts were becoming convinced ''that there are moderate and responsible groups which would be friendly towards the U.S. and could also govern. There is also a good probability that if a civilian successor regime came in with the blessings of most key oppositionists, including Khomeini, it would be greeted with relief by an Iranian public terribly fatigued by the turmoil of the past year.''[42]

The events of the following months proved this analysis to be wrong, since the moderate forces could not stand against Khomeini and the radicals. Ambassador Sullivan advocated a somewhat different approach: a deal between the army and Khomeini's more moderate supporters to accept the Islamic victory in exchange for preserving the military. By that time, the United States could do no better than those admittedly shaky options. A rapprochement with Khomeini would have been impossible—he had no intention of maintaining good relations with the United States—though his power and intransigence was not fully appreciated at State. Equally, neither the Iranian military nor the Shah had the resolution or organization necessary for an all-out ''iron fist'' confrontation with the opposition. In the summer of 1978, a policy of concession or of repression might have worked; by the winter of 1978 it was indeed too late.

Despite the Shah's departure from the country, the installation of a reformist cabinet, and frantic secret negotiations by the U.S. embassy to arrange a settlement, Khomeini's forces swept to power in February 1979. The Ayatollah and his followers erroneously identified the United States and its embassy as the true rulers during the Shah's regime and as the cause of all of Iran's problems. Khomeini, correctly understood, however, Washington's efforts to keep him and the radical fundamentalists from gaining power, but concluded that this policy would continue after the revolution's victory. In fact, the Carter administration wanted to get along with the new Islamic government.

Although the taking of diplomats as hostages in Iran was the most dramatic such event in U.S. history, there were precedents. During the Chinese Communist revolution in 1948–1949, for example, 22 employees of the U.S. consulate at Mukden were held captive for over a year. After a 1965 mob attack on the U.S. embassy in Moscow during an anti–Vietnam War demonstration, Secretary of State Rusk said, "We expect . . . that the host government will organize itself to provide full protection for our personnel and our official installations." Congress discussed cutting off aid to any country allowing such assaults on U.S. property, since it was well understood that no embassy could long hold out against an attack supported by the local authorities.[43]

The revolution's anti-Americanism and belief in a U.S. counterrevolutionary conspiracy led to a brief takeover of the embassy on February 14, 1979. On the same day, U.S. Ambassador to Afghanistan Adolph Dubs was kidnapped and killed. Nine months later, on November 4, after President Carter admitted the ailing Shah into the United States, the U.S. embassy and its occupants were again made hostage. Those who organized the seizure were determined to further radicalize the revolution and destroy moderate forces in Tehran who might, treasonously in the radicals' view, seek to rebuild relations with the United States. A few days later, the U.S. embassy in Pakistan was burned down by a mob while Pakistani troops were slow to respond; two U.S. servicemen died. In February 1980, U.S. Ambassador to Colombia Diego Ascencio was seized along with 15 other countries' ambassadors by terrorists and was released only after two months of negotiations.[44]

As the era of modern terrorism began, Washington adopted a policy of refusing to make political concessions to terrorists, although in practice it was often willing to negotiate or pay ransom for kidnapped Americans. Other governments' lack of cooperation weakened the fight against terrorism. In January 1977, France refused to hold Palestinian terrorist leader Abu Daoud for extradition. Similarly, Yugoslavia would not detain the terrorist "Carlos," who was traveling through that country. U.S. Ambassador Cleo Noel and Deputy Chief of Mission George Moore were murdered by Palestinian terrorists in the Sudan in 1973, but the killers were released from prison. Greek Cypriot gunmen who murdered U.S. Ambassador Rodger Davies in Cyprus in 1974 were freed there after minimal jail sentences. The assassins of U.S. Ambassador Francis Meloy Jr. and FSO Robert Waring in Beirut in 1976 were extradited but never prosecuted by Lebanese authorities.[45]

At State, such crises were monitored by the Operations Center. Ordinarily, that high-pressure, around-the-clock office receives 500 to 700 cables daily. The watch officers distribute incoming cables and keep track of the whereabouts of State's leadership in case they must be called for emergencies. There is a large, full-color world map with clocks set at different time zones, teletypes from commercial wire services, and special machines for making multiple copies simultaneously. "This job is like having constant jet lag," explained an FSO working there. But on this crisis the pressures were even far more intense than usual. On such fast-breaking problems, the Operations Center sits in the middle of a complex jumble of developments, avoiding any temptation to "cry wolf" but equally prepared to note the critical moment when events are changing and higher authority must be summoned.

Such distinctions were not difficult to make on February 14, 1979. After receiving a report from Afghanistan of Dubs's kidnapping at midnight Washington time, watch officers called Ambassador Anthony Quainton, director of State's Office for Combatting Terrorism, and others to form an Afghanistan task force. Top-priority cables arrived describing developments, while Quainton's group kept an open telephone line to the embassy in Kabul.[46]

Two hours later, the White House Situation Room's duty officer convened a conference call. "I hate to raise another problem area," he began, "but we have another serious problem." A stunned member of the Iran Working Group listened to a Marine security guard reporting, over a commercial telephone line from Tehran, the attack on that embassy. Rifle fire could be heard over a loudspeaker carrying the conversation to others in the room. Now word was passed to Vance, Newsom, and Saunders, who all rushed to the Operations Center.

That day had begun far more hopefully in Tehran. About 40,000 Americans had been successfully evacuated from Iran during the revolution, but now the end of the fighting brought people strolling peacefully in the streets. Some embassy staffers, tired of staying home during the final days of battle, had just gone back to work. An apparent return to normality was punctuated by the arrival of a message from Secretary Vance to the new Iranian foreign minister. Shortly after 9:30 A.M., however, rifles and machine guns were fired at the embassy compound. Marine guards scrambled for their posts, carrying tear-gas grenades and dressed in combat gear and gas masks, while embassy employees lay on the floor for safety. After about 20 minutes of shooting, armed leftists invaded the embassy; the Marines held their fire to avoid casualties and an even worse incident. The attackers vandalized offices and threatened the staff. Back in Washington all communications lines to Tehran went dead about 3:30 A.M.—the embassy had surrendered.

Tension mounted a few minutes later as one of the task force's members could be heard shouting over the poor connection to Kabul, "The ambassador is wounded? You say the ambassador is seriously wounded?" And finally: "The ambassador is dead." Soviet advisers, ignoring pleas by U.S. embassy officials to negotiate, told the Afghan army to attack the terrorists. Dubs was killed in the crossfire. The embassy staff in Tehran was luckier—the Iranian government intervened to obtain their release that evening—but that day must rank among the most difficult in State's history.

These tragedies foreshadowed the second seizure of the U.S. Tehran embassy, ten months later, on November 4, 1979, and the ensuing 444-day hostage crisis. The staff there had been augmented at the request of the Iranian government, itself under attack from nominally allied but more radical Khomeini supporters. These latter groups believed that the embassy was orchestrating rising anti-Khomeini opposition within the country. Friendship with the United States was poisonous to the revolution, they argued, and a break in relations was to the militants' political advantage.

Some of the FSOs taken hostage believe an additional factor provoking the embassy seizure was the long line of Iranian bureaucrats, businessmen, and even military officers outside its visa office. It provided visible evidence of U.S. willingness to give disgruntled Iranians sanctuary, and of Iranians' willingness to deal with the Americans. Yet even the most outspokenly anti-American officials were prone to seek special consideration for friends by shoving a few Iranian passports into the hands of any visiting U.S. embassy officer.

While the Shah's admission into the United States was the pretext for the embassy takeover, the action had deeper political roots. The Islamic radicals found the kidnappings achieved all of their aims at once: forcing the moderates out of power, destroying any chance of rapprochement with the United States, and removing the "nest of spies" from Iran's capital. The holding of the U.S. diplomats also became a rallying point for the revolution. Iranian politicians who opposed it were intimidated into silence.

Actually, the staff in Tehran and Precht in Washington had warned that the Shah's admission could bring retaliation against the embassy. The main advocates of admitting him to the United States were Henry Kissinger, New York banker David Rockefeller, and his aide, Joseph Read. Carter was aware of the dangers, but concluded that it would be dishonorable to turn away such a veteran ally and wrongly believed that the exiled monarch could receive needed medical treatment only in the United States.[47]

Nevertheless, while FSOs became the victim-heroes of the hostage crisis, State became a scapegoat. "The first thing we would do," said a *Wall Street Journal* editorial, "is to fire the whole lot of policymakers who have blundered us into such a mess."[48] The long, frustrating span of negotiations and confrontation with Iran was made inevitable by the time it took Tehran to form a decision-making authority as well as by constraining political considerations, particularly the U.S. desire not to push Iran into the USSR's arms.

Still, once again there were policy differences between State and the NSC staff. The State Department could only coordinate the U.S. response; the decisions were made in the White House. At a foreign policy breakfast shortly after the hostages were seized, Vance and Brzezinski took sharply contrasting positions. Brzezinski warned the president against allowing the crisis "to settle into a state of normalcy. . . . Yes, it is important that we get our people back. But your greater responsibility is to protect the honor and dignity of our country and its foreign policy interests." The implication was that military action to pressure Iran or a rescue mission was needed as soon as possible. Most of the hostages, of course, were more immediately Vance's "people." The secretary of state replied, "We're dealing with a volatile, chaotic situation in Iran, and negotiation is the only way to free them." The president and nation, he continued, "will ultimately be judged by our restraint in the face of provocation, and on the safe return of our hostages."[49]

During the first five months of the crisis, Carter followed Vance's advice. The president's decisions were implemented under the supervision of a special coordinating committee operating from the windowless, wood-paneled Situation Room in the White House basement, a far more spartan location than its furturistic image in various films. Undersecretary Newsom was in charge of the task force, with assistance from Assistant Secretary Saunders, Iran Office Director Precht, and others. At State itself, the Iran Working Group handled long, exhausting shifts facing endlessly ringing telephones, hundreds of inquiries from Congress and the hostages' relatives. It was often dependent for news on the mass media, given the paucity of direct contacts with the Iranian government.[50]

One constraint pushing the president toward diplomatic efforts was the

Soviet invasion of Afghanistan in December 1979. While the hostage crisis had a profound emotional and psychological effect on the United States, the dispatch of Soviet troops to install a servile pro-Moscow Communist Afghan government served to demonstrate Moscow's aggressive objectives. James Taylor, an embassy political officer in Kabul, was the first to discover the invasion. Driving to work on the evening of December 27, 1979, he saw vehicles heading straight toward him, ignoring his flashing high-beamed lights. Only 30 yards away, he realized that he was about to collide with Soviet armored vehicles. "An abrupt and bumpy left turn over the traffic island carried me past them," Taylor wrote, "but not before I saw their combat-ready infantry dismounting—moving off down the street behind the embassy toward Radio Afghanistan, about 500 yards away." Soviet bullets whizzed close to U.S. embassy employees, some of whom were busying themselves burning classified documents.[51]

Washington was frazzled by the frustrating hostage crisis—promising negotiations broke down in March 1980—and the Afghanistan invasion. When Iranian embassy representatives came to the State Department building to receive expulsion notices two months later, one of them said the hostages were being well treated and that some would prefer to remain in Iran. Precht replied with an obscenity, and the Iranian diplomats, claiming they had been insulted, stormed out of the meeting.[52]

This frustration with the diplomatic route helped lead to the White House decision to launch a rescue mission. Talks with Iran were going nowhere. Media coverage, the U.S. international image, and the ongoing presidential election all put pressure on Carter to achieve results.

Vance was doubtful about the military's ability to carry out a rescue operation—generals tell you they can accomplish a mission whether or not that is true, he told associates. His attitude can be traced to an earlier experience. As army secretary, Vance had to help decide whether to airdrop U.S. troops at night during the 1965 intervention in the Dominican Republic. The action was fortunately postponed, Vance recalled, "Because it turned out that our maps were faulty and . . . there would have been a substantial number of casualties."[53]

Vance had other concerns as well. State had just finished assuring European allies, during talks on sanctions against Iran, that if they imposed strong economic restrictions, the United States would not take military action. Vance worried that the U.S.'s closest friends would believe they had been deliberately deceived. Further, he held to a principle, quite contrary to Brzezinski's thinking, that the armed forces should not be used to achieve what might be done by diplomatic means. Finally, Vance reasoned that people would be killed unnecessarily and that Iran would simply round up other Americans in Tehran, including journalists, and hold them hostage.[54]

On April 10, Vance went to Florida for a weekend, and the following day, in his absence, an NSC meeting decided to go ahead with the rescue mission. Although Christopher represented State, Vance felt that he had been deprived of a personal opportunity to argue against the plan, that he no longer enjoyed the White House's confidence, and that Brzezinski had cut him out of a key decision. Vance decided to resign as soon as the mission was finished. "You would not be well served in the coming weeks and months," Vance declared, "by a secretary of state who could not offer you the public backing you need on an issue and decision of such extraordinary importance—no matter how firm I remain in my support on other issues, as I do, or how loyal I am to you as our leader."[55]

State's staff was dismayed by Vance's departure, which many blamed on Brzezinski's maneuvers. Veteran Democratic Senator Edmund Muskie, appointed by Carter to succeed Vance, emphasized his intention to be the administration's foreign policy spokesman,[56] but in his brief tenure he was never able to compete effectively with Brzezinski.

During the remaining months of Carter's term, Deputy Secretary Christopher managed the hostage negotiations while Carter himself handled day-to-day decisions, relying heavily on his closest aides. By the autumn of 1980, the Iranians had finally assembled a parliament empowered to set conditions for the hostages' release. The final defeat of President Abol-Hassan Bani Sadr's faction gave the Islamic Republican party radicals the total power they had been seeking, removing their domestic political incentive for continuing the crisis. Iraq's invasion of Iran in September heightened Tehran's need for assets impounded abroad and arms supplies embargoed by U.S. allies. This new war gave Iran another reason to settle the hostage crisis as quickly as possible.

With the Algerian government serving as intermediary, a U.S. diplomatic team led by Christopher spent 13 days in Algiers working under difficult conditions and well aware that disputes and mistrust might scuttle an agreement at any time. Finally, the hostages were freed on January 20, 1981, as Ronald Reagan was being inaugurated as U.S. president. Even when the hostages were put on a plane to leave Iran, there was worry about some last-minute hitch. The negotiators went into the U.S. embassy snack bar, exhausted but happy. Champagne was opened and passed around, and someone mused out loud, with typical diplomatic pessimism, "What would we do next if it hadn't worked?" In fact, the new Reagan administration ungratefully dropped the team's appointed members from government so quickly that their salaries stopped when they were still in Algiers. Their fares home were only paid after it was pointed out how disgracefully they were being treated.

The hostage issue monopolized the attention of U.S. policymakers longer than any crisis since Vietnam. Other problems, like Moscow's attempt to suppress the Solidarity union movement in Poland, which almost led to Soviet military intervention in the fall of 1980, had to be dealt with at the same time. But most other questions had to take a backseat, a reminder that while the system involves thousands of people, decisions on major issues are made by a very small group at the top.

In facing both the Nicaraguan and Iranian revolutions, Washington's perceptions and reactions often lagged behind events. By the time the government absorbed information from the field and made decisions, the situation had changed too much for these policies to be effective. The Iran problem encompassed both a failure of process and an internal political struggle. The debate over Nicaragua provides more examples of these factors.

There were two schools of thought in ARA. A liberal view saw pro-U.S. dictatorships as heading toward instability because internal problems and unrelieved popular disaffection would produce radicalization and revolt. The Cuban revolution and later guerrilla insurgencies elsewhere were interpreted as responses to domestic ills, attracting followers because of legitimate grievances. Timely political and social reforms were needed to defuse potential rebellions. The other viewpoint, defining itself as a realpolitik approach, thought tampering with existing regimes might only produce something worse for U.S. interests. The internal politics of friendly governments were none of the U.S.'s business, but this view was also quicker to see left-of-center reformist regimes as connected with Cuban and Soviet influence.

It is the job of all regional assistant secretaries to pull their subordinates behind U.S. policy despite clientism. At first, ARA's Central America office was protective of Nicaraguan dictator Anastasio Somoza; those dealing with dictatorships in Argentina, Chile, and Uruguay were reluctant to risk conflicts over human rights since bilateral relations were otherwise satisfactory. "I'm all for human rights," some FSOs would say, "but we shouldn't close our options with the existing governments." In contrast, the Andean, Mexican, and Caribbean offices, many of whose clients were democratic, were more positive about the directives.[57]

Part of the problem was with ARA's tactics. The East Asia bureau, for example, dealt with congressional human rights inquiries, even under Kissinger, by frank meetings. Just before Carter took office, however, a U.S. ambassador to Uruguay and an ARA deputy assistant secretary sent down to investigate told skeptical legislators that repression there was merely countering a Communist plot to overthrow the government. While Todman and other FSOs expressed distaste for the human rights policy, Todman's successor, Viron Vaky, knowledgeable and realistic about the region, believed the policy could be used to advance U.S. interests. But the caution of many ARA hands toward Carter's policy was also motivated by a desire to protect themselves. As one FSO said, "This is today's fashion, tomorrow it may change. Why should we burn our policy and career bridges?" To judge from what followed, they were right. Carter removed only two or three "conservatives," deemed too soft on Chile's dictatorship or human rights, but the Reagan administration made almost a clean sweep of the professionals.[58]

The fates of two individuals illustrate the problem. Lawrence Pezzullo joined the Foreign Service as a "mid-career" entry, and his experience outside of government made him bolder than most of his colleagues. After an assignment dealing with Capitol Hill, which sensitized him to Congress's human rights concerns, he became Carter's ambassador to Uruguay in 1977. Pezzullo was so well regarded that one ARA official went around saying the appointment "showed the system works." Another veteran officer responded, "If it worked so well you wouldn't be so surprised." By all accounts, Pezzullo did a superior job, mobilizing the embassy to press the junta for democratization. One ambassador explained, "The trick is in convincing the disparate agencies that there is one policy and that they have not just to deal with their own issue priorities but with a whole strategy." When a good man was needed for ambassador to Nicaragua, in the midst of the revolutionary crisis, Pezzullo was selected.

The career of another FSO, Frank Ortiz, was affected differently by political currents. In 1980, Ortiz was removed as ambassador to Guatemala and given a less important post for failing to support the human rights policy. When Reagan came into office, Ortiz flourished as envoy to Peru and later Argentina. ARA was hit harder than any other bureau in both White House transitions.

Anastasio Somoza's family had ruled Nicaragua for decades using U.S. support to bolster its authority; Washington never employed leverage to affect Somoza's policies. In 1972–1973, after a disastrous earthquake struck Nicaragua, some State officials wanted to use U.S. aid to force Somoza to broaden his government, but President Nixon and the U.S. ambassador, friendly to the dictator, refused. Reconstruction funds were largely stolen by Somoza and his cronies.

Opposition to Somoza grew in the following years. While the armed struggle was led by the Sandinistas' quarreling leftist factions, moderate middle-class elements were increasingly disenchanted as well. The assassination of respected dissident publisher Joaquin Chamorro on January 1, 1978, was a turning point as the Catholic church, business groups, labor unions, and others became active. These sectors were willing to negotiate with Somoza toward ending his dynasty and holding free elections, but the dictator stalled and the United States would not push him into a dialogue, confining itself to suggestions Somoza ignored.

In September 1978, the Sandinistas launched a military offensive producing a state of civil war; the moderate opposition demanded that Somoza leave office immediately. Assistant Secretary Vaky, who had just taken office, worked with support from State's leadership to encourage mediation between Somoza and the moderates, persuading the Dominican Republic and Guatemala to participate in multilateral efforts. Both Somoza and the moderates, now organized as the Broad Opposition Front, accepted the resulting international commission, which then produced a proposal for a peaceful change of government.

Within the U.S. government, the Defense Department thought Somoza could survive; the State Department believed he was likely to fall, but could not convince the NSC and CIA. Carter would periodically insert himself into the debate, but does not appear to have understand the issue very well.[59] At this critical juncture, Washington failed to pressure Somoza sufficiently for serious negotiations. The dictator produced a plan for a "plebiscite" administered by his government to demonstrate relative support for existing political parties. The opposition, which had no formal political party, gave up in disgust.

There was a split over the plebiscite within the U.S. government. Vaky and William Bowdler, the director of State's INR, who was serving as U.S. mediator, argued for rejecting Somoza's proposals, explaining that the plebiscite ploy could destroy chances for a negotiated solution. Deputy National Security Adviser David Aaron supported them, but the White House, along with Vance and Deputy Secretary Christopher, decided that the United States could not reject such an apparently democratic method.

So Bowdler tried to find some type of plebiscite acceptable to the opposition and, with his fellow Dominican and Guatemalan mediators, developed a formula for international oversight of a vote on whether Somoza should stay in office. Somoza again dug in his heels, realizing he could not win electoral support on retaining power, and the mediation collapsed in January 1979. The moderate front fell apart. Many of its leaders went into exile to work with the Sandinistas, who also gained support from Panama, Costa Rica, and Venezuela, while Cuba increased arms supplies to the guerrillas.

Meanwhile, Somoza consistently outmaneuvered the United States. When Carter sent him a letter with some praise, mainly intended to push him toward dialogue, Somoza published the positive sections. His plebiscite proposal put the U.S. government on the defensive while damaging cooperation between Washington and the moderates. Somoza also astutely used personal contacts, leaking all of State's proposals to congressmen John Murphy and Charles Wilson, who warned Carter against pressuring Somoza. This was an effective team, since Murphy chaired the subcommittee with jurisdiction over implementing the Panama Canal treaties and Wilson was the swing vote on foreign aid appropriations. The White House never rallied congressional supporters for anti-Somoza moves.

But Somoza's tactical shrewdness failed to save him from overwhelming opposition at home. By June 1979, when the Sandinistas launched their final offensive, Washington could only make a long-shot attempt to avoid a complete Sandinista victory. Vaky, Bowdler, and Ambassador Pezzullo, with the approval of Vance and Christopher and with the NSC's support, put together a plan to broaden the post-Somoza government. Somoza would leave the country, power would be turned over to the pluralist opposition front, a cease-fire would be implemented, and the army would be reformed and eventually merged with the insurgent forces. The Sandinistas accepted this proposal, which would have left a greater potential role for the moderates and army than would a Sandinista military victory, but the plan fell apart through Somoza's sabotage. When Somoza left the country in July 1979, his handpicked successor refused to abide by the commitments for a transfer of power. It was a suicidal, Samson-like act: The army collapsed, the moderates joined the Sandinistas, who now enjoyed support from a broad cross section of Latin American countries as well as Cuba, and a leftist-dominated government took power in Managua.[60]

Washington had forfeited four opportunities to prevent this worst-case outcome: the period after the assassination of moderate opposition leader Pedro Chamorro in January 1978; the initial mediation proposals in October 1978; the plebiscite compromise idea; and the doomed June 1979 minimalist formula. Somoza, of course, had no intention of leaving power until the end. By forcing the center to disintegrate, he hoped to polarize the situation, making Washington support him rather than accept a Sandinista victory. As the internal policy debate continued, half-measures and indecision at the top allowed chances to slip away.

Only U.S. pressure could have pushed him into an agreement. Vaky and Bowdler repeatedly attempted to gain such high-level support, but top decision makers would not consent. Given their own noninterventionist preferences, the White House, Vance, and Christopher felt hard put to contest the assertion that the United States should not "play God" by deciding Nicaragua's next government. As the crisis escalated, the White House would at best agree to the now insufficient measures proposed against Somoza during the preceding period. Preassuring Somoza to leave also held potential domestic political costs for the administration: If the radicals had still come to power in a post-Somoza era, it would have been blamed for "losing" Nicaragua to anti-U.S., pro-Soviet forces.

Each time Somoza made an initiative, Washington took two weeks to work out a response. As it proved unwilling to pressure Somoza, essential time was lost and U.S. credibility with the opposition eroded. Carter never provided needed leadership. Vance drifted in and out of the discussion, recommending caution. Brzezinski and his staffer on the region, Robert Pastor, opposed attempts to pressure Somoza into concessions, blocking any serious response to ARA's warnings.

The Carter administration was also ineffective in saving a reformist 1979 coup in El Salvador, which was soon recaptured by traditional hardliners. In January 1980, Bowdler, who had spent much of his career at ARA, replaced Vaky as assistant secretary. Two months later, another career man, Robert White, became ambassador to El Salvador. White's Senate confirmation was opposed by Senator Jesse Helms, who thought he was too liberal. Salvadoran rightists agreed and besieged White's residence in May 1980.[61]

The Reagan administration's quick ousting of Bowdler and White and other ARA personnel demonstrated one main lesson in the policy process's recent history: The metronome of ideological change was accelerating between the Nixon-Ford, Carter, and Reagan administrations, making life more difficult for the career staff. Those years also showed the staying power of Kissinger's revolution. The national security adviser and his staff were established as equal rivals of State. Yet without Kissinger's institutional and personal domination, competition was reaching dangerous heights. Both the Iran and Nicaragua crises showed how the policy process was slowing down to the point of paralysis and how inconsistent signals confused both friend and foe alike.

"We on the inside," Brzezinski later wrote about the infighting, "underestimated the political damage created by the public perception of Presidential indecision which the so-called Vance-Brzezinski split was generating." Brzezinski added, "Precisely because Carter did dominate the process, he did not feel troubled by the disagreement."[62] This analysis is wrong on two counts. First, although the two sides did agree on many issues, the conflict was not merely a matter of media manipulation or friction over "turf." These battles and philosophical differences were real, affecting the response to crises and foreign relations. Secondly, while Carter ultimately decided policy, he often did so too late to affect events for the better. This last problem was partly due to the system and partly to Carter's own shortcomings.

In Iran, accurate data only slowly rose to the top and there fell prey to disputes among policymakers. On Nicaragua, the lack of attention or comprehension by decision makers and the tendency to postpone or water down decisions produced a U.S. policy incapable of enough drive or consistency to succeed. Similar problems of information and understanding plagued Kissinger's responses to the crises of Bangladesh and southern Africa. But Kissinger, for better or worse, was decisive and held power to implement choices, while Carter and Reagan had several advisers presenting conflicting options and suggestions, a system presuming that the chief executive was able to transcend such divided counsel and develop a firm policy.

The NSC staff's form—small size, proximity to the president, and focus on a limited number of high-priority issues—as well as its leaders' personalities seemed to give it the advantage. Brzezinski's patronizing assessment portrayed Vance as "best when negotiating with decent parties in the world . . . at his worst in dealing with the thugs of this world. His deep aversion to the use of force was a most significant limitation on his stewardship in an age in which U.S. power was being threatened on a very broad front."[63]

Without overstating the point, however, these were characteristics of State as an institution in the business of diplomacy and mediation. As an instrument of the White House's drive for decisive success and a consistent strategy, the NSC was often equally characterized by a tendency toward conflict and confrontation. These attributes were visible in Brzezinski's own misestimates stemming from excessive eagerness to use force, exaggerate threats, and escalate conflicts.

The Carter administration's experience seemed to confirm a major structural shift in the U.S. policy system. "The idea that the Secretary of State is going to be chief policy spokesman isn't going to work," claimed political analyst Philip Odeen. Major foreign policy issues had too many economic and security implications for State to handle on its own. The secretary of state, Odeen argued, "cannot look at them from the White House point of view."[64] The Reagan administration set out to prove these assertions false, but may have succeeded in establishing them even more firmly as the new consensus.

9

On-The-Job Training: The Reagan Administration

Ronald Reagan wanted to alter dramatically U.S. foreign policy in a conservative direction and drastically revise management of the decision-making process. Reagan's team decried Kissinger's détente and arms control policy, which they saw as accepting Soviet superiority, as well as Carter's alleged weakness toward enemies and his undermining of friends. Strength and willpower, it claimed, would right the balance in the U.S.'s favor.

Rejecting Kissinger's solo dominance while also seeking to avoid the Carter administration's internal disputes, it concluded that the national security adviser's role must be reduced in favor of restoring the secretary of state's primacy. Since few of the Reaganaut right-wingers had foreign policy experience, however, the administration's two secretaries of state came from outside Reagan's circle of personal and ideological friends. For this reason—and also because State was considered a liberal bastion—the secretary was never given the authority necessary to lead the policy process. Without a strong, active presidential presence to resolve lower-level disputes, the previous two-sided battle became an even more confused melee, involving not only the secretary of state and national security adviser, but also the White House staff and secretary of defense. The decision-making process ended up more mangled than ever before.

Richard Allen, who had spent two decades in Washington without ever producing a notable idea, became national security adviser, with orders to maintain a low profile and smaller staff. Reagan chose Alexander Haig as secretary of state, partly on Nixon's recommendation. A career army officer assigned to the NSC in 1969 (Vance was one of those who recommended him), Haig rose to become Kissinger's deputy and NATO commander. Haig was a non-Reaganaut and a Kissinger disciple whose politics, style, and methods were unacceptable to the president's men. The arrangement was doomed from the start.[1]

As in any administration, the foreign policy system came to reflect the president's abilities and requirements. Neither Carter nor Reagan had prior international experience, but the two men were quite different in personality. Carter wanted to hear a variety of views and immerse himself in detail, attempting, often too slowly or uncertainly, to reconcile them. Reagan, in contrast, was not very interested in foreign affairs except where they touched on his ideological principles. He lacked the background and inclination either to run the process himself or to form a close partnership with someone who could do it for him. Consequently, his White House staff played a particularly important role, first as a counter to Haig and later by filling the vacuum left by Reagan's relative passivity.

Another complicating factor was the failure to establish clear decision-making channels. Haig's attempts to take the lead, regardless of his tactlessness, were unacceptable to the White House. Allen was incapable of playing a central role, and the administration had already decided not to have a strong national security adviser. Secretary of Defense Caspar Weinberger and CIA Director William Casey were more forward than most of their predecessors in advocating specific options, but they could hardly run policy. Nor would a weakened NSC staff or an inattentive president properly prepare for effective NSC meetings where advisers debated and the chief executive made choices. The result was confusion, rapidly shifting factions, and burgeoning suspicions. As one participant put it, "Everything was a fight."

Relations with the bureaucracy also began on a bad footing. Previous presidents had failed to protect the career staff, but the Reagan administration did not even safeguard its own appointees. Several assistant secretaries faced months of humiliating harassment from Senator Jesse Helms before they were finally confirmed. The spread of political appointees into lower positions, including special assistants as well as deputy assistant secretaries, increased. Never before had senior FSOs waited so long for new and often disappointing assignments.

The Reaganauts thought they were combatting resistance from a Foreign Service they saw as a hotbed of liberalism. Allen later wrote that the "counterrevolutionary power of the bureaucracy" and the "mainland China lobby" at State was wearing down Reagan's militancy. "The president was made to appear weaker than he really is on the issue of coping with . . . Soviet adventurism."[2] ARA underwent a particularly thorough purge; two deputy assistant secretaries at EA were forced out because Helms thought them too warm on U.S.-Peking ties. Morris Abramowitz, an FSO who had served with distinction as ambassador to Thailand, lost the post of ambassador to Djakarta because Defense Department enemies—he had properly forced the military to follow embassy priorities in Bangkok—slandered him to the Indonesians. Abramowitz became ambassador to military force reduction talks in Europe; the White House's next nominee for the Indonesia job had been involved in such questionable commercial dealings there that his nomination had to be withdrawn.

The adversaries in the State-NSC struggle for influence in the Reagan administration changed frequently: The original Haig-Allen combination only lasted one year. Allen was replaced by Reagan's friend, Deputy Secretary of State William Clark, in January 1982. Haig resigned after a number of policy disputes in July 1982, to be replaced by George Shultz. Clark restored the NSC's importance but was succeeded in October 1983 by his deputy, Robert McFarlane.

Haig's belligerence at bureaucratic infighting sprang from attempts to imitate Kissinger's foreign policy hegemony and frustration at not being able to do so. "He was always ready to do battle with 'the bastards' and the 'forces of darkness,'" said a former Nixon-era associate, "but now he's much more . . . super-sensitive."[3] This behavior was encouraged by the existing hostility and confusion, but it scarcely helped Haig's position.

Actually, Haig had experienced similar situations. At General Douglas MacArthur's headquarters in Japan, where Haig served in the early 1950s, and in the Nixon White House, a powerful leader had surrounded himself with loyalists who delighted in compiling enemies' lists. But now Haig himself was outside the magic circle. Once a bureaucratic samurai judged by fierceness in fighting his lord's battles, Haig had now become one of the bosses. Whereas he had once been rewarded for toughness in attacking other agencies and factions, his new role called for a cautious, deferential style.

Consequently, Haig's talk of saboteurs and his efforts to gain control of the process by acting as if he were already in charge only made matters worse. Kissinger had been Nixon's partner; such pretensions on Haig's part struck the Reaganauts as treasonous. Kissinger manipulated the media, while Haig's enemies devastated him with leaks. Finally, the administration's friction with the career staff, a secretary of state's natural institutional ally, transformed one of Haig's greatest potential strengths into an additional liability.

Originally, Haig proposed a policy structure placing most authority in his hands, intending to be, as in his much-quoted remark, "the vicar of foreign policy." Obviously, this pattern was based on Kissinger's 1969 coup preempting Secretary of State Rogers, but Kissinger was fulfilling Nixon's design while Haig was hardly accepted as Reagan's alter ego. Instead, the White House put Vice-President George Bush in charge of crisis coordination. Communication among decision makers was already breaking down; Haig let it be known that the decision was not discussed with him in advance.[4]

William Clark, the original number-two man at State, was the White House's choice to hold Haig to the Reaganaut line. Clark, a lawyer without foreign policy experience, had been Governor Reagan's adviser and appointee to the California Supreme Court. He got along with Haig, but played a relatively small role in his year at State. Clark's power over foreign policy would await his promotion as Allen's successor.

During his 18 months in office, Haig was ably helped by most of his own appointees at State. Assistant Secretary for European Affairs Lawrence Eagleburger, once Kissinger's expert on running the department, had been ambassador to Yugoslavia during the Carter years. He became Haig's trusted adviser and was soon promoted to undersecretary for political affairs, the highest career post. Eagleburger survived Haig's resignation and, until his own retirement in 1984, played the same role for George Shultz. In a city where few are without detractors, Eagleburger's ability and intelligence were widely respected.

Eagleburger's career illustrates one of the main FSO routes to success. He joined the Foreign Service in 1957 and first served as an economic officer in Honduras. Two years later he became political analyst on Cuba in the Bureau of Intelligence and Research. In 1961, he took Serbo-Croatian language training and was sent for four years as an economic officer in Yugoslavia. His next assignments were to the Secretariat, as special assistant to former Secretary of State Dean Acheson, who was advising President Johnson on NATO problems, and again back to the Secretariat as acting staff director. In October 1966, Eagleburger went to the NSC staff for a year to work on European affairs. He then served as a special assistant to the undersecretary for another year and was given the job of helping Kissinger with the transition, giving him a total of six straight plum Washington assignments. Kissinger kept him on as executive assistant until Eagleburger's health collapsed and he was given a relative rest as chief of the U.S. political mission to NATO and as a deputy assistant secretary of defense. When Kissinger became secretary of state in 1973, he brought Eagleburger back as an executive assistant, and in 1975, he made him undersecretary for management.[5]

Eagleburger's climb to the top was fostered by holding so many posts close to top policymakers. Also helpful was his range of experience with different institutions—State, NSC, the Defense Department, NATO—and with Europe, which has remained, albeit under increasing challenge, U.S. foreign policy's flagship region. He was in the right places at the right times and was not politically committed until the advantageous moment when Kissinger's star was on the rise. If Eagleburger had taken a political appointment or high-profile position at the end of Johnson's administration, for example, he would have been pushed aside by the incoming Nixon team.

Another FSO, Thomas Enders, a Haig discovery during the Cambodia crisis, had "peaked" too soon, making influential enemies in Congress and even at the Kissinger-era State Department. But he survived comfortably in ambassadorial posts until Haig returned him to Washington as assistant secretary for ARA. For the first two years of Reagan's term, Enders achieved a high degree of bureaucratic success, dominating the interagency process over Central America, the administration's most consistently important foreign issue. Enders combined a high degree of arrogance and ambition with a keen intelligence. Although very different in personality, Eagleburger and Enders were both hybrids, FSOs possessing many of the politicians' skills.

While Eagleburger and Enders were undoubtedly Haig's most important mid-level advisers, two noncareer men also played important roles. Richard Burt, a strategic analyst who had also worked as a *New York Times* correspondent, started as director of the Bureau of Politico-Military Affairs and then succeeded Eagleburger at the European bureau, becoming the first non-FSO at the latter post within memory. Eager to seize opportunities and credit, even by Washington standards, Burt made more than the average number of enemies. Haig's relationship with Paul Wolfowitz, head of the Policy Planning Staff, gave that group influence as the secretary of state's think tank. After Haig resigned, Wolfowitz became assistant secretary of state for East Asia.

African affairs during the Reagan administration were dominated by Chester Crocker, an area specialist who had been director of African studies at the Georgetown University Center for Strategic and International Studies. Crocker was sympathetic to the problems of black African states, but believed that a controversial "constructive engagement" toward South Africa would more likely encourage internal reforms and independence for its Namibia colony than would attempts to isolate the white minority regime.[6]

Haig was less satisfied with some other choices. He was barely on speaking terms with Undersecretary for Economic Affairs Myer Rashish and hardly utilized talented, respected Assistant Secretary for Economic Affairs Robert Hormats. Before he was finally let go, Rashish was kept for months in what one official called "the closest thing to purgatory that we have over here."[7] Senator Helms blocked Hormats's promotion to the job, leading to his resignation. Since no one effectively rescued State's leadership in international economic affairs, its position further declined toward a point of no return, even under George Shultz, the first economomist ever to serve as secretary of state. Haig was also unimpressed by his FSO assistant secretary for the Middle East, Nicholas Veliotes.

The Republican party's right wing was not pleased to see "establishment" figures, including Kissinger-era retreads, given the best positions. Helms was the appointees' tormentor. Partly as leverage to promote his own favorites, Helms held up confirmation of Crocker, Burt, Hormats, and others he considered too liberal. He wanted to ensure that State was neither too friendly with the Peoples Republic of China nor too unfriendly toward South Africa. The conservative lobby won a few mid-level posts, blocked a few disliked career people, and forced some transfers. After months of delay, Helms allowed Reagan's major appointees to go through unscathed, though embittered.[8]

Helms's chief aide, John Carbough, suggested, "We need an assistant secretary for human rights as much as we need an assistant secretary for motherhood." When Reagan's original ultrarightist candidate to head that bureau was rejected by the Senate Foreign Relations Committee in a conflict-of-interest controversy, Elliott Abrams was given the job. Abrams complained, with some justice, that many liberals had a double standard on human rights, criticizing right-wing regimes far more than leftist ones. But the Reagan administration had its own ideological slant on the matter and showed little interest in the issue, viewed as a distasteful Carter-era leftover.[9]

UN Ambassador Jeane Kirkpatrick, a Georgetown University professor, never got along with Haig; some of his subordinates, including Burt, leaked unfavorable stories about her. She grew more powerful after Haig's resignation and always maintained her own direct line to the White House.[10] With few exceptions, Reagan's other ambassadorial appointees were not outstanding. In theory, they were screened by a committee of State Department and White House officials, but the nominees included many whose sole qualification was their large contributions, political support, or personal connection to the Reagan campaign. About 44 percent of them were not FSOs, complained the American Foreign Service Association, the highest proportion of any modern administration.[11]

Haig was both victim and provocateur of the administration's internal conflicts. The government must speak with a single voice, he told State in his own inaugural message: "The president needs a single individual to serve as the general manager of American diplomacy. President-elect

Reagan believes that the secretary of state should play this role. As secretary of state, I would function as a member of the president's team, but one with clear responsibility for formulating and conducting foreign policy, and for explaining it to the Congress, the public and the world at large. The assistant to the president for national security would fill a staff role.''[12]

The tone of this statement reveals much about Haig's problem with the White House staff. Despite the lip service paid to the president's leading role, he made the chief executive seem a rather auxiliary figure. No one else in government, White House aides thought, would be left with much to say at all. They sarcastically dubbed Haig ''CINCWORLD,'' pseudo-military jargon for ''commander-in-chief of the world.''

Allen was another disappointment. His lack of energy and policy clout frustrated NSC staffers. In contrast to his predecessors, Allen was subordinate to other White House staffers. Allen failed in his first big assignment: piloting through Congress the controversial sale of AWACs early-warning planes to Saudi Arabia. The proposal barely squeaked through the Senate in October 1981, after Baker pushed aside Allen and took over responsibility. To make matters worse, even before he took office, *The Wall Street Journal* questioned his ''ability to distinguish between the affairs of government and his own personal business interests,'' citing documents showing his apparent use of White House connections to seek lucrative consulting contracts for himself and friends.[13] Mounting accusations of financial impropriety finally convinced the White House that Allen had to be replaced.

Yet even the weakness of this potential rival was more of a problem than a benefit for Haig. Rather than facing a more or less equal conflict with the national security adviser, he found himself clashing directly with presidential aides who thought Haig's attempt to portray himself as foreign policy chief a challenge to Reagan's own prestige. Chief of Staff James Baker, his deputy, Michael Deaver, and counselor Edwin Meese equated their own positions with the president's honor, considering themselves more representative of his wishes than any secretary of state could ever be. As political operatives, Haig later wrote, ''These men were intensely sensitive to the public mood and reluctant to take any action that might [result in] alienating public affections or creating controversy.'' Haig's high profile and strong rhetoric ran contrary to these objectives. If Allen could not handle Haig's perceived threat, the White House staff took the job on itself. They sat at the cabinet table at meetings, limited Haig's access to Reagan, and criticized him in the press.[14]

When Reagan was wounded in a March 1981 assassination attempt, Haig's ''I'm in charge'' statement was taken out of context by the White

House staff—and much of the public—as further evidence of an erratic and wildly ambitious character. Early acrimonious battles reinforced derision and fear of Haig. He collided with Budget Director David Stockman on foreign aid levels, with the Agriculture Department over continuing a grain embargo on the USSR, and with the Defense Department over a variety of issues. He may have been right on these points, but was outnumbered by increasingly hostile colleagues. Kissinger, too, had acted as if he could not afford to lose a single battle to rivals, but defeat was usually avoidable for him; Haig held a much weaker position.

Among other mishaps, Allen and Haig gave separate simultaneous briefings to reporters on arms control proposals. In March 1981, the White House and State had to disavow a statement by NSC staffer Richard Pipes that the United States and the USSR were heading for war; another NSC man, General Robert Schweitzer, was fired after an even more apocalyptic speech that he had failed to clear with Allen. The NSC staff had little influence on the interagency level. "From one meeting to another," said one official, "a different NSC representative will show up. No one seems to be assigned to bulldog an issue."

Haig and his supporters also had grievances against the White House staff. "Our problem has never been with the president," one appointee at State explained, but "with the clones." Yet the White House aides saw themselves as "clones" of Reagan, who held ultimate responsibility for the disorganization. Even Haig placed the blame at the president's door. Haig said that "greater discipline" was needed to overcome the "cacophony of voices" on foreign policy. Asked, "Who has to tighten up?" Haig replied, "The president."[15]

White House aides and the minions of Secretary of Defense Caspar Weinberger, with his own foreign policy agenda, launched a war of leaks reminiscent of the Carter administration. They criticized Haig's personality and policy in thinly veiled statements to the media. When columnist Jack Anderson cited "insiders" as saying that Haig was at the top of the president's "disappointment list" and "has one foot on a banana peel," Haig replied that someone in the White House was conducting a "guerrilla campaign" against him, thereby giving the remarks even wider circulation. There was more confusion to come. Trying to rescue Reagan from a statement that he could envision a limited nuclear war in Europe, Haig told a Senate hearing that Washington might launch a "demonstrative" nuclear blast in case of Soviet attack. The following day, Weinberger said NATO plans called for nothing "remotely resembling" such an operation.[16]

Through all this, Reagan blithely maintained there was "absolutely no foundation" to rumors of conflict, and that the media was only creating an impression of disarray. This was clearly untrue, and in November 1981, the president ordered Haig and Allen to end their feuding. While decrying such disorganization, the reaction of the press and the foreign policy community was quite different from their traditional approach. In the past, they would have judged this situation a result of NSC staff meddling and advocated placing the secretary of state in firm control. The conventional wisdom had now subtly shifted. *The New York Times* concluded, "Things won't work well with a strong national security adviser to the president [but] without a strong adviser, things won't work at all."[17]

Contrary to the Nixon and Carter administrations, State-dominated interagency and senior interagency groups operated without close NSC oversight. At the top, Reagan's planning group consisted of Bush, Haig, Weinberger, Casey, Meese, Baker, and Deaver, with Allen as note taker. But Allen's poor staff operation stymied effective meetings. Position papers were contradictory or not cleared with relevant bureaus; follow-up was rare. Reagan made up his mind on the basis of informal conversations with aides or cronies. One NSC member complained that decisions were "based on who talked more persuasively on a given day."[18]

This situation began to change after Clark replaced Allen. Derided for his obvious ignorance about international affairs, and more self-effacing than Kissinger or Brzezinski, Clark's unbeatable asset as national security adviser was a close relationship with Reagan. Clark quickly established himself as the center of authority over foreign policy, convincing the president to issue a seven-page directive defining the NSC staff's role and clear channels of responsibility. He reasserted White House control, issuing orders to Haig and Weinberger. Once again, the NSC staff produced policy studies, now called National Security Study Documents. Clark was well assisted by Robert "Bud" McFarlane, a Marine colonel and master organizer who had served in the Ford White House and as Haig's counselor at State.[19]

The Reaganauts now challenged Haig on each new appointment. When the job of assistant secretary for international organization affairs opened, Haig supported Alvin Drischler, a former aide to pro-Reagan Senator Paul Laxalt, while the White House preferred Gregory Newell, a second-level presidential aide who had no conceivable experience in the area. Newell won the position.[20]

As Clark asserted himself in the first half of 1982, Haig's power declined. The Reaganauts escalated attacks on Haig's "pragmatic" policies. When Haig warned that sanctions against a Soviet gas pipeline were alienating Western European allies, one White House aide commented that Haig "sometimes acts like Europe's ambassador in Washington."

The key meeting on the issue was held when Haig was out of town and, he claimed, Clark placed only the most hard-line option before Reagan. Facing a mix of media leaks, internal disputes, and apparent sabotage, Haig accused Clark of "conducting a second foreign policy." But Clark viewed Haig as the interloper, telling him, "You've won a lot of battles in this administration, Al, but you'd better understand that from now on it's going to be the *president's* foreign policy." Haig's hint at resignation in July was quickly accepted.[21]

Haig had forgotten, political writer Morton Kondracke noted, "the first rule of backroom bureaucratic politics: 'Maintain your base with the Boss.'" Ironically, Haig had fallen victim to the kind of national security adviser's behavior that he had promoted as a Kissinger aide. Clark had merely done to him what Kissinger had done to Rogers and what Brzezinski had done to Vance.[22]

Three other important factors maintained friction between the White House and State both before and after Haig's resignation. First, there was institutional mistrust of State as loyal to its own rather than administration goals. Reaganauts saw clouds of compromise and concession hanging over Foggy Bottom. Second, there were actual policy differences between White House advisers and State—appointees as well as career staff—over Central America, the Middle East, and arms control. Finally, Clark was unhappy with State's advice and policy implementation.

George Shultz, Haig's replacement, was not aggressive by nature and had a limited foreign policy background, although he had been Nixon administration treasury secretary and the president of Bechtel, a giant international construction company. His deputy secretary, law professor and economist Kenneth Dam, also lacked experience. Consequently, both men depended on Eagleburger's counsel.[23] "The whole secret of this administration is that the president is not to be humiliated," explained one official. "Whoever does this is dead. George Shultz never does it."[24] Unlike Haig, Shultz was a team player; but to avoid embarrassing or alienating Reagan, Shultz could not confront the Reaganauts too often or too directly, since they were personally, institutionally, and ideologically closer to the president.

Clark's inside track with Reagan was reinforced by daily briefings, one-page option papers, and his control of documents. Through McFarlane, Clark oversaw the interagency and senior interagency committees. Baker's Legislative Strategy Group often sought compromise to get the president's program through Congress; Clark favored more unyielding positions and public campaigns. When Baker expressed misgivings to reporters about one Clark policy, the usually soft-spoken national security adviser snapped, "No one else speaks for foreign policy in the White House except me."[25]

Against Baker and Shultz, Clark and such allies as Kirkpatrick, Undersecretary of Defense Fred Ikle, and others advocated a "full court press" against Moscow. In speeches not cleared with Baker or Shultz, the president inveighed against the USSR as an evil empire and advocated placing antimissile defensive weapons in space. On Central America, greater help was given the Nicaraguan "contras" while there was more talk of military victory in El Salvador.

Shultz and State became progressively less relevant in these debates. Even on economic matters, the secretary of state's personal interest, he had to accept Helms's favorite, Richard McCormack, as assistant secretary for economic and business affairs. The battle over pressing Japan to accept "voluntary" quotas on auto exports to the United States pitted the protectionists at the Commerce and Transportation departments, the eventual victors, against the Council of Economic Advisers, the Treasury Department, and the Office of Management and Budget. State was unimportant. Shultz was, however, able to convince Reagan to drop sanctions on European participation in the Soviet gas pipeline.[26]

Public opinion, becoming concerned over the dangers of nuclear war, developed a concomitant interest in arms control efforts. As usual, everything related to U.S.-Soviet relations was closely monitored by the White House. Negotiations on the military balance required close coordination between State and the Defense Department, the main task for which the NSC was originally established. But during 1981–1982, a weak NSC staff failed to cushion clashes between Haig and Weinberger.

Another complicating factor was the attempt of ACDA Director Eugene Rostow to take the leading role. The Reaganauts were so mistrustful of the Soviets and so determined to overcome Moscow's supposed military superiority as to be skeptical about arms control efforts. The hardliners lacked confidence in Rostow and arms control negotiator Paul Nitze, despite their strong defense records. Rostow was forced to resign in January 1983 due to disputes over his role and stance and was replaced by Ken Adelman, a young Reaganaut. The main battle, however, had already shifted to one between Assistant Secretary of Defense Richard Perle, who was extremely critical of arms control efforts, and Assistant Secretary of State Richard Burt, who favored at least cosmetic attempts to develop a proposal that might form the basis for talks.[27]

The most difficult internal struggles took place over the Middle East and Central America. In the former region, the Iran hostage crisis had been resolved on inauguration day, but the incoming president inherited stalled Arab-Israeli negotiations, a Lebanese civil war, and an Iran-Iraq conflict. The Iranian revolution and the Soviet invasion of Afghanistan made Washington fear instability in the Persian Gulf, an area whose massive oil production gave it great strategic importance.[28]

The Reagan administration's first-term Middle East policy went through three distinctive periods, each characterized by a different division of power among agencies and decision makers. During the first cycle, from January 1981 to June 1982, Haig and Weinberger presented different strategies to ensure Persian Gulf security. In the second period, from June 1982 to April 1983, Shultz launched a major initiative to resolve the Arab-Israeli conflict, and put the priority on removing foreign forces from Lebanon. The final phase, somewhat similar to Haig's earlier approach, allied Shultz and McFarlane against Weinberger and Casey to reestablish cooperation with Israel, combat Syrian hegemony in Lebanon, and continue lower-key efforts on the Gulf.

Conflict between Haig and Weinberger reflected distinctive institutional interests as well as personal views, since Weinberger's Defense Department benefited from large arms sales to the Arabs and needed their cooperation to coordinate the Gulf's defense. Although both put the main stress on Gulf security, Haig sought a "strategic consensus" to unite as many countries as possible against future Soviet aggression. Weinberger put the prime emphasis on the U.S.-Saudi connection, whose survival he saw as integrally tied to progress on the Palestinian question. Haig argued that the U.S.-Israel relationship did not really harm U.S.-Saudi links, based on the oil-rich kingdom's security needs, despite Saudi rhetoric to the contrary. Weinberger's unilateral actions often leaked into the media and made U.S. policy look foolish and disorganized. For example, he announced U.S. intentions to sell F-16 planes and Hawk antiaircraft missiles to Jordan, even though this decision had not been approved by State.

But the end of the first phase, and of Haig's tenure, came when Israeli forces moved into Lebanon in June 1982. Haig saw this as an opportunity—with the PLO and Syria so badly defeated—for a diplomatic breakthrough on Lebanon. U.S. pressure and mediation could help remove all foreign troops from the country, end the civil war, and reestablish Lebanese sovereignty. Yet, given the virtual civil war within the U.S. government, he did not clear either his strategy or immediate actions with colleagues. For Haig's rivals, this was the last straw.

While Haig was trying to lever the PLO out of Beirut by letting it worry about a possible Israeli attack, Clark, Bush, and Weinberger foiled the strategy by telling the Saudi ambassador to disregard that threat. Now Clark's personal influence with the president made the difference. Haig's resignation came in the midst of the crisis.

At his own confirmation hearings, Shultz stressed the Palestinian issue over all other international questions. The centerpiece of Shultz's program was a proposal that became known as the Reagan Plan, along lines prepared by NEA and the NSC's Middle East staff. In a speech on September 1, 1982, Reagan advocated a federated Jordanian-Palestinian state to extend King Hussein's rule to the West Bank and Gaza Strip, coupled with Arab recognition of Israel and border modifications in Israel's favor. The U.S. government would accept neither Israeli annexation of the territory nor a PLO-led Palestinian state. There were several reasons for the timing and shape of this proposal. Israel's invasion of Lebanon weakened Syria and the PLO, supposedly making them less able to veto such a settlement. It was already clear that the United States was the only factor that could promote serious negotiations, but to be a mediator Washington had to convince Jordan's King Hussein to enter negotiations. The speech was also intended to defuse domestic doubts about the administration's foreign policy competence.[29]

NEA's career staff, on whom Shultz was leaning heavily, considered themselves regional experts, but their analyses were often superficial and erroneous. Their conventional wisdom held that the Arab-Israeli conflict was such an overwhelming concern of Arabs that failure to reach a quick resolution in a manner acceptable to the Arab side would turn those countries toward pro-Soviet policies and leftist or Islamic fundamentalist revolt. While the PLO and Arab regimes were willing to negotiate, NEA argued, only Israel was blocking a breakthrough. U.S. pressure could deliver Israel to the bargaining table and pave the way for a solution to all problems. The region would become stable and pro-Western. If only decision makers listened to the professionals, FSOs moaned, the decline in U.S. influence would be halted. Politicians did not do so, they cynically commented to each other, only due to fear of losing Jewish votes. Some in NEA wanted to have the Reagan plan endorse Palestinian self-determination as a step toward a PLO-led state, but Shultz would not accept this idea.

Although many in NEA attributed such decisions to extraneous politics, decision makers had perfectly valid reasons for rejecting both the bureau's general analysis and its proposals for U.S. policy. If the politicians had their clientele, NEA's main concern was in reporting and soothing Arab grievances. Many of the best FSOs had become better rounded in recent years concerning their comprehension of the Arab-Israeli conflict, but earlier traditions remained potent. Their analysis was focused on Arab rhetoric rather than on the requirements, constraints, and actual behavior of Arab regimes.

While the Palestinian question is an extremely important and emotional one for Arab states, its centrality has often been exaggerated. Arab antagonisms and alliances with the West are the product of wider needs or complaints. The search for security in the Persian Gulf, for example, is far more significant in setting Saudi Arabia's policy toward Washington than the Arab-Israeli conflict. Furthermore, the nature of Islamic and nationalist views, as well as pressure from radical neighbors, would never allow U.S. bases in that country regardless of the alliance with Israel. Otherwise, Saudi interests, military purchases, and economic investments—even their desire to give Washington an incentive to limit support for Israel—set the basis for a strong bilateral relationship.

Making peace with Israel has been too risky in domestic and regional terms for most Arab regimes. Specific proposals are rejected out of mistrust for the PLO or jealousy of the Reagan Plan's favoritism toward Jordan. U.S. power to bring about a breakthrough is easily exaggerated since the obstacles and complexities are so great. Finally, NEA never convinced policymakers it could produce a settlement that would enhance such U.S. interests as limiting Soviet influence and strengthening regional stability.

On the contrary, in the case of the Reagan Plan and Lebanon negotiations, following NEA's advice caused the administration considerable embarrassment. Israel's opposition was only sealed by State's failure to consult it on the Reagan initiative. But contrary to confident predictions by Assistant Secretary Veliotes and his colleagues, the PLO and Jordan also rejected the proposal and the Saudis ignored the plan. Such factors as the PLO's intransigence, internal conflicts, and veto power over Jordanian participation, Syria's opposition to Amman's aggrandizement, and the Saudis' perpetually timid foreign policy had been left out of the equation. To make matters worse, almost up to the moment Jordan gave up on the plan in April 1983, NEA and the U.S. embassy in Amman were reporting that King Hussein would accept it.[30]

Providing the president with misleading estimates that put his name on an unsuccessful policy did not endear NEA to Shultz nor Shultz to the White House. Developments over Lebanon were even worse. Haig's original conception had followed Kissinger's style of realpolitik policy. But as Kissinger had shown, such a strategy of diplomacy, based on power leverage, requires careful coordination and timing—impossible requirements given the discontinuity and conflict in the Reagan administration. Personnel shifts and divided authority wasted months, and the failure to apply immediate pressure lost whatever opportunity existed for moving a stunned Syria and a shaken-up Lebanon.

Shultz had assigned former Undersecretary of State Philip Habib to negotiate the removal of foreign troops from Lebanon. He was assisted by

Deputy Assistant Secretary Morris Draper, who had studied in Lebanon two decades earlier. To gain time, Damascus led Habib to believe that Syria's withdrawal would be no problem after an Israeli pullback was negotiated. A Lebanon-Israel accord was reached in May 1983, but Syria hung on, rejecting the treaty and escalating its demands. The United States also had no coherent plan for an internal settlement to force compromise among warring Lebanese communities. The presence of U.S. Marines in Lebanon for month after month, in an attempt to intimidate the Syrians, could not succeed since Damascus knew Washington was bluffing and that domestic pressure would force a withdrawal. Heavy casualties from Syrian-aided terrorism heightened congressional and media criticism on the administration until the Marines were pulled out in March 1984.

By that time, Shultz's dissatisfaction with State's performance had already produced a thorough housecleaning. Habib suggested his own return to retirement since Syria refused to meet the architect of the Lebanon-Israel treaty. Veliotes and Draper were offered ambassadorships. It was too late, however, to save State's credibility with the White House. Clark stepped in, selecting his own deputy, McFarlane, as chief Lebanon negotiator. State had lost control of Middle East policy to the NSC staff.

Within weeks, however, McFarlane succeeded Clark. Now Middle East policy entered a third period. Chastened by his experience with the Reagan Plan, Shultz advocated close U.S.-Israel cooperation in Lebanon and elsewhere, following the advice of his Policy Planning Staff rather than that of NEA. Syrian double-dealing and PLO intransigence made Shultz and McFarlane hostile to those forces. While the U.S. presidential elections brought a hiatus on Middle East efforts, this new phase was primarily the result of events and experiences in the region itself.

The Reagan administration's evolving Middle East policy illustrated several types of procedural problems. As always, changes in the substance of policy went hand in hand with shifts in the policy structure and the distribution of power among individual decision makers. There had been friction over personalities and ideas between Haig and Weinberger, operational conflicts between Haig and Clark, State Department clientism and poor analysis on the Reagan Plan and Lebanon negotiations, and alternating periods of dominance between State and NSC. Overall, there were frequent, dramatic shifts of authority among agencies, individuals, and bureaus reacting to regional or internal struggles. In short, there was no long-term or intrinsic structure of authority; influence shifted depending on the specific issue and an individual's strength of personality or access to the president. Internal conflicts usually combined four different elements: substantive disagreements over policy, agencies' institutional viewpoints, bureaucratic ambitions and personal frictions.

Central America policy, the administration's most important ongoing problem, showed even more clearly the problems of such a mercurial system as policy conflicts among appointees intensified bureaucratic disputes. The administration first had to formulate an approach toward Central America when there were significant splits in its own ranks over the goals, timing, strategy, and tactics between a realpolitik group and the Reaganauts. Those favoring different policies competed by media leaks and contradicted each other's statements. Reagan's intervention made it possible for the Reaganauts finally to gain the upper hand, but even this victory took two years of effort and alliances among similarly minded people in different agencies.

These disputes overlay attitudes inspired by institutional responsibilities. For example, White House counselor James Baker was "moderate" because he worried that a strident policy might damage Reagan's domestic program in Congress and his reelection prospects. The Pentagon balked at direct U.S. armed intervention for fear of "another Vietnam" draining military prestige and resources. The Defense Department and CIA, as operational agencies, tend to play a smaller part in formulating policies, but once given a task, develop a vested interest in building up their role and funding. Earlier reservations or pessimistic reporting are squelched in their enthusiasm to get the job done.

The leftist Nicaraguan Sandinistas seized power in July 1979. Three months later, a reform-minded junta took over in El Salvador. The Carter administration was conciliatory toward the first and verbally supportive of the second. Secretary of State Cyrus Vance had said, "By extending our friendship and economic assistance, we enhance the prospects for democracy in Nicaragua. We cannot guarantee that democracy will take hold there. But if we turn our back on Nicaragua, we can almost guarantee that democracy will fail."[31]

By 1981, some of State's career staff still believed that Nicaragua, under pressure, could become nonaligned, if leftist. The guerrillas in El Salvador could not win the war, but the Salvadoran government could lose if it failed to strengthen its own political base by continued land reform, an end to rightist violence against moderates, and the co-optation of non-Marxist opposition forces. The belief by many political appointees outside State that ARA soft-liners were sabotaging Reagan's policy fed suspicions about Enders himself.[32]

But those holding such views had already been transferred, purged, or silenced by the new administration, which saw El Salvador's insurgency as a Soviet-Cuban effort to destabilize Central America, with Mexico and the Panama Canal as eventual objectives. Haig's realpolitik strategy envisioned that the insurgency could be defeated by getting Moscow and Havana to end support for the guerrillas, a strategy similar to the linkage policy unsuccessfully pursued by Kissinger over Vietnam. Nicaragua, the guerrillas' headquarters and rear area for training and regroupment, was itself considered beyond a point of no return on the road to dictatorship and alliance with Cuba.

To Reagan and his supporters from California and other western states, Latin America loomed larger than for establishment policymakers oriented toward Europe. The Reaganauts were in close contact with Latin American rightists just as liberal Democrats had bonds with the area's Christian Democratic and Social Democratic politicians. The salience of the Panama Canal for conservative Republicans—the battle over those treaties in 1978 had been the newly resurgent right's first foreign policy test—further sensitized them to Central America security issues.

Even before the inauguration, extreme conservatives were calling for a tough policy. The Council for Inter-American Security's Sante Fe Commission report, for example, advocated unremitting support for the Salvadoran military and for destabilizing Nicaragua. Three signers entered the Reagan administration: Lt. General Gordon Sumner became a consultant to State, Roger Fontaine joined the NSC staff, and Lewis Tambs became ambassador to Colombia.[33] The Republican party's 1980 platform accused Carter of standing by while Soviet-backed Cuba promoted revolution. It deplored the Sandinista takeover and U.S. aid to Nicaragua, opposed "Marxist attempts to destabilize El Salvador, Guatemala, and Honduras," and supported "the efforts of the Nicaraguan people to establish a free and independent government."[34] Vernon Walters, later ambassador-at-large, visited Central America for the Reagan transition team and promised support for conservative military regimes there.

UN Ambassador Jeane Kirkpatrick became the President's personal adviser on Latin America because Reagan felt her views articulated his preconceptions. President Carter, she wrote, thought "that the cold war was over, that concern with Communism should no longer 'overwhelm' other issues, that forceful intervention in the affairs of another nation is impractical and immoral, that we must never again put ourselves on the 'wrong side' of history by supporting a foreign autocrat against a 'popular movement,' and that we must try to make amends for our deeply flawed national character by modesty and restraint in the arenas of power and the councils of the world." Obviously, such a course would not be recommended to the Reagan administration. "The deterioration of the U.S. position in the hemisphere," Kirkpatrick warned, "has already created serious vulnerabilities where none previously existed, and threatens now to confront this country with the unprecedented need to defend itself against a ring of Soviet bases on and around our southern and eastern borders.''[35]

The Republican party's extreme right wing hoped to name one of its number as assistant secretary of state for the Bureau of Inter-American Affairs. Senator Helms's aide, John Carbaugh, headed the transition team on Latin America and quickly asserted himself. Asked about regional policy just after Reagan's inaugural, a State Department official responded, ''Why don't you ask John Carbaugh—he seems to be running things around here.''[36]

When Haig became secretary of state, however, he quickly dismissed Carbaugh and gave Enders the assistant secretary post. Helms reacted by delaying Enders's confirmation for several months and unsuccessfully insisted on the hiring of ultraconservatives as deputy assistant secretaries. Sumner was made a consultant to ARA, but Enders prevented his having much influence. The White House saw Helms and Carbaugh as nuisances, but it also felt uneasy with Haig and Enders.

While limiting Helms's influence, the administration also eliminated career people at ARA who disagreed with, or were not trusted to implement, its policy. Carter's last ARA assistant secretary, William Bowdler, was dismissed within 24 hours of Reagan's inaugural, though he could normally have expected an ambassadorial post. Ambassador to El Salvador Robert White, also an FSO, was ousted within 10 days and was offered no new assignment.[37] Other ARA officials, like Deputy Assistant Secretary James Cheek, were sent to posts outside the hemisphere; Ambassador Lawrence Pezzullo, after six more months in Nicaragua, became diplomat-in-residence at a U.S. university. Their replacements, like ambassadors Deane Hinton in El Salvador and John Negroponte in Honduras, as well as the newly installed ARA leadership in Washington, had no regional experience.[38]

Reagan's aides were not greatly interested in foreign policy and the NSC staff was too weak or preoccupied elsewhere to assert itself at first. Of the NSC staffers on the region, Roger Fontaine had little influence and Al Sapia-Bosch generally cooperated with Enders. Therefore, with Haig's support, Enders was able to dominate the policy process on Central America by chairing interdepartmental committees, controlling access to meetings, and regulating the flow of reports and option papers. As assistant secretaries often do, Enders brought his own friends and allies into the bureau, particularly L. Craig Johnstone, who had worked with him in Cambodia, as office director for Central American Affairs.

On the working level in other agencies, however, the beginning of a Reaganaut coalition was organized by Undersecretary of Defense Fred

Ikle, Deputy Assistant Secretary of Defense for Inter-American Affairs Nestor Sanchez, himself a CIA veteran, and Constantine Menges, the CIA's chief intelligence officer for Latin America, who all favored a tougher line.

Despite the presence of such activists, Weinberger had little interest in Central America, and the Joint Chiefs of Staff wanted to avoid a direct U.S. military role in the region. Remembering the Vietnam experience, they were reluctant to become involved in an unpopular war which would lack public support, hurt their budget requests in Congress, and subject them to political constraints. Deputy Secretary Frank Carlucci was sympathetic to the military's concerns. These objections surprised the Reaganauts, who complained that the Pentagon, like State, had its own policy priorities.

On the covert side, CIA Director William Casey lobbied for U.S. aid to anti-Sandinista guerrillas in Nicaragua—the "contras"—but the Agency's first task was to supply intelligence. The CIA had well-established sources in the region, but pressure to reach conclusions backing current policy influenced reporting. During the Carter administration, a congressional study noted, CIA reports played down Nicaraguan involvement in El Salvador. In contrast, with Reagan in office, the CIA tended to accept the Salvadoran regime's view of the situation and did not report well on rightist violence.[39]

Even more blatant was the manipulation of State's information. U.S. embassy reports of Salvadoran army atrocities and massacres of peasants were surpressed, or even attributed to the guerrillas, by officials speaking to Congress or the press. While the embassy repeatedly explained the involvement of leading Salvadoran officers in death squads, the administration, including Enders, spoke as if the killings were being done by extreme rightists who had no connection with the army. American diplomats on the scene knew that civilian moderates in the regime had no control; authority was monopolized by the hard-line, corrupt colonels. Back in Washington, the Salvadoran government was portrayed as democratically ruled. Embassy dispatches reported that the Salvadoran army was incompetently led and dangerously ineffective. Washington spoke of progress in the fighting. Once again, the reporting system was being corrupted. It was not merely a matter of misrepresenting the facts to the public, but also one of not taking these facts into account when making decisions. Ultimately, actions taken on the basis of such self-delusion could only end in tragedy for both El Salvador and U.S. interests.[40]

While Congress was an indirect actor in the policy process, it was a major consideration in gaining funds and public support. Yet it had neither enough dissidents nor sufficent motivation to force policy changes unless alienated to an unusual extent—a situation Baker and Enders sought to avoid. Consequently, while Congress made the president report on human rights in El Salvador, those who disbelieved his periodic "certifications" could not block continued aid. Some protested U.S. covert assistance to the Nicaraguan "contras," but lacked votes in the Senate to cut off funds. Of course, many members of Congress supported Reagan's policy, were indifferent to Central American issues, or feared being held responsible for "losing" El Salvador. Despite the apparent exercise of Congressional power in Lebanon and El Salvador, it was only a general constraint on the executive branch—rather than a maker of policy—and could usually be circumvented.

In the administration's first weeks, Haig attempted to make Central America a top priority. By doing so he could demonstrate to the Reaganauts his anti-Soviet toughness and establish public credentials as foreign policy "vicar." El Salvador would serve as a symbol of U.S. willingness to oppose Soviet-Cuban aggression and could put anti-U.S. forces on the defensive. A guerrilla "final offensive" had just been defeated in El Salvador, but the opposition had shown its military capacity in a graphic manner. Meanwhile, Nicaragua's authoritarian leadership disappointed those in Washington who hoped for a pluralist and nonaligned regime there.

Washington quickly suspended wheat sales and economic aid to Nicaragua and opposed its loan requests in the World Bank and Inter-American Development Bank. Haig warned Cuba against involvement in Central America, threatening to "go to the source" of the crisis, while Edwin Meese chimed in that the United States "could not rule out any means" of dealing with the issue. The campaign's centerpiece was State's February 1981 White Paper using materials captured by the Salvadoran military to argue that the USSR, Cuba, and Nicaragua were supplying arms and equipment to the guerrillas. The White Paper had little influence on public opinion, however, since several critiques challenging its interpretation of the source documents received wide publicity. Envoys were sent to Europe and Latin America to explain U.S. policy and present proof of Soviet and Cuban involvement. Haig even speculated that three American nuns killed and raped by Salvadoran soldiers might have been victims of "an exchange of fire" while running a roadblock. Kirkpatrick said the nuns were oppositionist "political activists."[41]

But this strategy ran the risk of stirring up public and congressional opposition by reinforcing the administration's "trigger-happy" image, worrying Congress and the voters about possible deployment of U.S. troops in Central America. Reagan rejected Haig's suggestions for blockading Cuba, while Baker and Deaver were anxious that the tough public statements alone might overshadow their economic programs and stir popular fears of "another Vietnam" in Central America. One White House official told reporters that the "train was going too fast." ARA officials also felt the escalation of rhetoric ruined chances for negotiation or maintaining congressional support. Acting Deputy Assistant Secretary of State John Bushnell warned privately in March 1981 that the mass media was overemphasizing Central America, claiming, "This story is running five times as big as it is." His stand was disavowed—the administration did not want to imply that Central America was unimportant—but for a number of months thereafter, Haig and the White House reduced public emphasis on the problem.[42]

Behind the scenes, however, they began to devote even more attention to the issue. Reagan secretly approved aid to the "contras" at a November 1981 NSC meeting. The White House introduced a Caribbean Basin Initiative for regional trade preferences, increased assistance, and provided incentives for U.S. investment. While Haig repeated warnings to Nicaragua, Enders went there in August 1981 to offer a nonaggression pact and curbs on anti-Sandinista military training camps in Florida if Managua stopped its military buildup and the arms shipments to El Salvador. The two sides failed to reach agreement.

The following March, while not ruling out efforts to overthrow the Sandinistas, Haig suggested U.S. financial aid and terminating support for the "contras" in exchange for an end to Nicaraguan involvement in the Salvadoran war. The Haig-Enders strategy, to intimidate Nicaragua from aiding the Salvadoran revolutionaries, was the origin of what became known as the "two-track" strategy. Washington would use military threats and economic pressure to gain Nicaraguan concessions while continuing military efforts to win the war in El Salvador.[43]

The Reaganauts did not block these efforts in 1981–1982 since they supported moves against Nicaragua, but were skeptical about—or even opposed—diplomatic initiatives. They estimated that the Nicaraguan government, and Cuba behind it, would stand firm and that negotiations would fail. Consequently, a much-expanded U.S. effort would be needed to defeat the Salvadoran guerrillas and to force Nicaragua's back against the wall. Reaganaut impatience was also prompted by pessimistic embassy reports that the Salvadoran government was performing poorly on the battlefield. Further, while officials at State viewed U.S. policy's domestic unpopularity as a response to excessive militancy, Reaganauts complained of media bias and the lack of a serious campaign to mobilize political backing. The large turnout for El Salvador's March 1982 elections brought a favorable echo in the United States, convincing the Reaganauts that a climate existed for winning that regime increased support. The failure to reach agreement with Nicaragua, its military buildup, and reports of the Sandinista's eroding base at home convinced the administration that the best opportunity to subvert or intimidate Nicaragua might soon be past. "The hours are growing rather short" to prevent "Managua from becoming another Havana," Haig warned.[44]

Haig's June 1982 resignation ushered in George Shultz, who had little interest in Central America. In the short run, Enders's power was unaffected, but Haig's departure exposed the assistant secretary to an increasingly active NSC staff, led by Clark. Haig's removal, Clark's gathering authority, Shultz's willingness to yield ground, and Kirkpatrick's ability to influence Reagan facilitated the Reaganauts' efforts to take control of Central American policy.[45]

From his exploratory August 1981 visit to Nicaragua until January 1983, Enders attempted to combine waging war in El Salvador with diplomatic efforts toward Nicaragua, but the opposition of the Reaganauts, who placed more hope than did State on the success of the U.S.-backed "contras" operating across the Honduran border, became stronger as the months passed. This covert CIA operation was ostensibly designed to interdict the arms flow from Nicaragua to El Salvador and to provide U.S. leverage toward Managua's agreement to cease intervention. But the Reaganauts saw the "contras" as a means of overthrowing the Nicaraguan regime.[46]

Events provided ammunition for both factions. Hinton reported that the Salvadoran government barely contained major guerrilla offensives in October 1982 and January 1983. This development apparently encouraged Enders to work harder on his two-track policy of negotiating as well as fighting, including a visit to Spain's Socialist Prime Minister Felipe Gonzalez, a potential intermediary. Clark, not informed of the meeting in advance, considered it an attempt to arrange talks behind his back and ordered that all future trips be cleared with him. When Hinton made speeches criticizing the rightist death squads as destroying El Salvador "every bit as much as the guerrillas," Clark, through the White House, ordered him to cease such statements. Yet Hinton's very pessimism was also used by Clark and Kirkpatrick to convince the president that State's policy was not working.[47]

White House aides saw the "two-track" option, embassy reporting, and State's criticisms of the Salvadoran regime as evidence of ideological softness. Some appointees accused the department of having an "obsession with perfecting the government of El Salvador" rather than prosecuting the battle against the Salvadoran guerrillas and Nicaragua. To mollify Congress, Enders made some conciliatory comments and calls for reform by the Salvadoran regime. Such tactical concessions further fed the

Reaganaut image of a State Department determined to dilute the struggle. Clark already felt that Enders was poorly implementing the President's orders; Kirkpatrick wanted the administration to go on the offensive in the public opinion battle, arguing that appeasing Congress only increased its obstructionism. Bureaucratic jealousies provided the last element undermining Enders. Enders's arrogance and his procedural domination moved his key potential White House ally—Baker—at least temporarily into Clark's camp.

All these developments gave the Reaganauts a chance to displace Enders and push for a more aggressive strategy. Clark convinced the president in January 1983 to institute a high-level policy review and to send Kirkpatrick on a 10-day trip to the region. She returned with the conclusion that the Salvadoran government and army were demoralized by uncertainty over U.S. support and that the other countries also wanted a firmer U.S. policy, blaming their loss of confidence on congressional criticism and State's talk about negotiations.[48]

To remedy these perceived shortcomings, Kirkpatrick proposed emergency aid increases and urged the president to buoy the Salvadoran regime by announcing that Washington would not back down. Casey called for expanding the "contras'" operations, claiming the Sandinistas might be overthrown by the end of 1983. Reagan accepted these recommendations and also appointed former Senator Richard Stone as special Central America negotiator, further eroding Enders's authority. One reason for Reagan's intervention was the NSC staff's February leak of Enders' "two-track" memorandum. The president was unhappy with being confronted publicly by a State Department position seemingly at variance with his views. With Reagan finally motivated to support Clark, Kirkpatrick, and their allies, two years of leadership by State and Enders on Central American policy was at an end.[49]

Enders's reassignment as ambassador to Spain and Hinton's recall in June 1983 confirmed a power shift that had already taken place. Insult was added as well when a White House official claimed Hinton was "burned out." But the feisty ambassador warned that a U.S. military commitment for victory would be "so massive that it's not even worth discussing."[50]

Planning to keep control of the issue himself, Clark wanted to replace Enders and Hinton with men ideologically compatible to the Reaganauts. He considered Ambassador Negroponte in Honduras and U.S. Ambassador to Guyana Thomas Gerald, a retired admiral, to succeed Enders. A White House staff member voiced the cruder aspect of prevailing anti–State Department views by saying, "You don't handle Central American politics with tea and crumpets on the diplomatic circuit." Angry FSOs responded with criticism of "foreign policy amateurs" who ignored expert opinion and their own embassies' on-the-spot evaluations.[51]

Undersecretary Eagleburger convinced Shultz, however, that State should not surrender too much authority over these appointments. Ambassador to Brazil H. Langhorne Motley, a political appointee close to Clark, became the compromise choice as ARA assistant secretary, the eighth man to hold that post in nine years. State managed to save the ambassadorship to El Salvador for a career FSO, Ambassador to Nigeria Thomas Pickering. Still, Shultz had lost practical control over U.S. policy in the region.[52]

A series of dramatic measures followed, making the issue the administration's number-one priority. McFarlane, Clark's deputy, was put in charge of a special Central America working group. In March, Reagan declared the defense of the region against Marxism-Leninism as vital for U.S. national security. Congress was cowed by the president's powerful, nationally televised April speech to a joint session. If El Salvador was taken over by pro-Soviet forces, he hinted, Congress would be held responsible for having failed to provide enough aid. Politicians, aware of precedents for being blamed with the "loss" of a country to Communism, dampened their dissent.[53]

The president made similar speeches in the conservative, electorally important South: in Texas, where the administration hoped Chicano voters would be supportive; in Florida, with its many strongly anti-Communist Cuban émigrés; and in Mississippi. The White House, NSC, State, and the U.S. Information Agency (USIA) developed an active public program, including courses to prepare government officials to defend administration Central American policy. The White House operation, under presidential assistant Faith Whittlesay, held regular Wednesday meetings to rally conservative groups. At one such gathering in June, they planned ways to counter a Washington march protesting U.S. involvement in Central America. Otto Reich, AID's Cuban-born deputy director, was named coordinator for public diplomacy on Central America at State. USIA worked on presenting Washington's policy abroad, particularly in Western Europe, where polls showed great concern over the issue.[54]

The Reaganaut coalition was now dominant. Clark coordinated and the Defense Department handled military aid and training, with Deputy Assistant Secretary Sanchez serving as liaison to Clark. The CIA ran covert operations against Nicaragua from neighboring Honduras; Negroponte directed action on the scene. Kirkpatrick monitored State's performance and advised Reagan. Constantine Menges, who had long argued that Mexico was the real target of Soviet-Cuban subversion, came from the CIA to the NSC staff. Fontaine and Sapia-Bosch both left.

In July 1983, Reagan named Kissinger head of a bipartisan group to suggest options for policy in the region. The administration hoped the commission would undermine congressional criticism by recommending a program along essentially the same lines it had been pursuing. The months necessary for a commission study would buy time for entrenching the post-Enders policy.

However, any immediate political gains were undermined the next day when news leaked out about a large U.S. military exercise in Honduras and the Caribbean, Big Pine II, planned by the Joint Chiefs of Staff at Clark's behest. State, U.S. ambassadors, and Central American governments had not yet been told of the final decision. An angry Shultz protested directly to Reagan about Clark's failure to keep him informed. The administration was seriously embarrassed and Congress was again aroused about the possibility of direct U.S. military involvement.

Congressional demands for cutting off U.S. aid to the "contras" were reinforced by uncertainty over administration objectives in Nicaragua. Was the goal to overthrow the Sandinistas or simply to intimidate them from helping Salvadoran guerrillas? Kirkpatrick denied the former objective: "We have minimal and maximal goals in Nicaragua. And I truly believe that they are not identical with the 'contras.'"[55] Skepticism of such assurances in the House of Representatives produced votes to sever aid to the "contras." These results showed the Reaganaut's failure to quiet congressional dissent, but the Republican majority in the Senate blocked any practical effect.

On one level, Clark continued Enders's strategy: encouragement for the Salvadoran government, intimidation for Nicaragua and Cuba. As one DOD official put it: "We're playing a little cat-and-mouse game with them, putting a little squeeze on, making them wonder what's going to happen next. Ultimately, the idea is to convince them that allowing the Salvadoran guerrillas to use Nicaragua as their headquarters for revolution is not a good idea if they want to keep their own damn revolution."[56]

Nonetheless, the administration had gone beyond that approach. Ikle provided the best definition of the new objectives in a September 1983 speech cleared by Clark. "We must prevent consolidation of a Sandinista regime in Nicaragua that would become an arsenal for insurgency." Congressional opposition to covert operations was creating "a sanctuary" for the rebels, who would never settle for a fair democratic process. Consequently, he concluded, "We do not seek a military defeat for our friends. We do not seek a military stalemate. We seek victory for the forces of democracy."[57]

The Reaganauts doubted the possibility of negotiating with the left, whether in El Salvador or in Nicaragua. Both Reaganauts and "pragmatists'' went along with most of the 1981–1983 policy of pressuring Nicaragua into negotiating, although the Reaganauts put the emphasis on pressure, while State stressed compromise. The Reaganauts were more interested in affecting the internal situation in Nicaragua, while State limited its concern more strictly to Managua's involvement in El Salvador. The department talked more about curbing the ultraright's activity in El Salvador and seeking reforms; the Reaganauts thought that efforts to strengthen social change and centrist forces might divide the regime and damage the war effort. So the administration covered up U.S. embassy reports about continued military control, corruption, and repression. If left undisturbed, the Reaganauts thought, their policy would bring military victory through the Salvadoran army and the ''contras.''

Reagan's lack of interest in foreign policy or procedure fostered conflict among subordinates. When given a choice between the Reaganaut approach—represented by his old friend Clark and his new friend Kirkpatrick—and the distant ARA bureau, whose ideological views and loyalty he suspected, Reagan's decision could not be in doubt. But another Reagan whim, the transfer of Clark in October 1983, soon upset this new policy structure by removing the key man ensuring the Reaganaut alliance's rule. McFarlane, Clark's successor, had less personal influence with the president and seemed a compromise choice between Kirkpatrick, the Reaganauts' candidate, and Baker, the choice of ''pragmatic'' political operatives on the White House staff. Shultz made a small comeback in reviving Enders's original two-track plan, but the consensus against any accommodation with Nicaragua was now too strong to reverse.

The Reaganaut coalition was, therefore, able to continue its control of Central America policy even without Clark. State now played a relatively minor role as the Defense Department and CIA implemented growing programs of military and financial support to the Salvadoran military and Nicaraguan ''contras.'' A new major figure was General Paul Gorman, head of the U.S. Southern Command, who oversaw aid to the Salvadoran army, organized regional security, directed maneuvers, and constructed facilities. State was annoyed by the fact that Gorman was now more powerful than any of the U.S. ambassadors on the scene.

Given the Reagan administration's reading of U.S. interests, it never seriously contemplated accepting a leftist revolution in El Salvador. Since any compromise with the Salvadoran opposition was seen as inevitably producing a Marxist-dominated government, this option was also rejected. Consequently, the U.S. government backed the Salvadoran regime in the war. In this context, the internal debate revolved around whether to encourage a moderate civilian, reform-oriented regime that could muster broader support or to leave the traditional military and its power untouched, supposedly allowing better prosecution of the war. Relatively successful 1984 elections in El Salvador and the strengthening of moderate President Jose Napoleon Duarte's hand, along with a stabilization of the military situation, allowed some temporary compromise between the Reaganaut and State Department approaches.

There was a wider range of options and debate about Nicaragua. At State, some officials felt that negotiations and pressure might persuade Nicaragua away from trying to spread revolution, and that Washington could coexist with a Managua that changed its foreign policy. The Reaganauts rejected this idea because they considered Nicaragua an intrinsically unacceptable extension of Soviet-Cuban influence, doubted whether any Sandinista government could abandon revolutionary goals, and wanted to change the country's internal political system. The Reaganauts inevitably concluded that only the Sandinista regime's fall could remove the threat to El Salvador and the region as a whole.

Regardless of their views on the ultimate outcome of U.S. pressure on Nicaragua, all factions agreed that intimidation was needed to discourage Nicaragua from involvement in El Salvador. The Reaganauts, however, would countenance much higher levels of escalation, as their policy after February 1983 demonstrated. The October 1983 invasion of Grenada, underlining willingness to use force in the area, posed obvious potential parallels for Nicaragua.[58] Yet it was difficult to assess the actual objective of Reaganaut strategy, since economic, political, and military pressure to change Nicaragua's policy could not easily be separated from efforts to bring down the government. Indeed, the strategy's crux, at least originally, was that Managua's doubt about this distinction would be more likely to produce concessions.

State Department critics argued that the Reaganaut policy of threats without incentives would only toughen Nicaragua's stand, pushing it deeper into the Soviet bloc by offering a fight to the finish as its only option. Further, the Reaganauts had no program for resolving the problems with El Salvador and Nicaragua. The former situation seemed to follow a worrisome and not unfamiliar pattern, requiring larger and larger amounts of U.S. aid and involvement.

The case of Central America shows that the president's control of the foreign policy process is constrained by several factors, including major internal battles, mistrust between political appointees and the Foreign Service, and escalating interagency competition. Ideological struggles and accelerating personnel shifts made it more difficult to build a consensus even within the executive branch. The roles of Congress, the media, and public opinion intensified and complicated the struggle while providing new fronts for rivalry. Intelligence and reporting from the field became weapons in such battles rather than means for adjusting policy to reality. Even presidential backing was no guarantee of success, since both Carter and Reagan vacillated over which subordinates to support.

The chief executive must constantly decide what he wants, how well officials are providing it, whose advice he will heed, and to whom he should delegate authority. Changing fortunes of battle and personnel shifts shortened the lives of each policy and every decision-making framework. An analysis of the administration in January 1981 would show State as reducing past NSC dominance, while a study at the end of that year would have noted a "typical" conflict between the secretary of state and the national security adviser. Similarly, reviews in 1982 claimed that the experience of exercising power had moderated the Reagan team's original views. Another look, twelve months later, would conclude that the national security adviser and the ideologues were in control. Yet Reagan himself then damaged that faction by moving Clark to a domestic post, partly due to Shultz's complaints.[59] To understand the substance and direction of U.S. policy, separate—and equally short-lived—models on the distribution of power, alliances, and viewpoints among top decision makers would have to be drawn up for every issue.

Brzezinski suggested the Shultz-Clark friction showed the "problem of who makes foreign policy is not the product of a conflict of personalities," since these two men were not known for "ego trips" or flamboyance. Rather, the culprit was a built-in NSC-State rivalry.[60] Indeed, events did demonstrate that the traditional primacy of the secretary of state and his department had become fictional. The quick abandonment of Reagan's early experiment with a weak NSC staff indicated White House refusal—despite its own original intentions—to return to a pre-Kissinger policy system. The national security adviser's challenge to State was an entrenched practice.

Even more dramatic were the accelerating shifts among policymakers themselves. After eight years, the Kennedy-Johnson and Nixon-Ford administrations each had only three people serving as secretary of state or national security adviser. During Carter's single four-year term, the number was still three while, well before his first term was completed, Reagan had five different people in these positions.[61]

In Haig's colorful phrase, the administration's foreign policy was "a permanent case of a struggle for power around the president. . . . You can't have a troika, and then a quadriped, which is now down to a dynamic duo."[62] The consistency of decision making, already fragile by foreign standards, was further called into question, and the two-sided battle between State and the NSC staff threatened to become only a small part of broader institutional and individual power plays involving the

White House staff, the Defense Department, and even the CIA as near-equal participants.

Not surprisingly, Brzezinski proposed that the national security adviser take first place, since foreign policy decisions were so closely related to national survival and to domestic issues that the White House would no longer delegate authority to a department viewed as outside its immediate control.[63] The advantage of this solution was its apparent efficiency and responsiveness to a president's particular world view and wishes.

At the same time, this very insulation from foreign perceptions and actual issues removed decisions further from the facts and causes of problems. For example, it was easier for the NSC staff or White House aides to ignore embassy reports from El Salvador warning that policies were failing. A review of some of the—to be blunt—dangerous and reckless schemes sometimes thought up by national security advisers Kissinger and Brzezinski shows State's important restraining role. When one looks at the department's slow pace and clientist obsessions, the NSC staff's potential dynamism seems more attractive. In short, any particular organizational framework has its price. Structure must also correspond to the needs and abilities of different presidents and subordinates. Such trade-offs make books or studies proposing some "solution" to policymaking problems useless and ignored.

While the trend toward leadership by any particular individual or agency creates shortcomings, an even worse problem is the trend toward anarchy—the absence of a clear chain of command and the persistence of internal conflict among decision makers. The result is a growing discontinuity of policy, not just on a quadrennial but even on a monthly basis, and an inability to pursue any consistent strategy. The intellectual vision and apparent unity of purpose each incoming administration tries to impose is lost in the complexity of events, the need for quick responses to developments, and differences among proud and opinionated people at the top with their own jurisdictions and views.

Clearly, then, the machinery involved in the foreign policy process actively shapes the outcome. Even the most treasured and established ideas in the international affairs lexicon—credibility and security, linkage and deterrence, Cold War and détente, containment and negotiation, human rights and realpolitik—lie in the realm of abstraction. Even the most passionately held ideological theories or cool pragmatism only have meaning if applied to specific situations that do not neatly fit their expectations. At some point, decisions must be made and implemented. No ideology or organizational scheme can ever take for granted the manner in which that is done.

10

State Department People

Working for the State Department is, in theory, a glamorous and exciting job, too glamorous for American public opinion, which exaggerates the department's elitist and snobbish propensities. Of course, knowing the deepest secrets, being involved in fast-moving and world-shaking events, enjoying a high living standard in exotic locales, and dealing with the powerful has its rewards. Yet all this is only a small part of State's work and affects fewer employees than outsiders think.

More accurate than the romantic stereotype is the view of State as an enclosed and limited world, a bureaucratic hierarchy in which the tedious has as much place as the glorious. Some sections are overstaffed backwaters producing little of consequence; others constantly struggle to keep up with the issues making daily headlines, and the stressful, unrelenting pace can actually burn people out. Everything must be done quickly, mistakes are not permissible, and knowledge is quickly outdated. In crisis situations, portions of the department become like military units under enemy shelling.

While trying to maintain a sense of urgency and importance, officials know that much of what they are doing is futile. Reports or policy papers whose every word has been fought over for days or weeks may go unread by those at the top. Most initiatives will not bring results; hundreds of events are monitored before a significant development turns up, perhaps a government's subtle signal in a newspaper article or speech. Briefings are prepared for those too preoccupied to listen. Energy and effort must be exercised over and over again to avoid an undesirable incident or to press for a long-sought objective. Patience is for diplomats what courage is to soldiers, but all too often the dissonance between urgent perfectionism and disappointing reality breeds cynicism and spiritual exhaustion.

As veteran FSO Charles Bohlen describes it, "An awful lot of the work—I'd say fully 50 percent . . . is really routine work. . . . You've got to have high requirements to enter because any one of the young people coming in could rise to positions of responsibility. And when you're abroad, even a vice consul can do things that can harm or help the reputation and standing of the United States in a given country. But then when you get this high degree of qualifications, education, and so forth, and then you have to apply them to really routine tasks, you get a certain amount of discontent. It's bad for morale."[1]

Political ideals undergo the same process. Patriotism must be a real factor in anyone's decision to join State. After all, there are many better-paying and less demanding jobs. A strong dose of idealism—a passion to make the world a better place and a belief in the intrinsic importance of accurately understanding foreign developments—is often part of an FSO's makeup. But exposure to so many different points of view throughout the world and discovery of the dirty and amoral actions that often mark politics (including office politics) can erode these sentiments. The prescription for promoting the national interest shifts from one administration to the next, friends and enemies constantly change places, and U.S. policy is inevitably inconsistent or hypocritical. It is hard to maintain a sense of righteousness. Indeed, the code of diplomacy stresses emotional detachment and pragmatism, expediency over morality.

These two characteristics—pessimistic skepticism and an uncommitted aloofness—contrast sharply with the politician's energetic confidence and partisan commitment. This difference makes political leaders distrust the career people and even puts the State Department staff somewhat out of step with American culture itself.

The concept of bureaucracy is that of impersonal regulation: clearly defined standards and rules applied equally to all, with rewards based on performance, ensuring an effective and fair system. This is easier envisioned than done. In State, as elsewhere, personal relations play a major role. Those who rise are often less able to do the job well; they are merely better equipped to appear more qualified. Inequities and frustrations may induce the resignation of the able and conscientious, who have less time and skill for office politics, flattery, blandness, and the acquisition of patronage. A half-century has passed since one FSO wrote, "I have lost all my illusions. I told myself bitterly that a man who entered the diplomatic service with a view to making it a career was a hopeless optimist, a determined sycophant, or a congenital idiot." More than a few contemporary FSOs hold similarly gloomy feelings.[2]

In the post-Kissinger era, with increasingly ideological administrations, the pressure for loyalty and the number of non-FSOs appointed to upper- and mid-level jobs have increased. "Many water-walkers have become hall-walkers," commented one depressed FSO, meaning that even those whose careers once seemed to rise rapidly found it difficult to obtain a new or satisfactory assignment.

The State Department understandably requires a great deal of conformity, given the delicate nature of its tasks. When one represents his country—and policies with which he may disagree or even believe harmful—there is little room for individuality. Emphasis on this quality molds FSOs' personalities to discourage creativity. Instead of interpreting facts and events on their own merits, officials watch superiors to determine acceptable conclusions. Talented people may find conscientious patrons to reconcile these competing pulls, but it is common to see important posts filled by those who survive and accumulate seniority by avoiding anything that could possibly turn out to be erroneous or controversial. Other FSOs are highly ambitious and competitive, even ruthless, in their pursuit of advancement. Cliques and alliances protect the interests of both types; the mediocre benefit by discouraging others from being too different or energetic. The best FSOs are victims of such defects, and State carries these additional burdens in performing its already difficult tasks.

The institution usually does an adequate job, but diplomacy is a field where even small mistakes can be very costly. Even 100 percent proper performance by the staff must be filtered through the competence of political appointees and the overall policy structure. There are far fewer intelligence failures than there are analytical failures, and fewer of those than there are policy errors. Good information, analysis, and advice make bad decisions less likely, but certainly do not bar them. The ultimate frustration is that foreign countries simply do not behave as Washington predicts or desires.

There are also some differences between FSOs and the civil service specialists in intelligence, public affairs, and other areas, who serve in Washington at the same type of job throughout their careers. The wider historic split was well illustrated when U.S. Ambassador to Japan Joseph Grew described a long Memorial Day weekend in the 1930s: "We played golf daily . . . for four days." His opposite number at home penciled in, "We worked."[3]

In the mid-1950s, the number of civil servants at State was sharply reduced and many of them joined the Foreign Service. This process went too far, since many positions require a degree of experience and expertise that only long tenure and specialized training can provide. Appointees are inherently transient, and the constant rotation of FSO generalists disrupts operations and discourages efforts to master knowledge about a specific country or subject. When FSOs are moved, there is rarely any overlap allowing outgoing officials to pass on knowledge to replacements. Several Yugoslavs told George Kennan when he finished his tour as ambassador there, "It's quite discouraging. . . . Just as we feel we're getting to know" an American diplomat, he is "suddenly transferred to another country."[4]

A generalized form of specialization survives only with State's political, economic, consular, and administrative "cones." In theory, the system gives them all an equal chance for advancement and the opportunity to become ambassadors; in practice, political officers look down on the other groups, a partial explanation for State's chronic weakness on economic issues.

State is almost an institution without an organizational memory. One scholar, explaining the lack of research, concluded that the career staff "believe in making policy through some kind of intuitive and antenna-like process." FSOs consider their most important ability is to make "mature and balanced judgments about complex situations," a skill that can only come from inner resources and long experience; it cannot be taught. Consequently, as one researcher put it, FSOs "operate with that which is given and they rarely consider whether the level of insight and understanding might be raised as a result of deliberate effort."[5]

The unique aspects of the Foreign Service personnel system are a further source of strain. FSOs must start laboring to obtain desired assignments as much as a year ahead of time, just as they are settling into their current position. Choosing one's next job is of great importance for career advancement, since some bureaus and tasks are considered dead ends. Embassy slots are rated on the post's morale and living conditions, importance and activity, and the presence of personal allies or patrons. The officer must also consider family preferences. Growing numbers of husband and wife FSO teams add to assignment complexities.

"Worldwide availability" is a basic Foreign Service principle, but any energetic officer can maneuver to gain one of his preferred slots. Those eager for promotion may want to stay in Washington, though foreign assignments can only be avoided for so long. Various exchange programs have been developed, allowing FSOs to take a stint at the Defense Department, Congress, a corporation, or as a university teacher or student. The best jobs for those on the way up include assignments to a country desk, the Secretariat, Operations Center, Policy Planning Staff, or as special assistant to an appointed official (the higher, the better). Positions in the political section or as an assistant to the ambassador are the most desirable embassy posts.

While traveling the cycle of assignments, FSOs must be concerned about annual efficiency reports and the danger of "selection out," removal from the service for receiving a low rating or failing to reach a higher grade. But sympathy promotions allow many to avoid expulsion. Being human, superiors also grade their staff on grounds other than competence. Such mercy or favoritism contributes to another problem. As officers reach higher ranks, there are often no commensurate jobs for them to fill; they must haunt the corridors, nervously seeking a suitable position.

While some crave to be at the center of action in Foggy Bottom, others joined the service largely because they wanted to live abroad, sampling different cultures, or because of their attraction to a particular part of the world. In the past, FSOs hated to serve in Washington, where higher living costs and fewer benefits put a significant dent in living standards. Staffers who live well abroad, with housing and servants beyond anything they could hope to afford at home, used to call Washington "the worst hardship post."

Today, however, working spouses and the threat of terrorism make stateside service more popular. Overseas assignments can take a toll on families. The psychological stress for children having no real home (or studying back in the United States, separated from parents), difficult cultural adjustment, boredom, isolation, and language barriers have in recent years brought alcoholism, psychological difficulties, and even drug problems among FSOs and dependents.

Working at State demands far more than the usual 40-hour week for many, while others have little to do. Officers stationed overseas are almost never off duty. Not only may they be called on at any hour of the day or night, they also represent the government in every aspect of their lives and personal encounters. Even socializing is work. Attending parties, seemingly an attractive way of making a living, pales after weeks of mandatory and boring appearances following an intensive workday.

In a sense, the FSO signs over his personality to the government. Every word must be carefully weighed. Changes in policy sometimes require dramatic reversals in position. Since they are representing the views of the U.S. government rather than their own, FSOs are supposed to become vessels of communication, without personal views. Many of them learn to radiate blandness and to censor their own opinions. An ideal pose is to give the impression of great knowledge while revealing little of substance.

By necessity, FSOs must represent U.S. policy in formal conversations or where their remarks will be interpreted as official statements. According to State's regulations, "An employee may be held accountable for deliberate and unauthorized public expressions whether written or spoken, which by violating the confidentiality of privileged information, impede the efficiency of the Service."[6] When Secretary of State Vance told a group of officials in 1980 that he was resigning because he could no longer defend administration positions before Congress, a career staffer responded that he had to do it all the time, whether he agreed with policy or not.

State's officials also live with the painful knowledge of the department's poor reputation among the general public. Seldom do these public servants take on the aura of heroes. A more common expression of the average citizen's attitude can be found in a guide's lecture overheard as his busload of tourists passed State's offices. There, he told them, "The United States supports thousands of people in ease, idleness and luxury." In response, FSOs defensively turn inward, seeing foreign policy problems as inherently intractable no matter what the department does about them. Consequently, they think State functions as well as can be expected and that critics cannot comprehend such matters because they are ignorant of the problems it faces.[7]

"The fact is," wrote an FSO, "that no other branch of officialdom is so often berated in public, so frequently accused of pursuing the wrong policy, so roundly scolded by demagogues from the grass roots, so thoroughly misunderstood by the masses in whose behalf it labors. Badgered by Congress, kicked around by the ill-informed, tackled by the politicians like a dummy at football practice, the department which bears the chief burden for our safety in a predatory and tricky world is everybody's whipping boy." John Paton Davies, a China specialist purged during the McCarthy era, later wrote bitterly, "We have a long and assiduously cultivated tradition of disrespect for officialdom. The public gets pretty much what it invites: a bulk of depressing mediocrity. To get a large body of talent, you have to pay for it in cash, prestige, and tenure."[8] But despite all the criticism, the service is still attractive to young people, as the large number of applications show yearly.

FSOs are unique people. By design and experience, they are not given to enthusiasms; their dramas are internal, rarely showing on the surface. Each career can be summarized by a list of foreign cities, with intermittent mentions of Washington. Friends and acquaintances are scattered around the world, and it is common to encounter someone met or worked with in Buenos Aires, Oslo, Tehran, or Tokyo. Informal regional clubs and alliance networks built up over the years among FSOs have great power over careers and transfers.

Far from luxurious, working conditions in Washington—in terms of space, decor, secretarial assistance, and even office equipment—are below those of the average business and well beneath the demands of any junior corporate executive. Expense accounts are hardly lavish; overseas, many must dig into their own pockets to cover work-related "entertainment," the cocktail parties, dinners, etc., that enhance contacts with important foreigners to gather information, explain U.S. actions, and gain goodwill.

On top of all their other worries, FSOs must be security conscious. Spies do exist. The efforts of foreign secret police and intelligence agencies, using electronic surveillance and other means, can be extensive. Documents have to be protected; secrets must be kept. Dealings with journalists are necessarily cautious, although those who provide leaks are rarely caught or punished. Each administration swears it will discover the culprits, but deliberate leaks, often from the top, have become an integral part of the policy system. Officials use them to gain the attention of superiors, launch trial balloons, scuttle rivals' plans, advocate personal or institutional proposals, and influence foreign governments.

There are additional frustrations for the conscientious. Reporting the truth is not always rewarded, as many since the China hands have discovered. The career staff is not supposed to make policy but rather to provide the information for higher-ups to do so. This is one of the first lessons taught incoming FSOs. Despite influence in shaping the way issues and options appear to superiors, FSOs cannot escape the recurrent feeling that no one is listening to them and that policy is unconnected with what they perceive as reality. After the frustrations, bureaucratic barriers, anxiety, and low morale, someone may only gain a position of authority when he or she is too worn down to use it. As one FSO comments sadly, "I have spent my whole career trying to prevent the disasters I have predicted from happening."

Disasters can happen to careers, too, and one never knows who may be determining one's future. The department is a whisper mill in which an individual's "corridor reputation"—what others think of him—affects his status. People call up their own friends and allies to determine who would be good at a particular post. Despite periodic assessments on the performance of FSOs and civil service employees, one political appointee comments, "Academics are judged by their publications, lawyers by their victories in cases, but how are people here evaluated? It's never really clear." Such problems contribute to low morale. Many FSOs, as George Kennan put it, "have the feeling that their fate is determined by people who neither know them nor care about them personally or have any personal experience that would qualify them to understand and to judge Foreign Service performance.''[9]

Former Undersecretary of State David Newsom lists as an FSO's necessary qualities: ''An understanding of our own nation; a balanced sensitivity to other societies and peoples; a firm grasp of . . . international relations; and the skill to bring this knowledge together in advancing both the interests of our country and the establishment of working understandings with others.'' Historian Arthur Schlesinger writes with equal accuracy, ''At times it almost looked as if the [Foreign] Service inducted a collection of spirited young Americans at the age of 25 and transmuted them in 20 years into bland and homologous denizens of a conservative men's club.'' Kennan, a defender of the corps, says, ''I have seen, over the decades, an unduly high percentage of older men in this Service who prematurely lost physical and intellectual tone, who became, at best, empty bundles of good manners and, at worst, rousing stuffed shirts.'' Roger Morris comments, ''It was one of the whispered little secrets of younger aides . . . that the boss, whatever his respected name, reputation and apparent success, was not really all that bright or informed.''[10]

Again, this does not negate the courage and abilities of many at State, but the grind is more powerful than the glorious. There are as many mediocre timeservers and ferociously ambitious manipulators as there are bold and imaginative people advancing U.S. interests in the best possible manner. The system creates precisely what one would expect, so that those successfully playing the game while retaining their energy and creativity are particularly remarkable people. One of them, John Franklin Campbell, explained that ''the part of the machine that recruits and hires and fires and promotes people can soon control the entire shape of the institution.'' Attempts to reform State fail because they rarely consider the atmosphere and incentives that shape the behavior of those working there.[11]

Professor Christopher Argyris's extensive interviews with FSOs produced similar conclusions. They placed high priority on avoiding open conflict with others, since State's norms ''inhibit open confrontation of difficult issues and penalize people who take risks.'' One FSO stated, ''To make real changes you have to be a wavemaker and that's dangerous. It could harm your career.'' Said another, ''I think that one reason I have succeeded is that I have learned not to be open: not to be candid.'' Requesting frankness, he thought, ''was like asking us to commit organizational suicide.'' Such behavior produces mutual mistrust and FSOs soon learn to maintain a careful facade. These characteristics, Argyris concludes, produces feelings of bitterness and powerlessness.[12]

A bureaucratic and promotion system based on peer review and uncertain standards teaches the career service to prize caution. The importance of getting along is reinforced by the difficulty of many veteran FSOs in finding satisfactory jobs outside State. Having lived in State's closed system and overseas so long, they are somewhat insecure with American society and the culture shock experienced on returning home.

Nevertheless, one constantly meets men and women who do not conform to these limitations. Some have a particular concern over the quality of their work and a relatively low level of ambition; others—extroverts, able politicians, and those well studied on the issues—are able to succeed given their substantive and personality skills. Quality is often appreciated, and most talented people prefer to surround themselves with others, who will make them look better, in contrast to the insecure, who feel safe only with mediocrity.

While foreign assignments, a bond of secrecy and tradecraft, remnants of past esprit de corps, and a sense of being besieged tie the service together, broader recruitment has introduced more variation. The traditional FSO's profile—upper-class, Ivy League, white Anglo-Saxon male —has been altered by changes in American society and government policy. Decades ago, the service became open to a wider geographical, educational, and ethnic cross section. Beginning in the 1960s, more women and blacks were encouraged to join the ranks. The mid-career entry program permitted a greater variety of people with more diverse experiences to participate. Although women and blacks may still find acceptance far from complete, State is much more heterogeneous than in the past.

Still, if the Foreign Service is no longer a smug men's club, it is more like one than any other part of the U.S. government. White male FSOs are often bitterly critical in private about the alleged quality of female and minority colleagues, arguing they were hired or promoted to fill quotas and not on ability. The best officials in these categories generally quit, they say, to take better opportunities outside the service, leaving behind those with fewer options. It is obvious that the shrinkage of available top jobs has created resentment among senior male officers, making them overstate the number of such promotions. Women and black FSOs argue that these private comments reflect bias at State. They attribute resignations to poor assignments, bad treatment, and slow promotions. There is also another problem. The utility of diversifying the recruitment pool is precisely to broaden department culture and introduce people who may be more inclined to question, hold different perspectives, and stress other priorities. Yet these characteristics are often punished rather than valued at State.

In the past, wives of FSOs dutifully followed spouses from post to post. Their social performance was rated as part of the husband's evaluation: Typically, "Mrs. Jones was an asset to her husband and to the American Foreign Service." Only in the 1970s were regulations changed to make it easier for wives to obtain embassy jobs abroad (a common practice by other countries) and to allow married women to remain in the service.

During the 15 years after 1957, the percentage of women in the Foreign Service actually decreased, but between 1970 and 1980 it rose from 5.3 to 11.5 percent of the corps, from 174 of 3304 to 413 of 3581. The proportion is higher among the professional civil service, USIA, and AID personnel. Men still receive more easily the boss's confidence and support, kinder evaluations, and faster promotions. The situation is far from satisfactory; even sexual harassment remains a problem.[13]

In fact, only about 16 percent of FSO positions are held by women. A 1983 U.S. Commission of Civil Rights report noted, "Minorities and women currently are almost totally absent from top appointed positions, other than ambassador, at the State Department." President Carter appointed 8.8 percent minorities and 7.5 percent women to ambassadorial posts; President Reagan's figures were, respectively, 8 percent and 5.6 percent. The Carter administration chose more, although still a relatively small number, to higher slots at State. Ironically, a department whose very purpose is to deal with other cultures and peoples, places a remarkably high emphasis on homogeneity and tends to regard the different as inferior.[14]

The relationship between appointees and career staff is equally complex. FSOs tend to be culturally conservative, but some patterns of professional experience encourage aspects of liberal thinking: a preference for political over military means and greater awareness of foreign perspectives. In addition, many FSOs do not share the appointee's tendency to see the United States as always in the right. Career people are aware of, and sometimes exaggerate, limits on U.S. power; appointees are more ambitious in their expectations of accomplishment. Many staffers have a special, institutional interest in the Third World, a stress on indigenous roots of problems over the East-West aspect, and sympathy with foreign aid.

At the same time, the careerist's world view is skeptical about a prime ingredient of liberal foreign policy thinking, namely, applying morality to international relations. The staff's belief in the status quo's staying power is an attribute that makes it moderately conservative. The principle that foreign policy should be above domestic politics and beyond ideology leads it to distrust all camps.

Each appointee must decide his or her attitude toward the career service. Working arrangements emerge gradually, with many administration loyalists consciously determined not to be co-opted by the bureaucracy. If their experience is limited, policymakers are dependent on the permanent staff for understanding procedures and obtaining information. The newcomers frequently believe that facts are being withheld or that options are excessively predetermined. Eager to set a fresh course, they want to ensure they are invited to relevant meetings, asked to clear orders, and provided with important cables. They will inevitably discover that papers are late or that requests are not filled to their satisfaction. Yet their demands may be contradictory: They want the career staff to be detached, but accuse it of being bland; they demand discipline, but can brand this as lack of imagination; they require experienced judgment, but may call this negativism. One FSO complains, "Presidents and their aides need scapegoats. They can't blame the administration so they blame the secretary of state and if they can't blame the secretary of state they criticize the department's staff."

Recent administrations have been particularly determined to preserve ideological purity. This factor and the desire of individual appointees to accumulate power for themselves make them reluctant to delegate authority. Those placed in office by voters or the president do not relish being told that there are things they cannot do or that circuitous procedures must be followed to obtain results. When appointees do accept a subordinate's recommendations, this can put them at odds with administration colleagues who counterpose what they see as White House objectives or who are listening to the competing portions of the bureaucracy over which they rule. The president's staff often starts to hint that the appointees at State have sold out to the eternal, immobile, and allegedly hostile career staff.

Those appointees accustomed to making candid public statements as private individuals must adjust to the scrutiny and delicacy of international affairs. In many cases, they have to learn about complex issues from scratch. Scuffles over relative power will occupy an administration in its early days and sometimes all the way through its term. "If the Foreign Service spent as much energy against the Soviets and Cubans as they did toward advancing their careers and protecting their turf, those countries would have ceased to threaten the United States long ago," exclaimed one frustrated Reagan appointee. But the same tongue-in-cheek complaint can be said of the appointees as well.

The most skillful appointees at State will labor assiduously to develop their own links to the White House staff, NSC, other departments, and even Capitol Hill. A successful assistant secretary must build such alliances and play a strong role in the interagency coordination process. This is by no means an easy accomplishment. Only two ARA assistant secretaries—Thomas Mann under Johnson and, before his firing, Thomas Enders under Reagan—gained this kind of predominance. No assistant secretary for European affairs has built such a position in recent decades.

The appointee is likely to worry more about his connections up the ladder than about his relations with the bureau he heads. Only rarely can an assistant secretary, particularly one in charge of a front-line regional bureau, do a good job in both directions. This requires both substantive knowledge and political ability. Perhaps the best examples have been several recent NEA assistant secretaries—Joe Sisco and Alfred Atherton under Nixon, and Harold Saunders in the Carter administration—and Chester Crocker of African Affairs in the Reagan administration. All of them, it is interesting to note, had experience in lower positions within the bureaucracy.

Ambitious staffers aim to join the ranks of those holding a presidential appointment as undersecretary, assistant secretary, or ambassador; the last-mentioned is still an FSO's main career goal. Appointees, however, may correctly view the working level as lacking a framework for analyzing events. The politicals see policy proposals as partisan, arising from a clear-cut taking of sides in internal debates. In contrast, the staffers tend to lack a sense of strategy and an appreciation of other agencies' positions and of political factors. FSOs are professionally, if not always personally or bureaucratically, neutral—that is the central principle of a career service. When they are not seen as such by superiors, they are usually perceived as unfriendly.

In a few cases, an FSO or civil servant may declare himself militantly on the current administration's side, a step which, depending on timing, can lead to rapid promotion or an end to his career. More often, unless elevated to mid-level posts, they are content to allow the politicals to take responsibility. After all, as one appointee put it, "The expert may be right or he may be wrong, but the risks of being wrong loom much larger at the top than at the bottom. Bureaucrats could talk all they wanted about taking risks and managing the consequences of defeat; political leaders had to take the risks and suffer the consequences." [15]

Many FSOs see this problem in different terms: "Expertise is not merely ignored it is often resented. The policymakers' attitude is that I'm higher than you and I'm the one who has been chosen to make the decisions." Career people know that those who zealously served their country have often done worse than those who put primary emphasis on energetically, though cautiously, pleasing the incumbent administration.

Another general difficulty is staying in tune with the White House. This was no problem for Kissinger, who lacked any serious competitor for Nixon's ear. When, in contrast, the president is open to different, disagreeing advisers, there are going to be conflicting interpretations of his will. In the words of one former high official, "There just naturally comes a series of mutual resentments and recriminations in which the enthusiasts say that the cautious people are undercutting the president and the cautious people say that the enthusiasts are getting ahead of the president."[16]

Whether or not the president's people and the career people work well together, one of the primary complaints within State itself is the department's internal management. Like many of its contemporary problems, this is hardly a new issue. As early as 1946, an administrator noted, "The people doing the clerical end of the work there don't have the faintest idea of the standards prevailing in the well-run agencies."[17] Almost every employee has firsthand stories about misplaced files, wasted money, and irrational personnel policies. The number and variety of reforms and reorganizations has often made matters worse, destroying consistency and inhibiting experience. Poor management also occurs because officers promoted as good writers or diplomats find running an office requires a different sort of skill.

Much conversation among FSOs centers on analyzing the complex, shifting regulations that govern promotion, perquisites, status, and security. A law passed in 1980, for example, reshuffled rankings and created a Senior Foreign Service at the apex. Some officers try to plan a career by meticulously choosing each post for what it can add to their record. During particular years, the true connoisseur realizes, those who are economic officers, study Spanish, or switch to ARA rise more rapidly due to crises or statistical flows. All these administrative, behavioral, and personnel problems have been discussed for years and seem no closer to resolution today than in the past.[18]

Yet two factors about the career staff maintain an overriding importance. First, any given job in Washington or in an embassy can be made or broken by the competence of the person who holds it. His or her performance at key moments, no matter how tedious the normal run of business, can determine the success of U.S. policy and even the fate of nations. Second, people will more likely perform effectively when they believe that they and their work are treated fairly by the government. With declining living conditions and greater dangers overseas (some FSOs go to work in bulletproof cars), wages inferior to private industry, and limited career prospects, the Foreign Service is having considerable difficulty retaining good people. A Bureau of Personnel official commented, "If you found a Foreign Service career successful only if it reaches the senior grades, you stand a significant chance of being disappointed."[19]

In evaluating the work itself, many insiders and observers of State find overstaffing and excessive paperwork as major causes of sluggishness. The growing complexity of issues has usually been dealt with by inflating titles and adding new layers of bureaucracy. Certainly, State has to deal with more foreign countries—over 150 today, as compared to only about 50 in 1945—and whole new sets of problems like terrorism, the environment, drug-smuggling, and nuclear proliferation.

Yet each new office means one more special interest with its own turf and clients to protect. Conflicting jurisdictions inevitably develop over any matter. A cable concerning U.S.-French relations in Africa could involve the Africa, European, Politico-Military, Economic, and International Organization bureaus, which all must coordinate positions and initial documents before recommendations can go up the hierarchy. This requires more time while diluting responsibility and decisiveness.[20] It is hard to see how this process can be avoided except by a tight monopolization of power at the top or by moving decision making altogether outside of State, which is precisely what has happened in recent years.

The proliferation of information produces a flow of cables and memoranda so great that many officers spend much of their day just reading them. One official commission found Washington did not sufficiently guide embassies to provide what it needed: There were too many reports with too little analysis. It approvingly quoted Kissinger's complaint that "mere reportage of events which have already taken place and about which in many cases we can do little is not sufficient. I require not only information on what is happening, but your most thoughtful and careful analysis of why it is happening, what it means for U.S. policy and the directions in which you see events going." Other studies reached similar conclusions. Even when computerized, the accumulation of two million reports after less than three years made "conducting a search on any subject . . . likely to bring forth a much longer listing of documents than can possibly be applied," noted one evaluation.[21]

Internal studies suggested "that the 'useless' or 'minimal usefulness' component is as high as one-third of all telegraphic reporting from some larger posts" and "that as much as one-sixth to one-third of all reports are only vaguely related to policy, and as many as two-thirds of all political and economic reports provide no interpretation or analysis of the event reported on." Yet "high reporting output is almost always commented on favorably in efficiency reports and inspection reports while low output—

if mentioned at all—is usually cited as a sign of lack of ambition or imagination.''[22]

U.S. missions are also challenged by the mass media's speed and comprehensiveness. When foreign correspondents start writing about something, the diplomats had better not be far behind. Given this alternative channel of information and the fact that U.S. missions are often not well briefed from Washington about the latest policy developments, the embassy can lose credibility with the host country, further impairing its function.[23] But FSOs themselves do not always have a good understanding of the local country's politics and culture. Washington or the ambassador may distort dispatches to reflect favorably on their predilections and efforts. Examples include the censoring of pessimistic reports about Saudi responses to U.S. policies, Iran's revolutionary upheaval, events in El Salvador, and predictions on foreign elections.

The main problem of maintaining objectivity is not outright political bias but bureaucratic interests. A diplomat writing a memorandum of conversation after an official meeting, one FSO explains, ''is under an almost unbearable . . . pressure to report that he expressed his government's point of view just a little more skillfully, a little more forcefully and effectively, than he actually did. . . . It is natural for the human to want to make himself appear to good advantage.''

Another temptation is when ''The facts, and the interpretation of them which he intends to submit, will be unwelcome to his own government.'' When a foreign leader rejects a U.S. position, the diplomat can feel, ''Officials at home may attribute this to his own lack of finesse.'' But if temptations are not resisted and the truth is not accurately presented, ''He would be . . . misleading or confusing his own statesman and policymakers. (An enemy agent could do not more.)''[24]

Embassies have a vested interest in upbeat reporting. At one orientation lecture the deputy chief of mission explained that, to maximize U.S. support and aid, the political officer should show that the host government ''is honest, efficient, popular, and a true friend of the United States.'' The economic officer should report ''regularly on the real progress the government is making toward balancing its budget and utilizing foreign aid effectively. Don't let the team in Washington down. . . . Be boosters, not knockers.'' This attitude was unusually explicit but not unusual because the ''reward'' for accurate assessment can be that Washington, facing endless demands on limited resources, will cut back assistance to the unfortunate country with accurate U.S. embassy reporting. If relations deteriorate, the honest diplomats will appear to have failed in their mission.[25]

The best are also the sharpest critics of the reporting process. A young

FSO known for his brilliant, and unheeded, dispatches comments, "Imagine someone living abroad responsible for reporting on a country where he doesn't know the language, never rides the bus nor deals with average people except for maids or chauffeurs. He only stays in the best hotels or most expensive residential districts and talks to other Americans, foreign diplomats or government officials." Yet, he added, "We do as well as most other foreign ministries and better than the journalists."[26]

Many FSOs also become convinced by close contact that their host country's government is an asset for U.S. interests. They come to sympathize with its political positions and its case against rivals. "Clientism" breeds conflict within the department and mistrust in the White House. The results can vary from the humorous to the tragic. "Tell Madame Gandhi how lucky she is," Lyndon Johnson told a startled Indian ambassador as he left a White House meeting in 1968. "She's got two ambassadors workin' for her . . . you here and [U.S. Ambassador Chester] Bowles out there."[27]

Roger Morris, then an NSC Africa specialist, bitterly recounted that when Biafra seceded from Nigeria in 1967, the wife of a U.S. diplomat toasted to its destruction. Embassy officers told visitors that pictures of starving Biafran children were a publicity stunt. Attempts were made to alter or suppress reports unfavorable to Nigeria, including eyewitness accounts of atrocities. Dissent was punished by unfavorable performance ratings. Irritated by such wild clientism, Washington sent letters and then visitors to urge more complete reporting, but State itself was reluctant to offend victorious Nigeria as Biafra's collapse became imminent.[28] This was, of course, an extreme case, but other examples include battles between China and Japan hands in the 1930s, conflicts pitting the East Asia and African bureaus against the European bureau responsible for relations with colonial powers in the 1950s and 1960s, and partisanship today between U.S. embassies in India and Pakistan, the Arab states and Israel, and anywhere else where countries are at odds.

In these intramural and other maneuvers, successful FSOs and appointees must be good bureaucratic politicians. The key ability is to convince others to do as you wish, using skills ranging from personal charm to writing options so that the one you support will be adopted. This last exercise often follows the "Goldilocks principle"—one option is too soft, one is too hard, and the policymaker predictably selects the one that is just right. Flexibility is another requirement. As one FSO put it, "You have to be a good diplomat to work for Carter one week and Reagan the next." Winning bureaucratic successes, meeting extremely short deadlines, gaining clearances from other bureaus, and keeping one's superiors and staff reasonably happy also require the diplomatic virtues.

Yet while such adventures offer opportunities for exhilaration and resourcefulness, there is much bitterness among FSOs. They are supposed to be, in the words of one of them, "pro-nothing except pro-U.S.," but many feel the sentiment is not reciprocated by their country. Kennan despairingly calls diplomacy a "thankless, disillusioning and physically exhausting profession" whose acolytes are "professionally condemned to tinker with ill-designed parts like a mechanic with a badly built and decrepit car, aware that his function is not to question the design or to grumble over the decrepitude, but to keep the confounded contraption running, some way or other."[29]

Once again, such attitudes are alien to the political appointee, who views his administration as a new model of automobile he has helped design with all the latest features. If the career staffer complains about the policy system, an appointee considers it as the president's system. Such world-weariness seems like lassitude and a self-fulfilling prophecy drowning hope in routine, effectiveness in pessimism.

Administrations have their own pattern of behavior. During the election campaign, candidates promise to improve the policymaking process, assemble a team that speaks with one voice on foreign policy, and name more ambassadors on merit. When elected, the new president pledges primacy to the secretary of state and asks the department for new ideas. When it fails or comes up with proposals the White House does not like—though these will usually be self-censored out—disillusion begins to set in. About 12 to 18 months later, a shake-up will remove appointees who do not fit or who are on the losing side of policy disputes.

There will also be a mellowing process as new personnel acquire experience and closer working relations with the career staff, and as the administration's distinctive views are worn down, particularly since many of the issues and crises they face will be different from those they foresaw. The president will continue to complain about State's performance and be upset about leaks. Most of the latter, however, come from his own appointees. As powerful and aggressive leaders collide, the president has a choice: settle the disputes decisively, back up a chief adviser to make decisions, or allow matters to drift. Meanwhile, the staff will continue to report developments and the positions, inquiries, and initiatives of foreign governments, monitoring U.S. relations with various countries, helping to formulate and implement policy, coordinating with other government agencies, promoting U.S. interests and commerce, and protecting the rights of Americans abroad.

Decision making is a complex process involving a large number of factors and people. Involvement in foreign policy creates a skeptical attitude toward longer-range perspectives, a narrow perception of what is possible, and an element of cynicism over the difference between public pronouncements and behind-the-scenes realities. The need to choose between equally problematic options as well as the speed and obduracy of events encourages feelings of passivity. As for the creativity so often called for by the politicals, the staffers respond that there are many good ideas but that they are not relevant to the current situation or leverage of the United States.

The strain between political appointees and career staff is ultimately unresolvable and even partly useful by providing a variety of perspectives. Foreign policy is rarely subject to neat organization. Its complexity and rapidly changing profile make it inherently disorderly. While outwardly the system demands conformity and consensus, it is always driven by individual willpowers and objectives.

In the battle to convince the president, an idea or proposal is only as strong as the bureaucratic backing mustered and the competing alliances built by supporters and opponents. The information on which decisions are made is dependent on the knowledge and sensitivities of those who supply it. The options selected are based on the skill and political ideology of those making the choices. In U.S. foreign policy, the process and people involved have a predominant role in shaping the nation's positions and actions.

11

The Policymakers

The United States is the world's most powerful single nation. This would, in theory, make the U.S. State Department the globe's most significant foreign ministry. But the White House staff, NSC staff, CIA, Defense Department, Congress, and other institutions have divided and disputed control of diplomatic decision making and operations. These conflicts produced the resignation of three secretaries of state in the last three administrations: William Rogers in 1973, Cyrus Vance in 1980, and Alexander Haig in 1982. The struggle for power and influence among individuals, viewpoints, and agencies has often been as dramatic as the world events it mirrored.

Most studies of foreign policy chase after the minor secrets of the diplomatic process: the exact date and contents of meetings, memoranda, and decisions. But the greatest secret of state—how decisions are made and implemented—applies to all circumstances. This mystery is understood by most of those engaged in the policy process, but it is not comprehended by many outsiders.

U.S. foreign policy is currently in a serious crisis of both form and substance. The public at home and abroad is well aware that rising internal conflict and confusion within the U.S. government has led to poor performance in handling international crises. These problems have reached a point where they interfere with America's ability to cope with the serious problems it faces. The pattern of bitter internal disputes and poor coordination has been surprisingly similar in the Carter and Reagan administrations despite the sharp political differences between the two presidents.

At the root of U.S. policymaking's unique style is a distinctive American attitude toward international affairs and those who practice diplomacy. "Foreign policies demand scarcely any of those qualities which are peculiar to a democracy," wrote Alexis de Tocqueville, that shrewdest of commentators on America, in 1835. "They require, on the contrary, the perfect use of almost all those in which it is deficient." More recently, veteran diplomat Charles Bohlen claimed, "There's no doubt about it, the American system of separation of powers was not designed for the conduct of foreign affairs."[1]

The philosophy of de Tocqueville sprang from a European view of realpolitik, an unsentimental, amoral view of states, power, and the pursuit of national interest that has never fully penetrated the American psyche. Henry Kissinger championed this approach and it is strongly entrenched among those following a foreign policy career. Yet such ideas remain quite suspect for the general public and most politicians. Both the Carter and Reagan administrations rejected them as elitist and immoral, failing to fight actively either for human rights or against Communism. After two centuries of life as a nation, Americans are still seeking some satisfactory way to conduct and to comprehend their relations with the rest of the world.

A country's foreign policy must blend actions necessary for its survival and prosperity with goals arising from the nation's values. Throughout history, most states have had to emphasize the former; the United States has had an almost unparalleled opportunity to stress the latter. A people whose enemies, and even neighbors, were distant and who were long protected by two ocean moats considered international relations a very marginal concern.

Until the 1940s, the American people and their government also had little experience with political instability, foreign threats, invasion, totalitarianism, underdevelopment, and other phenomena unfortunately common in the outside world. This pleasant cultural and historical legacy nevertheless created handicaps for the functioning of the United States as a great power. "Nowhere," writes historian Walter Laqueur, "has there been so little understanding of how a dictatorship works or so little appreciation of the importance of ideology (or religion or nationalism) in politics. In no other country has there been so much good will—which is to say willingness to ignore or at least belittle the existence of genuine conflicts among nations, ideologies, and political systems." A British diplomat notes, "Americans are not good at the observation of subtle graduations, the long-term calculations, the patient endurance of irremediable inconveniences. . . . Patience is not a typically American quality but it is one of the greatest diplomatic virtues.''[2]

American culture also teaches that vigorous and determined action can master problems and achieve goals. This idea presupposes some domination over circumstances, but the world is largely ruled by forces outside U.S. control and by people holding viewpoints quite different from those prevailing in Washington. Consequently, foreign policy is very different from the previous experience of politicians trained in domestic politics. The need to cope with other governments requires both an understanding of their situations and adjustments in U.S. behavior. Obviously, presidents and their senior advisers would prefer to avoid the frustrating delays, compromises, and dead ends that often occur in diplomacy. They approach the State Department deaf to warnings of problems or constraints limiting their power, like the nineteenth-century rogue millionaire who told his lawyer, ''Don't tell me what's legal, tell me how to do what I want to do.''

The White House's low regard for State is exceeded by the department's poor reputation among the public, media, and Congress. This problem has remained remarkably consistent over decades. The conclusions of a 1950 article, entitled ''Why Americans Hate the State Department,'' remain timely: ''The secretary of state exists only to recognize the existence of a world which Congress would rather ignore; of obligations which Congress distrusts and tries to turn to its advantage or to reject. . . . The people distrust this institution because it is our chief link with the outside world, and Americans are uneasily aware that our contacts with the outside world have brought us much more pain than pleasure.''[3]

U.S. political traditions also make the nation's policy structure unique in structure and spirit. In contrast to Washington's bureaucratic pluralism, most other countries have a single line of foreign policy authority: The president, prime minister, or chief general decides policy himself or delegates as much power as he wishes to the foreign minister. The foreign ministry is highly professionalized, with far fewer appointees among its officials or ambassadors. There is often a high career official—a secretary-general or permanent undersecretary—to maintain continuity and the links between political leaders and staff.

Parliamentary systems eliminate much executive/legislative conflict. Dictatorships diminish the role of both press and legislature, reducing the pressure of domestic politics on foreign policy. The media, controlled or curbed, does not leak secrets or details of internal power struggles. The American system remains unique, seeming like a free-for-all in contrast to more staid or controlled arrangements. In addition, the global scope of

U.S. policymaking—most countries focus attention on their own region and relations with the superpowers—adds additional complexities.

Obviously, the U.S. policy system must be in accord with national traditions and institutions, but it could be far better even within this framework. The scope of U.S. responsibilities does not excuse shortcomings. On the contrary, the inability to handle serious problems, the tendency to act only in the face of crisis, and the lack of policy coordination, along with other problems, are only made more worrisome by such awesome power.

Understandably, in light of these fundamental difficulties, the State Department faces unnerving contradictions. There has always been the strain of representing an inward-looking country to the world. During most of U.S. history, State administered the largely minor and routine matters involving U.S. foreign relations, excepting brief periods when strong presidents like Theodore Roosevelt and Woodrow Wilson personally took the helm. With the rising importance of international affairs after 1945, foreign policy became a prime area of concern on a daily basis. In a world facing constant crisis, U.S. actions were crucial in maintaining peace and shaping events. New agencies were created to deal with aspects of foreign affairs: the Department of Defense, the CIA, and the NSC. The CIA and Defense Department staked out a role in the 1950s, but it was the buildup of the NSC staff as the White House's own foreign affairs team in the 1960s that truly provided the chief executive with an alternative policymaking center to State.

The NSC was originally designed as a committee of all the involved agencies. But in the Nixon years, with Henry Kissinger as virtuoso conductor of this new power center, the NSC staff became executive manager of foreign policy. If Kissinger's successors could not match his degree of control over foreign policy, they created enough divergence of authority to prevent anyone from being in command. It became harder for administrations to create a united and consistent policy and took them longer to learn how to operate the levers of power, until half of each term or more was spent in these tasks.

The growth of a small, efficient NSC staff allowed a president to circumvent the seemingly slow and uncertain State Department. White House advisers have the advantage of proximity to the president and a detailed knowledge of domestic politics. The NSC staff's foreign policy expertise and access to embassy cables and policy papers permits it to intervene more frequently and in greater detail and also allows the president's political aides to gain increasing influence over foreign policy. At its most extreme eclipse during Kissinger's NSC years, State was described as "a collection of desk officers who answer the mail, compose and receive telegrams, and carry on relations with foreign governments at the level of the routine and pedestrian."[4]

It is not surprising that the president would favor his chosen colleagues over the "alien" State Department. As one scholar accurately explains, "Concerned with direction and results, presidents are usually predisposed to cut through the rigidities of complex bureaucratic systems and the cautions of the foreign policy regulars."[5] This pattern was clearly visible in the Nixon, Carter, and Reagan administrations. Freeing themselves from State's bureaucratic web, Nixon and Kissinger were able to maneuver quickly and dramatically to open relations with China, negotiate a comprehensive U.S.-USSR arms control agreement, and implement a patient, shrewd Middle East policy. But the same process of circumventing State also gave birth to badly flawed policies in southern Africa, southern Asia, and Cambodia, which staff experts could have improved if they had been allowed to do so. After Kissinger, the national security adviser's office and staff could never be the same.

State's shortcomings had much to do with its decline, but a system dominated by the national security adviser or, more commonly in the last decade, a power vacuum has problems of its own. Effectiveness was undermined by the disruptive effects of State-NSC competition. When the NSC staff, designed to reconcile interest groups, becomes one of them in its own right, there is no trusted, neutral body to perform this function.

Both the Carter and Reagan administrations had many criticisms of Kissinger's style and policies, but they could not escape his innovations. Carter chose the bureaucratically aggressive Brzezinski as national security adviser and the mild-mannered Vance as secretary of state. The president thus diversified advice but also built a split of personality and politics into the heart of his administration. Nominally, Carter reacted against Kissinger's process by reestablishing State's preeminence, but in practice the NSC staff was again soon challenging it. When the two sides cooperated—on the Panama Canal treaties, strategic arms talks with Moscow, and the Camp David negotiations—things went well. But, inexorably, the Carter administration gained a reputation for ineptness and vacillation because the two groups contradicted each other in public and in private.

When Carter, acting as referee, was slow to intervene in disputes or to make up his own mind, policy faltered; timing, as the crises over revolutions in Iran and Nicaragua showed, was crucial. The Iran hostage crisis and the Soviet invasion of Afghanistan pushed Carter in Brzezinski's direction. Vance resigned because he failed to stop the ill-fated Iran rescue mission, but that last debate also symbolized his declining stature within the government. A White House whose self-image was that of innovative, antiestablishment outsiders saw the Foreign Service as the advocate of business as usual. This image was most accurate in relation to the administration's human rights policy, which the career staff tended to see as unwarranted interference in other countries' affairs. Carter, like Kissinger, also distrusted State as the fountainhead of media leaks.

The Reagan administration, like the Carter White House, was eager to do everything differently from its predecessors. It went even further in expressing determination to cut the NSC staff's influence, choosing the unenergetic Allen as national security adviser and reducing the size of his staff. Secretary of State Haig, a veteran of Kissinger's team, took this as a signal that he would dominate the process, as "vicar" of foreign policy. The White House quickly decided that it did not like this idea, seeing Haig as ideologically suspect and personally distasteful. He was forced to resign after less than 18 months in office. Within State, relations between political appointees and career staff were worse than in the Carter years, as hard-line Reaganauts deemed FSOs disloyal. The national security adviser again tried to dominate the scene. Bureaucratic battles shook the process over U.S.-Soviet relations, El Salvador, and the Middle East. The Lebanon crisis provided a case study of a policy wrecked by problems of process: internal bickering, inconsistent strategies, changing personnel, and lack of coordination.

The secretary of defense and CIA director achieved an almost unprecedented role in decision making on foreign policy issues. The Defense Department's job is to supply, organize, and direct the U.S. armed forces, but this task involves it in many issues overlapping State's operations. Defense receives its own political analysis and reporting from its own bureaus, the Defense Intelligence Agency, as well as from military attachés and advisory teams located in U.S. embassies. Secretaries of defense have been strong adovcates on issues including the Vietnam War, arms control, U.S.-Europe relations, and Persian Gulf security.

Policymakers at Defense also have the advantage of almost unlimited financial resources and greater unity compared to poverty-stricken and often fragmented State. The military services are, by training, reluctant to become too involved in political matters; the career bureaucracy is more deferential toward appointees than are its counterparts at State. Although the Defense Department sometimes sees its institutional interest as preferring military means for solving problems, it can also be very cautious. Generals are more aware than civilians about the difficulties of success. Once they are given a mission, however, they single-mindedly pursue its fulfillment, rarely reevaluating costs or prospects.

In contrast to the White House and NSC staffs, the CIA is an information-gathering, rather than a policymaking, agency. While, ironically, covert operations monopolize public attention, most of the CIA is a huge research staff using scholarly methods to analyze a wide range of secret and open sources. The Agency often has far more detailed information and expertise than State, since its employees specialize to a far greater degree. Yet the CIA plays a smaller role in making political decisions than is generally understood. Comparing the working atmospheres at State and CIA, reality is also the reverse of what one might expect. Since CIA employees are so carefully screened and enjoy greater job protection, they are far more relaxed and—within their own circle—can be more outspokenly critical of policy than colleagues at State.

Congress is another major participant in foreign policy, although State's staff decries its sometimes ham-handed interventions. "Whatever the shortcomings of the average FSO, he is far smarter than the average congressmen," is a typical remark. While some legislators win respect, FSOs' experience with congressional junkets overseas—every diplomat has horror stories on the subject—further stirs their distaste. Members of Congress share the public's unfavorable stereotypes of State, which has no lobby or special interest group to protect its image. Yet State's declining ability to restrain decision makers and transmit congressional concerns to them, beginning in the Nixon years, has inspired Congress to play a more direct and productive role in foreign policy.

Senators and representatives have to spend so much time on electioneering, politics, and domestic issues that relatively few of them develop the background or time necessary for a sophisticated understanding of foreign countries and international issues. In recent years, however, the growth of its own staffs and the Congressional Research Service provide Capitol Hill with expertise and improve the capacity of legislators to deal with foreign affairs through speeches, resolutions, hearings, and laws, as well as budget and appropriations bills.

FSOs view themselves as the proper custodians of foreign policy and begin to think, almost subconsciously, of Congress and the White House as interlopers. But the president and his chosen advisers hold responsibility for exercising power and must consider all factors affecting an issue. Congress also represents legitimate interests and is sometimes more correct than the White House or State on a particular point. By ignoring considerations outside its own jurisdiction—the responsibilities of other bureaus, domestic policics, budgetary, and military needs—State's staff can be guilty of provincialism, unimaginative reliance on precedent, and an overweening desire for bureaucratic peace.

No one institution is always right, which is the benefit of having a pluralistic structure. At the same time, this system is extremely taxed as elected officials, appointees, and career staffs struggle together to keep up with a complex and rapidly changing set of alliances, enmities, wars, conflicts, friendly and unfriendly governments, political parties, revolutions, coups, economic vicissitudes, leaders, and issues. This maelstrom is difficult enough to daunt a combination of all the world's computers. The resulting burden isolates those involved in a round-the-clock world of their own, a society with its own rules and culture.

When Americans and foreigners think about U.S. foreign policy's decision-making process, they usually assume that it functions in a combination of two ways. The first, the civics textbook model, is based on the apparently reasonable belief that the president sets out with a clearly defined concept of ideology, strategy, and national interest. The administration attempts to accomplish his objectives by marshaling assets in a concerted manner. Every action and statement provides a clue to the projected plan and goals.

A second variety of perception sees American policymaking as either remarkably incompetent or shrewdly conspiratorial. The former approach focuses on mistakes or failures—the Vietnam War and the Iranian revolution, for example—and evinces despair and amazement that the United States does so badly. Shortcomings are attributed to blindness on the part of a misguided government or a complacent, mediocre bureaucracy. The latter school believes that the appearance of disarray only covers a hidden agenda born out of cleverness or bad faith: Elected officials may think the career staff sabotages their initiatives; Senator Joseph McCarthy and his followers went so far as to attribute poor performance to treason; and political extremists and Third World leaders see a plot in every jot of department activity. But whatever cause to which U.S. diplomatic behavior is attributed, the conclusion is that there is an extraneous reason why it does not follow a logical or consistent pattern.

These interpretations do not successfully account for the pervasive ambiguity and conflict in the U.S. government. A better explanation is that intentions and ideologies are redefined by institutions, factions, and individuals divided over goals, methods, interpretation of facts, and personal ambitions.

The resulting diplomatic style is subject to the power of the process. By *process,* those involved in policymaking mean the ways in which information and options are analyzed and passed up the ladder from below and decisions are then made and sent back down the hierarchy. Power springs not only from an individual's place, but also from his ability to manipulate the system and to ally with similarly minded officials. Success usually necessitates compromises with other interested parties or the ability to bypass bureaucratic rituals by writing the key cable, composing the options, or using one's own direct channel to get things done.

Process filters and deflects political views and original objectives, determining the accuracy of data that reach leaders, the nature of decisions, and the effectiveness of implementation. Policy does not move in a straight line but is the outcome of numerous compromises and battles often quite extraneous to the issue at stake. The variation among the behavior and degree of success of administrations is determined not merely by whether they are dominated by liberals or conservatives, hawks or doves, or by their choice of priorities, but also to a large degree by the way they organize the process.

Giving process its due, however, is not the same as the shortsighted model of bureaucratic primacy often adopted by those caught up in the daily whirl of Washington policymaking. Events in international politics and foreign capitals obviously have tremendous weight in defining U.S. actions as do, increasingly, domestic factors, the media, and public opinion. Top policymakers enter office with their own perspectives and ideology, not easily swayed by subordinates or internal compromises. Process, like an individual's perception of reality, is a lens, defracting the world view of the president and subordinate decision makers. Their minds still determine the interpretation of the problems, their choices still decide the direction of any U.S. response. Each administration even alters the structure, up to a point, to fit its own needs and personnel.

The key figures are more than 600 high-level decision makers, appointees, and civil servants at the top of the foreign policy pyramid. These include secretaries, undersecretaries, assistant secretaries and their deputies at State and Defense as well as equivalent officials at the White House, the NSC staff, the CIA, AID and USIA, plus important ambassadors abroad and ranking military officers.[6] A former FSO rightly claims that the idea, "Propagated by image-building presidents and the personality-oriented media, that 'the buck stops here' ignores the reality, well known to everyone in government, that every 'buck' is forged, shaped and framed by immediate subordinates." Proposals and information—"pieces of paper," as they are called in government—pass up the line to condition the views of decision makers. For every highly publicized decision the president or secretary of state makes, dozens of smaller ones are made by those lower down.[7]

At the same time, however, most of these individuals change from one administration to the next, and in this manner the style and substance of U.S. foreign policy has also shifted drastically from one president or secretary of state to another. Discontinuity at the top lends some unpredictability, and hence unreliability, to U.S. policy, and the turnover is built into both the appointee and the electoral systems. Walking past offices emptied of departing officials in the White House, the Executive

Office Building, or at State itself on a new president's inauguration day is an unforgettable experience with that fact.

Consequently, the traditional gap between political appointees and career staff is only one of three factors straining the policy process. In addition, the president's immediate subordinates are increasingly appointees from outside the bureaucracy and represent administration factions. Rather than compete with the career staff, administrations in the post-Kissinger era battle within their own ranks. Friction with the permanent bureaucracy remains important, but that sector is more easily dominated than are other appointees, with an equal claim to represent the president.

Finally, while policymakers are not deaf, they are hard of hearing. Carter, Vance, and Brzezinski already knew what they thought about realpolitik, human rights, and Iran's revolution; and, similarly, Reagan, Haig, and Clark did not need some junior official to provide them with views on the USSR, arms control, or Central America. Presidents and their top decision makers hold primary responsibility for the priorities and substance of U.S. foreign policy.

Since State is at the core of the appointee/career rift, it is at an institutional disadvantage in an era of growing conflicts among appointees and strong ideological conviction among leaders. Precisely because it represents a link transcending appointees and ideologies, the career service is viewed with suspicion by each new incumbent. The New Deal thought State reactionary, Eisenhower Republicans believed it full of leftist New Dealers, the Kennedy and Carter people thought it hostile to progressive policies, while Nixon and Reagan were equally convinced that State was tainted with liberalism. To liberals, State appears hidebound, fusty, and uncreative; to conservatives, it seems filled with idealistic liberals of doubtful loyalty or defective common sense.

The preferences and personnel selections of a president and secretary of state make different agencies and bureaus rise or fall in importance. The distribution of power and influence differs with each administration, reflecting the personalities and abilities of its leaders. The debate over management and tables of organization, which has so dominated discussion on the system's shortcomings, is sterile because it ignores this vitally important personal element.

Despite changing structures and people, the difficulties remain remarkably consistent. Hundreds of articles, books, and reports have recommended solutions. Shuffling of bureaus, responsibilities, and titles has been a constant feature of department life. "Let's Re- Re- Reorganize the State Department" was the sarcastic title of one article on the subject. This parade of reforms became a problem itself, since constant revisions of organization and procedure only added to the confusion. As one veteran FSO complained, "There's too much tinkering with the machine. Nine times out of ten, there was nothing particularly wrong with the machine except making it work by getting the right people in the right jobs. That's what decides things."[8]

The number of institutions and people involved in policymaking makes it possible to organize the process in many ways, determined by, among other factors, the secretary of state's and national security adviser's personal relationship with the president, the president's willingness to intervene in foreign affairs, State's conflicts with other agencies, and the secretary's arrangements with the department bureaucracy.

Obviously, there is no sense in having a secretary of state whom the president does not allow to perform his function. When the president and his secretary of state can no longer work together, as happened with Reagan and Haig in 1982, the latter will not long remain in office. A president unwilling to support or trust his secretary of state, as has happened with increasing frequency in recent years, must find other sources of advice and leadership. Further, while each administration has its share of talent, important jobs can be filled in a remarkably casual way. The new president or secretary of state may be unfamiliar with his own selections for subordinate posts, in which case they can be politically or personally incompatible, or they can be unqualified people gaining positions through connections and misleadingly impressive résumés.

If the president is willing to back one person as foreign policy chief, whether the national security adviser or secretary of state, relations between agencies can be put on some sort of ordered basis through consultation and coordination, with the president or his chosen instrument as referee, or by depriving some departments of real power. State has been eclipsed as authority gravitated to the White House and as other agencies staked out influence of their own—the CIA under Eisenhower; the NSC staff under Kennedy, Nixon, Carter, and Reagan; the Defense Department under Kennedy, Johnson, and Reagan.

In short, the U.S. policy system is defined by three alternatives: It can have a clear leadership or bog down in confused struggle; it can have a center of authority in the State Department, the NSC, or nowhere at all; or it can provide for State to play an important role or be limited to diplomatic housekeeping. In all three areas the tendency over time has been toward the least desirable and most anarchic arrangements.

Good organization does not have to entail good policy, but a well-functioning process can ensure more accurate information and a more careful consideration of options, obstacles, or consequences by whomever the American people elect as president. Other problems—despite various administrations' confidence in solving them—have proven remarkably intractable. The lack of time to think, the difficulty of planning for the future or anticipating crises, the impossibility of finding some magic formula for structuring, simplifying, or smoothing the process have all been constant complaints of those involved with State over the last 40 years. The world changes too quickly to permit much time to think. Planning is stymied by the difficulty in predicting the future behavior of dozens of countries and hundreds of issues, a situation whose complexity is only exceeded by the weather and in which a single error can lead to catastrophe.

Endless lists of suggestions for change have also never succeeded because they tend to minimize the inherent problems of policymaking and ignore the special interests that benefit from the status quo. "The critics are right," wrote former FSO John Campbell, "but they have not been able to change the organization they criticize." It is equally inevitable that, in Campbell's words, "The more sensitive the issue, and the higher it rises in the bureaucracy, the more completely the experts are excluded while the harassed senior generalists take over." If the career staff is pessimistic about a certain project, it will be replaced by loyal, energetic fixers who assure their bosses that the job can be done. Knowing this, individual officials at State feel the pressure and opportunity to become yes-men themselves.[9]

The State Department's institutional culture is partly adapted to the dilemmas of foreign policymaking and partly a stubborn entrenchment of unimaginative careerism. Some habits and internal agendas prevent the organization from achieving its supposed functions of coping with international problems and promoting U.S. interests. Such perverseness is the disease of bureaucracies, and State is the archetypical bureaucracy. Kissinger's theory was that the cautious and knotted department must be circumvented to make major policy changes or diplomatic breakthroughs. "I discovered that it was a herculean effort even for someone who had made foreign policy his life's work to dominate the State Department cable machine," he wrote. "Woe to the uninitiated at the mercy of that extraordinary band of experts."[10]

"As a citadel of foreign service professionalism," writes one scholarly observer, "the State Department is an inhospitable refuge for ideas and initiatives blown in from the cold. 'It's all been tried before' is a refrain . . . at the heart of the department's perceived unresponsiveness." As another study puts it, "Political appointees seem to want to accomplish goals quickly while careerists opt to accomplish things carefully."[11]

Kissinger spoke of the "blindness in which bureaucrats . . . measure success by the degree to which they fulfill their own norms, without being in a position to judge whether the norms make any sense to begin with."

Yet political appointees would be shocked and angered if the career staff decided to set priorities on its own instead of implementing superiors' decisions. On one hand, they fault State for having a mind of its own, favoring what one NSC official called "only one option, the preferred policy." The appointees sense State sees them, and even the president, as transient meddlers in its business. On the other hand, they are annoyed that State does not have enough strength or determination to take the lead. White House officials conclude that power must be removed from State's hands and placed in their own. In short, policy analyst I. M. Destler concludes, "State tends to end up with the worst of both worlds—neither the lead role nor a secure piece of turf."[12]

More than 20 years ago, Senator Henry Jackson proposed sound principles that have still not been applied to any appreciable extent. "Our best hope," Jackson wrote, "lies in making our traditional policy machinery work better [by] getting our best people into key foreign policy and defense posts." "No task," he concluded, "is more urgent than improving the effectiveness of the Department of State."[13]

The process's own recent history best explains why all these tasks remain unaccomplished. Some useful principles to keep in mind in trying to understand what has gone wrong with the foreign policy system:

1. The president must actively use his prestige and power to end disputes and mobilize the slow-moving bureaucracy; in many cases, nothing else will do. Even if he delegates authority to a secretary of state or national security adviser, the president can still intervene when he wishes and obtain alternative views from others. The worst case is a president who neither makes choices himself nor efficiently settles the battles below him, a situation prevailing in the Carter and Reagan administrations.

2. There must be a leading figure below the president who has a large measure of operational authority over policy. Otherwise, there will be an unbridled battle between competing forces, including the White House aides, NSC staff, State Department, and Defense Department. Conflicts arising from personality clashes and substantive disagreements will produce ineffectiveness and public embarrassment. The battle will take up an enormous amount of energy better devoted to other tasks.

3. In Washington, as a much-used phrase has it, "Where you sit is where you stand." World view is largely shaped by an individual's or agency's responsibilities. Consequently, the State Department, NSC staff, White House aides, Defense Department, CIA, and Treasury and Commerce departments each represent a portion of reality which must be brought together to make or evaluate decisions. One of the State Department's main jobs is to explain constraints on U.S. power and foreign

states' views. Leaders often rightly discount suggestions based on these principles; despite what many FSOs think, White House refusal to follow their advice is neither sinful nor illegal. But the dangers of ignoring State can be high-risk activism and unilateralism. While diplomacy may emphasize mere words, arguments and explanations can often have their role in convincing other governments of U.S. determination, reliability, and correctness, avoiding the need for stronger steps.

4. Within the State Department, a small number of administration appointees have to deal with a far larger number of career people. The politicals may mistrust the bureaucracy's loyalty, and the permanent staff sometimes questions the appointees' abilities. The career people attempt to "educate" or "indoctrinate" appointees to accept their priorities and proposals, making it more difficult to break with past policies. Different sections of State have special constituencies or objectives and insist on protecting their "clients." The White House worries, therefore, that the secretary of state is being co-opted by the bureaucracy.

5. Given this resistance and competition, the secretary of state has to use a great deal of time and energy to get anything done. He must negotiate not only with foreign countries, but also with other sectors of his own government. In Kissinger's words, "The nightmare of the modern state is the hugeness of the bureaucracy, and the problem is how to get cohesion and decision in it." Former Secretary of State Dean Acheson joked that "to columnists, correspondents, legislators, some academicians, and most New York lawyers over forty, foreign affairs are an open book," chiding them for failing to understand how bureaucratic conflicts complicate diplomacy.[14]

6. Overstaffing, excessive paperwork, and endless meetings tie up time and waste resources. The most important single factor is the need for competent, knowledgeable people in key positions. The best-qualified should be promoted and the best-informed should be heeded, but it is remarkably difficult to devise a system that produces these results. Administrations and their individual appointees need around two years to learn how to perform their jobs, but by then their tenure is growing short while repeated reorganizations and personnel changes further reduce skills and continuity.

The conflict of individuals and institutions in the policy process means that neither the president nor other high officials can merely give an order and wait for things to happen. They must induce or manipulate others to do their bidding. Experience is needed, therefore, not only in dealing with the substance of foreign policy, but in coping with the mechanics of bureaucratic warfare as well. This talent, supposedly the policymaker's special skill, is as necessary as knowledge of foreign politics and issues.[15]

Although they must analyze entire societies, American diplomats inevitably still spend most of their time dealing with other countries' leaders. After all, like it or not, what those in power think is usually more important for international relations than are the views of peasants or students. But on those increasingly frequent occasions when political upheaval is about to overturn this pattern, State does far worse. As an FSO who served in Iran explained, "Any ambassador sitting down to write up a report would prefer to begin, 'During my dinner with the king . . ,' with an eye on his memoirs, rather than 'During my 27 cups of coffee with 27 local notables. . . .'"[16]

Burning ambition can mean an obsession with pleasing superiors and carefully tailoring one's life, which can make people both individually unpleasant and narrow in dealing with the actual issues at stake. "There are a lot of ambitious guys who are neither bright nor able," explains Richard Cooper, Carter's undersecretary of state for economic affairs, "yet they do well. The guy who's willing to cancel his wife's birthday party [to work instead] is the guy who's likely to become an assistant secretary of state." Out of office, the would-be policymaker is in a similar situation. Of a reporter who later became a high official, Cooper noted, "As far as he was concerned, every potential secretary of state was a sterling character. He's in a holding pattern and he doesn't want any of them to say, 'You SOB, look what you said about me.'"[17]

Given these traits and the increasingly complex nature of the policy system itself, there is a growing danger that those on top will react more and more on the basis of instinct and internal struggle and less and less in response to the facts. More than once, American leaders have created a world of illusion—the Bay of Pigs invasion, Vietnam, and Iran—in which ignorance led to choices remote from reality. Policies that have looked so promising on paper have repeatedly done poorly in history.

Still, while it is possible to find plenty of defeats, errors, and even absurdities in the modern history of U.S. foreign policy, State's overall job with limited resources has been a reasonably good one. Its record is worthy of respect, particularly in light of its heavy burdens. Only rarely and briefly, as in the Iranian hostage crisis of 1979–1981, is there recognition for the dedication, skill, and even heroism of many of its personnel.

The history of U.S. foreign policymaking falls into three distinct periods. From the founding of the United States up to the 1940s, diplomacy usually remained a low priority, involving mostly routine work. Decisions were made informally by a very small group of people. After 1945, the system was subject to great strain and opportunity as the United States became a major world power. State forfeited the opportunity to play the leading role because of its intrinsic problems, bad decisions, and the onslaught of McCarthyism. Presidents looked for alternatives. The third, and current, era began in 1969 with the rise to power of the NSC staff under Kissinger. The policy system became more complex and conflict-ridden than ever before. This turmoil brought the process increasingly into the public eye and seriously damaged the Carter and Reagan presidencies.

One leading figure in the State Department's recent history correctly points out that it would be wrong to give the "impression that people engaged in foreign policy spent most of their time in internecine scraps." These conflicts are often the most interesting part of the process, but "they absorbed ten percent of time and energy, while ninety percent went into constructive accomplishment. The problem for the historian is how to reflect that factual truth and still record the . . . strains and stresses."[18] In the history of U.S. diplomacy, differences over the interpretation of developments and conflicts for power usually ended with cooperation toward a common end. The critical issue is how they have affected the quality and content of the policy actually produced.

Presidents and Secretaries of State

Franklin Roosevelt 1933–1945

Cordell Hull	1933–1944
Edward Stettinius	1944–1945

Harry S. Truman 1945–1953

James Byrnes	1945–1947
George Marshall	1947–1948
Dean Acheson	1948–1953

Dwight Eisenhower 1953–1961

John Foster Dulles	1953–1959
Christian Herter	1959–1961

John F. Kennedy 1961–1963
Lyndon Johnson 1963–1969

Dean Rusk	1961–1969

Richard Nixon 1969–1974
Gerald Ford 1974–1977

William Rogers	1969–1973
Henry Kissinger	1973–1977

Jimmy Carter 1977–1981

Cyrus Vance	1977–1980
Edmund Muskie	1980–1981

Ronald Reagan 1981–

Alexander Haig	1981–1982
George Shultz	1982–

Notes

1. *Foundations of State*

1. Text in Richard Morris, *Great Presidential Decisions* (Greenwich, Conn., 1966), pp. 43–44.
2. Edward Tatum, *The United States and Europe 1815–1823* (Berkeley, 1936), p. 219.
3. Ibid.
4. Arthur Schlesinger Jr., "America Experiment or Destiny?" *American Historical Review,* Vol. 82, No. 3 (June 1977), pp. 512–13. For example, Woodrow Wilson in 1919: "America is the only idealist nation in the world" and has "a spiritual energy in her which no other nation can contribute to the liberty of mankind." Cited on p. 517. These statements may sound like July 4 rhetoric, but they reflect sincerely held beliefs that often shape policy.
5. Tatum, op. cit., pp. 240–43.
6. John Gaddis, *Russia, the Soviet Union and the United States* (New York, 1980), p. 14. This analysis has often been heard in recent times, in commentaries on "the imperial presidency" and over Watergate.
7. Norman Graebner, "John Quincy Adams," in Frank Merlie and Theodore Wilson, *Makers of American Diplomacy* (New York, 1974), p. 112.
8. Text in Morris, op. cit., p. 97.
9. Joseph Jones, *The Fifteen Weeks* (New York, 1955), p. 151.
10. Felix Gilbert, *To the Farewell Address: Ideas of Early American Foreign Policy* (Princeton, 1961), p. 114.
11. Article II, section 2, of the Constitution, on the powers of the president, stipulated: "He may require the Opinion, in writing, of the principle Officer in each of the executive Departments, upon any subject relating to the duties of their respective offices."
12. For example, in the Constitutional Convention, Alexander Hamilton's plan of gov-

ernment contemplated a supreme executive, "to have the sole appointment of the heads or chief officers of the Departments of Finance, War, and Foreign Affairs." Gaillard Hunt, ed., *Writings of Madison,* Vol. 3 (New York, 1903), p. 195.

13. See Gaillard Hunt, *The Department of State of the United States* (New Haven, 1914), pp. 1ff.
14. For sources and discussion, see ibid., Chapters 2 and 3.
15. Henry Johnston, ed., *Correspondence and Papers of John Jay,* Vol. 1 (New York, 1890), p. 440.
16. For sources, see Hunt, op. cit., pp. 82–83.
17. See "Jefferson's Opinion on the Powers of the Senate respecting Diplomatic Appointments," in Julian Boyd et al., *Papers of Thomas Jefferson,* Vol. 17 (Princeton, 1961), p. 379: "the transaction of business with foreign nations is Executive altogether."
18. See "Letter to Thomas Jefferson," January 21, 1790, in John Fitzpatrick, ed., *The Writings of Goerge Washington,* Vol. 30 (Washington, D.C., 1944), pp. 510–511.
19. See D. White, *The Federalists: A Study in Administrative History* (New York, 1948) Chapters 2–3. The act creating the Department of Foreign Affairs in 1789 said that the Secretary must perform "such duties as shall from time to time, be enjoined or intrusted to him by the President of the United States, agreeable to the Constitution." The act establishing the Department of State in the same year stipulated that he was to receive from the president bills, orders, and resolutions of Congress.
20. See "Letter to Comte de Moustier," May 25, 1789, in Fitzpatrick, op. cit., Vol. 30, pp. 333–335.
21. For Jefferson's opinion of courts, see Boyd, op. cit., Vol. 8, pp. 269–270.
22. Ibid., Vol. 17, pp. 216–217.
23. See, for example, "Letter to Elbridge Gerry," January 26, 1799, in M. Peterson, *The Portable Thomas Jefferson* (New York, 1975), p. 478: "I am for free commerce with all nations; political connections with none; and little or no diplomatic establishment."
24. See "Plans and Estimates for the Diplomatic Establishment," in Boyd, op. cit., Vol. 17, pp. 218–231; idem, "The Consular Establishment," Vol. 17. pp. 244–256.
25. "Circular to American Consuls," ibid., p. 423, on the written communications now required formally of consuls: "that you give to me from time to time information of all military preparations, and other indications of war which may take place in your ports; . . . and in general that you communicate to me such political and commercial intelligence, as you may think interesting to the United States."
26. George Kennan, *Realities in American Foreign Policy* (Princeton, 1954), pp. 13ff.
27. U.S. Department of State, *A Short History of the U.S. Department of State 1781–1981* (Washington, 1981), p. 6 (hereafter, Department of State). In 1937, President Roosevelt issued an executive order forbidding uniforms or official costumes not previously authorized by Congress.
28. Rachel West, *The Department of State on the Eve of the First World War* (Athens, Ga., 1978), p. 6.
29. Ibid. In 1814, British troops burned Washington, including the State Department building and its library. Original manuscripts of the Declaration of Independence and the Constitution were saved.
30. Department of State, op. cit., pp. 9–10.

31. The other sections were: translations, archives, laws and commissions, pardons, remissions, copyrights and libraries, disbursing, and superintending.
32. Arthur Frost, "Nathaniel Hawthorne, Consul at Liverpool," *Foreign Service Journal* (August 1958), pp. 10–13.
33. Department of State, op. cit.
34. S. F. Bemis, ed., *The American Secretaries of State and their Diplomacy*, Vol. VII (New York, 1933), pp. 3–70; F. Owsley, *King Cotton Diplomacy* (Chicago, 1931); E. Adams, *Great Britain and the Civil War* (London, 1925).
35. Martin Duberman, *Charles Francis Adams* (Boston, 1961), Chapter 20.
36. West, op. cit., pp. 7–8.
37. Warren Ilchman, *Professional Diplomacy in the US, 1779–1939* (Chicago, 1961), p. 20.
38. Department of State, op. cit., pp. 17–18. Adee's transfer to Washington came after he impressed department officials in handling the extradition from Madrid of Boss William Tweed, the head of the notorious Tammany Hall ring.
39. West, op. cit., pp. 9–11. Adee bought clothing in lots of thirty-six and numbered them to check the material's longevity and deter thieves. He had a dozen pairs of identical shoes, and carried several spoons lest restaurant silverware not fit his mouth. Although a witty speaker, he rarely appeared at social functions. His will ordered the burning of all his papers. His favorite book, perhaps appropriate for his vocation, was *Alice in Wonderland*. See also R. Gordon Arneson, "Anchor Man of the Department: Alvey Augustus Adee," *FSJ* (August 1971), pp. 26–28.
40. Department of State, op. cit., p. 19.
41. Graham Stuart, *The Department of State* (New York, 1949), p. 200.
42. Howard Beale, *Theodore Roosevelt and the Rise of America to World Power* (Baltimore, 1956), pp. 7–12.
43. West, op. cit., p. 236.
44. Ibid., p. 1. In fairness, it can be argued that European foreign offices also significantly misestimated events.
45. George Kennan, *Russia Leaves the War* (Princeton, 1956), pp. 41–45.
46. Ibid., pp. 48–52; Edgar Sisson, *One Hundred Red Days: A Personal Chronicle of the Bolshevik Revolution* (New Haven, 1931).
47. Kennan, op. cit., pp. 63, 378–379; R. Bruce Lockhart, *British Agent* (New York, 1933), p. 220.
48. Ilichman, op. cit., pp. 138–139.
49. James Bamford, *The Puzzle Palace* (New York, 1982), pp. 20–81.
50. Ilichman, op. cit. Only two women were selected by 1925; by 1939 the service was virtually closed to them. Only eight blacks were accepted by 1942. Yet there was some broadening over the 1898–1914 era when one-third of the secretaries serving in Europe had gone to Harvard and another third went to Yale, Princeton, or some foreign university. Waldo Heinrichs, Jr., *American Ambassador: Joseph C. Grew and the Development of the U.S. Diplomatic Tradition* (Boston, 1966), pp. 95–98.
51. Robert Schulzinger, *The Making of the Diplomatic Mind* (Middletown, Conn., 1975), pp. 88–90. Hugh de Santis, *The Diplomacy of Silence: The American Foreign Service, the Soviet Union, and the Cold War, 1933–1947* (Chicago, 1980) pp. 16–20.
52. MacMurray had tried for a year to sell the piano, which was finally given away to an American professor in Istanbul. MacMurray to Nathaniel P. Davis, May 14, 1941, MacMurray papers, Box 188 Princeton University. I am grateful to Professor John

De Novo for this letter. It took months of begging before the embassy was sent sufficient secretarial help. In the meantime, one young FSO was forced to resign in disgrace for trying to stretch funds by changing his paycheck on the black market, a common practice among diplomats.

53. Elting E. Morison, *Turmoil and Tradition: A Study of the Life and Times of Henry L. Stimson* (New York, 1964), pp. 252–55.
54. Betty Glad, *Charles Evans Hughes and the Illusion of Innocence* (Urbana, Ill., 1966) pp. 132–140.
55. de Santis, op. cit., p. 36.
56. In his introduction to Lewis Einstein, *A Diplomat Looks Back* (Yale, 1968), pp. x–xii.

2. *The Challenge of Global War, 1933–1945*

1. Cited in Dean Acheson, "The Eclipse of the State Department," *Foreign Affairs* (July 1971), p. 600.
2. Charles Bohlen, *Witness to History 1929–1969* (New York, 1972), p. 129; Robert Dallek, *Franklin D. Roosevelt and Foreign Policy* (New York, 1979), p. 532.
3. Beatrice Berle and Travis Jacobs, eds., *Navigating the Rapids, 1918–1971: From the Papers of Adolph Berle* (New York, 1973), p. 206. Roosevelt's original leadership group at State included veteran officials William Phillips as undersecretary and Carr in charge of administration. After their retirement, Welles moved up to undersecretary, chosen by Roosevelt over Hull's old friend, R. Walton Moore, who became counselor.
4. Dean Acheson, *Present at the Creation: My Years in the State Department* (New York, 1969), p. 3.
5. Ibid., pp. 10–17.
6. Berle and Jacobs, op. cit., p. 7. Roosevelt appointed Ruth Bryan Owen, William Jennings Bryan's daughter, as first woman ambassador, to Denmark.
7. Robert Bendiner, *The Riddle of the State Department* (New York, 1942), pp. 109–110.
8. See Arnold Offner, *American Appeasement: U.S. Foreign Policy and Germany 1933–1938* (Cambridge, 1968), pp. 94ff. Messersmith joined the Foreign Service in 1914 and served as consul-general in Berlin from 1930 to 1934. Undersecretary Phillips described Messersmith as "probably better informed than anyone else about Nazi programs and activities."
9. Department of State, *Foreign Relations of the United States, 1933* (hereafter *FRUS*), Vol. 2 (Washington, D.C., 1949), p. 217, Gordon to Hull, April 9, 1933.
10. See William Dodd Jr. and Martha Dodd, eds., *Ambassador Dodd's Diary, 1933–1938* (New York, 1941), entries for December 16, 1933, and July 22, 1934.
11. For example, Department of State, *FRUS, 1934,* Vol. 2 (Washington, D.C., 1951), pp. 229–243.
12. Department of State, *Peace and War: United States Foreign Policy 1931–1941* (Washington, 1943), n. 44, Messersmith Memorandum, March 22, 1935, noting Dodd's concurrence in a warning to Washington that absolutely no trust could be placed in Nazi promises and that Hitler's aim was "unlimited territorial expansion." Of Hitler, Dodd wrote in 1934: "I have a sense of horror when I look at the man." Dodd and Dodd, op. cit., p. 126. See also Department of State, *FRUS, 1936,* Vol. 2 (Washington, D.C., 1954), p. 149, Dodd to Hull, September 18, 1936.

13. Dodd tended to view Nazism as the work of Hitler and his close associates alone. Thus Dodd was unable to understand the significance of the anti-Semitic persecutions except as a manifestation of Hitler's personality. See Sumner Welles, *Seven Decisions that Shaped History* (New York, 1951), p. 216; William Langer and S. Everett Gleason, *Challenge to Isolation,* Vol. 1 (New York, 1952), p. 8.
14. See Offner, op. cit., pp. 68–69; Dodd and Dodd, op. cit., pp. 4–6, entry for June 16, 1933.
15. James Compton, *The Swastika and the Eagle* (Boston, 1967), p. 69.
16. See, for example, Department of State, *FRUS, 1933,* Vol. 2, pp. 226–228, 251–252, 360–365; Department of State, *FRUS, 1936,* Vol. 2, pp. 159ff; and Offner, op. cit., p. 166.
17. Department of State, Division of Western European Affairs, memorandum, February 16, 1937, cited in Offner, op. cit., p. 177.
18. See Hugh Wilson, *A Career Diplomat* (Westport, 1973), pp. 38–39, Wilson to Sumner Welles, June 20, 1938: "I have frequently felt that the hostile state of mind in the United States is of the greatest danger to our staying aloof from any conflict. . . . The older I grow, the deeper is my conviction that we have nothing to gain by entering a European conflict. . . ." See also Department of State, *FRUS, 1938,* Vol. 1 (Washington, D.C., 1955), p. 490, Wilson to Hull, August 28, 1938; William Kaufmann, "Two Ambassadors," and Franklin Ford, "Three Observers in Berlin: Rumboldt, Dodd, and Francois-Poncet," in Gordon Craig and Felix Gilbert, *The Diplomats 1919–1939* (Princeton, 1933); Robert Dallek, *Democrat and Diplomat. The Life of William E. Dodd* (New York, 1968), pp. 199–202.
19. Berle and Jacobs, op. cit., pp. 169, 200.
20. Dallek, *Franklin D. Roosevelt. . . ,* p. 192.
21. George V. Allen to Laurence Steinhardt, January 16, 1943, Steinhardt Papers, Box 41, Library of Congress.
22. See also John Kenneth Galbraith, *Annals of an Abiding Liberal* (Boston, 1979), pp. 157, 307. Michael Straight, a wealthy young American recruited as a Soviet agent at Cambridge by Anthony Blunt, worked briefly at State as an unpaid volunteer, though he apparently never supplied much information to his Soviet contact. See Michael Straight, *After Long Silence* (New York, 1983).
23. See William Stevenson, *A Man Called Intrepid* (New York, 1976), pp. 84–96; David Kahn, *Hitler's Spies* (New York, 1978) p. 96; Warren Kimball and Bruce Bartlett, "Roosevelt and Prewar Commitments to Churchill: The Tyler Kent Affair," *Diplomatic History,* Vol. V, No. 4 (Fall 1981), pp. 291–312; Richard Whalen, *The Founding Father: The Story of Joseph P. Kennedy* (New York, 1964), pp. 303–314; *Documents on German Foreign Policy 1918–1945, Series D,* Vol. 9, (Washington, D.C., 1953) p. 417.
24. Charles Thomas, *Allies of a Kind: The United States, Britain, and the War Against Japan: 1941–1945* (New York, 1978), cited p. 92.
25. Bohlen, op. cit., p. 122; see also Robert Sherwood, *Roosevelt and Hopkins* (New York, 1948), p. 774–775.
26. See Chapters 3 and 4, below.
27. Sherwood, op. cit., p. 227.
28. Ibid., p. 270.
29. W. Averell Harriman and Elie Abel, *Special Envoy to Churchill and Stalin, 1941—1946* (New York, 1975), p. 230.
30. Acheson, op. cit., p. 38; see also Hull, Vol. II, op. cit., p. 1227.

31. Thomas Campbell and George Herring, *The Diaries of Edward Stettinius, Jr. 1943–46* (New York, 1975), p. 7.
32. John Blum, ed., *The Price of Vision: The Diaries of Henry Wallace, 1942–1946* (Boston, 1973), p. 246.
33. Sumner Welles, op. cit., p. 119.
34. Waldo Heinrichs, Jr., op. cit., p. 235. For those working on the Far East, even getting a paper to Hornbeck was a minor victory. His lieutenant, Maxwell Hamilton, "would fix on every phrase, every word, and every comma to achieve the perfection that could pass the Hornbeckian judgment. He often failed." Hornbeck had taught in China and became chief of the Division of Far East Affairs in 1928. Emmerson, op. cit., pp. 78–81, 104. On U.S.-Chinese relations during the war, see Michael Schaller, *The U.S. and China, 1939–1945* (New York, 1979).
35. Acheson, op. cit., p. 240; see also William Roger Louis, *Imperialism at Bay* (New York, 1978), p. 158.
36. James MacGregor Burns, *Roosevelt: Soldier of Freedom* (New York, 1970), p. 129.
37. Francis Loewenheim et al., *Roosevelt and Churchill: Their Secret Wartime Correspondence* (New York, 1975), p. 198.
38. Berle, op. cit., p. 401.
39. Acheson, op. cit., p. 3.
40. Stettinius Papers, Box 218, Folder NEA-Murray, University of Virginia Library.
41. Sherwood, op. cit., pp. 754–56.
42. Despite these frustrations, a strong-willed U.S. ambassador could have remarkable influence in smaller countries. When the Cuban military organized a coup against the elected president in 1944, the incumbent requested Ambassador Spruille Braden's intervention. Braden quickly visited the generals, warning them that Washington would not recognize any seizure of power. The would-be junta quickly disbanded. Sometimes, Braden later explained, an ambassador had to act on his own: "There are other ways of diplomacy than the Departmental cable and instruction." Braden, op. cit., pp. 458–459.
43. *The New York Times,* August 6, 1943.
44. Dallek, *Franklin D. Roosevelt . . .*, p. 421. Berle claims the FBI intercepted Communist party orders in December 1942 ordering attacks on Dunn, Atherton, and later on Berle himself. Shortly before Welles resigned, the party was called on to attack both Hull and Welles, but when the latter quit, the party decided to defend him, ironically destroying his chances to go to the Moscow conference.
45. Sherwood, op. cit., p. 757. "It was clear that Mr. Stettinius was not a Disraeli, Metternich or a Machiavelli, but . . . neither had I seen anyone else on the horizon who had such abilities." Walter Millis, *The Forrestal Diaries* (New York, 1957), p. 54.
46. Berle and Jacobs, op. cit., pp. 455–456.
47. Acheson, op. cit., p. 88.
48. Arthur Krock in *The New York Times,* January 4, 1945. These new appointments included Joseph Grew, a Foreign Service career man since 1904, as undersecretary. The new assistant secretaries were the following: William Clayton, a Texas cotton magnate and successful government administrator respected by Roosevelt, headed economic affairs, replacing Acheson, who shifted to congressional relations. Clayton's group of bright young economists from the wartime agencies included John Kenneth Galbraith, future ambassador and assistant secretary George McGhee, Walter Rostow, and talented Harvard professor Edward Mason. Nelson Rockefeller,

who had been White House coordinator of Inter-American affairs, took that regional assistant secretary position, while Dunn was responsible for the other geographic bureaus. Archibald MacLeish, the great poet, was charged with public and cultural affairs.

49. Text in James Bishop, *FDR's Last Year* (New York, 1974), p. 202.
50. Bohlen, op. cit., p. 125.
51. Martin Weil, *A Pretty Good Club* (New York, 1978), pp. 155–156.
52. George F. Kennan, *Memoirs, 1950–1963* (hereafter Vol. 2) (Boston, 1972), p. 48.
53. Ibid., pp. 166, 230–231.
54. Arthur Schlesinger, "Origins of the Cold War," *Foreign Affairs* (October 1967), p. 39.
55. Harriman and Abel, op. cit., pp. 345–347.
56. Berle and Jacobs, op. cit., pp. 462–468; see also U.S. Department of State Record Group 59, 861.9111/1-1045, Harriman to Stettinius, January 10, 1945, National Archives.
57. Weil, op. cit., p. 210. Secretary of the Navy James Forrestal held Roosevelt responsible for permitting "an organizational set up that was fundamentally unsound." Millis, op. cit., p. 283.
58. Harriman and Abel, op. cit., pp. 455–456.
59. Berle and Jacobs, op. cit., p. 477.
60. Bohlen, op. cit., p. 222.
61. Harriman and Abel, op. cit., p. 449.
62. de Santis, op. cit., pp. 156–157, 167.
63. Weil, op. cit., p. 314.
64. Heinrichs, op. cit., pp. 366–390.
65. Campbell and Herring, op. cit., pp. 317–318.
66. Sherwood, op. cit., p. 757.
67. Campbell and Herring, op. cit., p. 184. Stettinius's reorganization had sought to relieve assistant secretaries of administrative chores so that they could serve as the secretary's advisers, but this required too many operating officers to report directly to the overworked secretary. As time went on, the system eroded the reforms and worked its way back to the old patterns.

3. *State Takes Command, 1945–1952*

1. Francis Heller, *The Truman White House: The Administration of the Presidency 1945–1953* (Lawrence, Kansas, 1960), pp. 122–123; Harry Truman, *Year of Decision: 1945* (New York, 1955), pp. 14–17.
2. Heller, op. cit., p. 124.
3. Ibid., p. 153; Ernest May, *The Ultimate Decision: The President as Commander-in-Chief* (New York, 1960), p. 137.
4. Ibid., pp. 197–202, 220. The foreign aid program perhaps provides the most vivid example of the bureaucratic belief that changing the name of a program accomplishes something.
5. Alexander de Conde, *The American Secretary of State: An Interpretation* (New York, 1962), pp. 164–165.
6. Arthur Krock, "Washington Hasn't Enough Time to Think," *The New York Times Magazine*, December 9, 1945.

7. Richard Burns, "James Byrnes," in Norman Graebner, ed., *An Uncertain Tradition: American Secretaries of State in the Twentieth Century* (New York, 1961).
8. *DOSB*, July 8, 1945, p. 45.
9. When Forrestal, a hard-liner, suggested in October 1945 that Truman explain the situation with the Soviets to the American people, Byrnes demurred as this would "give the Russians an excuse for claiming that we had furnished provocation which justified their actions." Millis, op. cit., p. 102.
10. See Truman, op. cit., pp. 22–23; John Gaddis, *The United States and the Origins of the Cold War 1941–1947* (New York, 1972), pp. 285–286; Harriman, op. cit., pp. 524–530; Robert Donovan, *Conflict and Crisis* (New York, 1977), pp. 155–162; James F. Byrnes, *Speaking Frankly* (New York, 1947) and *All in One Lifetime* (New York, 1958); Robert Messmer, *The End of An Alliance: James F. Byrnes, Roosevelt, Truman and the Origins of the Cold War* (Chapel Hill, 1982). Turner Catledge, "Secretary Byrnes: Portrait of a Realist," *The New York Times Magazine*, July 8, 1945. Truman had assured nervous senators that nothing would be done at Moscow without his clearance, which made him particularly sensitive to Byrnes's trespass. When Byrnes returned to Washington, he scheduled a radio speech before consulting with the president. Truman remembered bawling him out in a meeting on the presidential yacht, but there are many different versions of the story. Acheson's description of Byrnes is particularly apt: "He was an independent operator using half a dozen close associates upon those problems. People in the Department and any problems upon which he was not working personally hardly existed." Acheson, op. cit., p. 163.
11. Donovan, op. cit., p. 161.
12. *The New York Herald Tribune*, November 30, 1945.
13. *DOSB*, December 21, 1945. Heinrichs, op. cit., pp. 372–380; Emmerson, op. cit., pp. 236–239. In May 1945, Acheson referred to Grew as "Prince of Appeasers," but later admitted Grew had been right. David McLellan and David Acheson, *Among Friends: Personal Letters of Dean Acheson* (New York, 1980), p. 55.
14. *DOSB*, December 2, 1945, pp. 893–894. Byrnes asked Walter Lippman to head Public Affairs, but the columnist preferred to maintain his independence. Signs of attempts at "news management" appeared early. During the Greek civil war, when Washington supported the Athens government against Communist guerrillas, State officials worked to place supportive articles and block publication of critical ones. Lawrence Wittner, "The Truman Doctrine and the Defense of Freedom," *Diplomatic History*, Vol. 4, No. 2 (Spring 1980), pp. 178–179.
15. H. Stuart Hughes, "The Second Year of the Cold War," *Commentary* (August 1969), pp. 27–31.
16. William Maddox, "The Foreign Service in Transition," *Foreign Affairs* (January 1947), pp. 306–307.
17. *DOSB*, December 23, 1945, pp. 992–993.
18. Acheson, op. cit., pp. 135–163; Spruille Braden, *Diplomats and Demagogues* (New Rochelle, N.Y., 1971), pp. 345–351. Acheson described Braden as "a bull of a man physically and with the temperament and tactics of one, dealing with the objects of his prejudices with blind charges, preceded by pawing up a great deal of dust."
19. Donovan, op. cit., pp. 151–152.
20. *The New York Times*, September 21, 23, and 25, 1945. MacArthur had said only 200,000 U.S. troops were needed to occupy Japan and the rest could go home, a

statement horrifying State, which was trying to counter congressional pressure for quick demobilization.

21. *DOSB,* December 9, 1945, pp. 930–933; Earl Latham, *The Communist Controversy in Washington* (Cambridge, 1966), pp. 205–210, 262–263. *The New York Times,* November 28, 29, December 8, 10, and 11, 1945; *DOSB,* December 2, 1945, pp. 882–883; *Time,* December 7, 1945, pp. 18–19.
22. Robert Griffith, *The Politics of Fear* (Lexington, Ky., 1970), pp. 39–57; Latham, op. cit., p. 94. More security leaks apparently occurred from OSS, the Treasury Department, the War Production Board, and the Board of Economic Warfare than from State, which became the scapegoat for such problems.
23. Latham, op. cit., p. 179.
24. Allen Weinstein, *Perjury* (New York, 1978), pp. 346–369.
25. Undesirable employees were encouraged to resign or seek transfers to the UN. They were assigned dead-end, routine jobs to encourage departures. Byrnes, *All in One Lifetime,* pp. 321–325. Vincent recalled 1945–1946 as the time "everybody was spying on everybody else." Weil, op. cit., pp. 49, 262–263. Marzani and others complained that OSS knew of their Communist ties when they were recruited. Maurice Isserman, *Which Side Were You On?* (Middletown, Conn., 1982), pp. 182–184.
26. U.S. Senate Foreign Relations Committee, "State Department Employee Loyalty Investigation," 81st Congress (1950), p. 1778, prints the House study. Case No. 12, though not identified, is obviously Hiss.
27. Ibid., Case 14, pp. 1778–1779.
28. Ibid., Case 50, p. 1787.
29. Ibid., Case 51, p. 1787.
30. Ibid., Peurifoy's testimony, pp. 1248–1250. *Newsweek,* March 3, 1947.
31. *The New York Times,* February 10 and 15, March 13, 1946. Moscow also increased harassment of American diplomats. A vice-consul was beaten and an embassy clerk was threatened with trial for "hooliganism." *The New York Times,* March 12, May 5 and 16, 1946.
32. *The New York Times,* March 15, 16, 21, 26, and 27, April 10, 14, 24, 26, and 27; Acheson, July 3, 1946; Krock, August 2, 1946.
33. James Reston in *The New York Times,* June 20, 1946. Even Acheson felt the economic pressure, *The New York Times,* May 21, 1947. Henry Villard, *Affairs at State* (New York, 1965), p. 123.
34. Letter to the editor, *The New York Times,* July 30, 1946. See also *The New York Times,* July 21, 27, and November 14, 1946. Before passage of the new law, the U.S. ambassador in London was paid less than half of his British counterpart's salary and faced a far worse tax situation.
35. *The New York Times,* May 15, 1946. So great were the pressures that the head of the Office of Foreign Service had a breakdown. On the struggle over reform, see Smith Simpson, "Perspectives of Reform," Part Two, *FSJ* (September 1971), pp. 21–23.
36. Ibid., December 29, 1946.
37. As *The New York Times* put it in a December 30, 1946, editorial: "Authority was simply not on the job. . . . Sometimes it has seemed that one sector . . . was deliberately pulling against another." Poor administration meant inconsistency and "general confusion." Joseph and Stewart Alsop, in a column, "We Have No Russian Policy," *The Washington Post,* January 4, 1946, saw the failure to develop a consistent policy toward the USSR as springing from these problems, but a position

was developing behind the scenes and would emerge in the following year. The transitional nature of the period might have led to some exaggeration on the extent of disorganization.

38. Gaddis, op. cit., p. 324; text of speech in *The New York Times,* September 7, 1946; *U.S. News,* September 13, 1946; *The Nation,* September 14, 1946.
39. Byrnes, *Speaking Frankly,* op. cit., pp. 239–242; *DOSB,* September 29, 1946, p. 577. Wallace first tried to make a deal: He would remain silent and stay out of foreign political affairs if State stayed out of foreign economic work. John Blum, op. cit., pp. 577, 620.
40. Braden, op. cit., pp. 356–370; Acheson, op. cit., pp. 187–190; Roger Trask, "The Impact of the Cold War on United States-Latin America Relations 1945–1949," *Diplomatic History,* Vol. 1, No. 3 (Summer 1977).
41. George Kennan, *Memoirs, 1925–1950* (hereafter, Vol. 1) (Boston, 1967), pp. 292–295.
42. Gaddis, op. cit., p. 315.
43. *DOSB,* June 16, 1946, pp. 1045–1047.
44. David Lilienthal, *The Journal of David E. Lilienthal, Vol. 2.* (New York, 1964) pp. 158–159. McLellan and Acheson, op. cit., pp. 64–66; Bohlen, op. cit., p. 259. On liberal reaction to the appointment, see Alonzo Hamby, *Beyond the New Deal, Harry S. Truman and American Liberalism* (New York, 1973), pp. 173 and 255. Marshall's willingness to listen to his own officials, James Reston noted ironically, "is regarded by many . . . in the department as a remarkably interesting experiment." *The New York Times,* February 5, 1947.
45. Joseph Jones, op. cit., p. 109. Dean Acheson, *Sketches From Life of Men I Have Known* (Westport, Conn. 1961) pp. 162–163. Alexander de Conde, "George Marshall," in Graebner, ed., op. cit., p. 250; *The New Republic,* January 20, 1947, pp. 5–6; Robert Elson, "The New Strategy in Foreign Policy," *Fortune* (December 1947), pp. 80–85.
46. Villard, op. cit., p. 20.
47. Jones, op. cit., pp. 117ff.
48. Lilienthal, op. cit., p. 267.
49. Elson, op. cit., p. 219.
50. Millis, op. cit., pp. 131, 263, and 316. Thomas Etzold and John L. Gaddis, *Containment: Documents on American Policy and Strategy 1945–1950* (New York, 1978), p. 17; Anne Kasten Nelson, "National Security I: Inventing a Process," in Hugh Heclo and Lester Salamon, eds., *The Illusion of Presidential Government* (New York, 1981), pp. 233–235.
51. See Acheson et al., "Planning in the Department," *FSJ,* March 1961, p. 22. Original members of the Staff included Carlton Savage, a 20-year department veteran and the group's executive secretary; John Paton Davies Jr., who had spent most of his life in China; Jacques Reinstein, the economist; George H. Butler, a Latin Americanist; Henry Villard, who worked on Africa and the Middle East; and Issac Stokes, a lawyer with special knowledge of UN and international organization issues. "Brain Trust," *U.S. News,* October 10, 1947, p. 52.
52. Nelson, op. cit., p. 238.
53. Donovan, op. cit., pp. 306–307.
54. Harold Graves, "State Department, New Model," *The Reporter,* November 8, 1949. On Acheson's hiring, see Acheson, "The President and the Secretary of State," op. cit., pp. 34ff.

55. Weil, op. cit., p. 143n.
56. In Acheson's words, "If a President will make decisions, you're in luck. That is the essential quality. And if he has a high batting average in the correctness of his decisions, then you're in clover." Cited in Cabell Phillips, *The Truman Presidency* (New York, 1966) p. 433. "It is immensely important that . . . the President and the Secretary of State understand each other completely and that they know what their respective roles are," wrote Truman. Harry S. Truman, op. cit., p. 255.
57. Lilienthal, op. cit., p. 520.
58. Ronald Stupak, *The Shaping of Foreign Policy* (Miami, Oh., 1969), pp. 36–41.
59. Graebner, op. cit., p. 270. Republican conservative stalwarts John Foster Dulles and Senator Robert Taft endorsed Acheson's nomination.
60. During the first three years of the Truman Loyalty Program the FBI investigated over 16,000 employees. Only 2 were removed as security risks, but an additional 202, who were under suspicion on security grounds, resigned or were eliminated in employment cutbacks. Weinstein, op. cit., pp. 384–385, 506–509, 620; *The New York Times,* January 8, 9, and 14, 1949; Hamby, op. cit., p. 394.
61. Marshall never understood domestic politics, and his few conflicts with Truman came on this point. He opposed Truman's idea to send Chief Justice Fred Vinson to Moscow on a peace mission in autumn 1948 and criticized some of Truman's Palestine policy, attributing it to political motives. Bohlen, op. cit., pp. 268–271.
62. Kennan, Vol. 1, op. cit., pp. 345–347, 365–367, 426–427; Etzold and Gaddis, op. cit., p. 135. Kennan faced further difficulties when Republican foreign policy adviser John Foster Dulles told the press that Kennan was dangerous and wanted to let the Peoples Republic of China into the UN. Concluding he had little future in the department, Kennan left in July 1953.
63. Frances Heller, op. cit., pp. 50–51.
64. Christopher Sykes, *Crossroads to Israel* (Bloomington, Ind., 1973); J. C. Hurewitz, *The Struggle for Palestine* (New York, 1968); Barry Rubin, *The Arab States and the Palestine Conflict* (Syracuse, 1981); Philip Baram, *The State Department in the Middle East, 1918–1945* (Philadelphia, 1978); John De Novo, *American Interests and Policies in the Middle East 1900–1939* (Minneapolis, 1978).
65. *The New York Times,* August 22, September 1, and October 29, 1946.
66. Donovan, op. cit., pp. 312–331, 369–387; Millis, op. cit., p. 362.
67. Jonathan Daniels, *The Man of Independence* (New York, 1950); de Conde, op. cit., pp. 256–257.
68. Philip Jessup, *The Birth of Nations* (New York, 1974), pp. 272–293.
69. Ibid., pp. 51–68.
70. George Herring, op. cit., p. 115.
71. Ibid., pp. 97–110. However, the Near East bureau succeeded in avoiding any American commitment to France's North African colonies as part of NATO agreements in 1951–1952. Jessup, op. cit., pp. 103–135.
72. Gary Hess, "The First American Commitment in Indochina," *Diplomatic History,* Vol. 2, No. 4 (Fall 1978), pp. 334–339.
73. Lorraine Lees, "The American Decision to Assist Tito, 1948–1949," *Diplomatic History,* Vol. 2, No. 4 (Fall 1978). Secretary of Defense Louis Johnson opposed State's efforts to help Tito.
74. Millis, op. cit., p. 534.
75. Tydings Committee hearings, op. cit. pp. 1759–1763. State's detailed refutations are in *DOSB,* June 12, 1950, pp. 963–973, and June 19, 1950, pp. 1012–1017.

Joseph Esherick, *Lost Chance in China: The World War Two Dispatches of John S. Service* (New York, 1974); Gary May, *China Scapegoat: The Diplomatic Ordeal of John Carter Vincent* (Washington, 1979). As early as January 1947, the pro-Chiang China lobby tried to block Vincent's promotion to career minister. Acheson and Byrnes finally gained confirmation of this appointment by promising he would not be sent to any important post in Europe or Asia.

76. Dean Acheson, "Crisis in Asia—An Examination of U.S. Policy," *DOSB*, January 23, 1950, pp. 111–115.
77. Kennan, Vol. 2, op. cit., p. 197; Acheson, op. cit., p. 355; Griffith, op. cit., p. 48.
78. John Allison, *Ambassador from the Prairie* (Boston, 1973), p. 115.
79. O. Edmund Clubb, *The Witness and I* (New York, 1974) p. 282; see also *DOSB*, February 27, 1950, pp. 327–328.
80. Interviews; Tyler Abell, *Drew Pearson's Diaries 1949–1959* (New York, 1974) pp. 115–120.
81. *DOSB*, March 27, 1950, pp. 479–481, April 24, 1950, p. 655, and May 15, 1950, pp. 752–753; John Service, "Pertinent Excerpts," *FSJ* (October 1951), pp. 22–23.
82. When Truman tried to put Jessup, one of Acheson's closest colleagues, on the U.S. UN delegation in 1951, the Senate subcommittee refused to recommend approval and the president gave Jessup a recess appointment. Griffith, op. cit., pp. 146–151; *DOSB*, November 5, 1951, pp. 736–737.
83. *The New York Times*, March 27 and April 22, 1950.
84. *DOSB*, May 8, 1950, pp. 711–716.
85. Allison, op. cit., pp. 120–121, 149–150.
86. Kennan, Vol. 2, pp. 26–32; Merle Miller, *Plain Speaking: An Oral Biography of Harry S. Truman* (New York, 1974) pp. 265–285; Phillips, op. cit., p. 288–291; Warren Cohen, *Dean Rusk* (Totowa, N.J., 1980), pp. 45–59.
87. Dean Acheson, "The President and the Secretary of State," op. cit., p. 40.
88. Kennan, Vol. 2, op. cit., p. 30; Heller, op. cit., pp. 39–40; *The New York Times*, August 17, 1950.
89. Seyom Brown, *The Faces of Power* (New York, 1968), pp. 47–51; Bohlen, op. cit. p. 292; Acheson, *Present* . . . , op. cit., p. 373; McGeorge Bundy, "The H-Bomb: The Missed Chance," *The New York Review of Books*, May 13, 1982. Bohlen and Kennan disagreed with Acheson that Moscow was planning further aggression, but the Soviet specialists had little influence in the department.
90. *The New York Times*, March 12, April 27 and 28, May 4, 5, 20, and 25, June 1, 3, 5, and 8, 1951. Almost every aspect of the department came under attack. The passport bureau was criticized for having issued 18 passports to Communist party members at a time when it was receiving 1500 to 2000 applications a day.
91. Ibid., June 3, July 29, 1951, February 12 and 20, 1952. For Kennan's views of these and other cases, see Kennan, Vol. 2, op. cit., pp. 205–221.
92. *The New York Times*, January 30, July 11, 12, and 24, August 3, November 4, and December 9, 1952.
93. Cited in *The New York Times*, August 3, 1952.
94. Cited in Acheson, op. cit., p. 719; see also *FSJ* (February 1953), pp. 26–27, 51–52.

4. *A Horseless Rider, 1953–1961*

1. Don Price, ed., *The Secretary of State* (Englewood Cliffs, N.J., 1960), p. 57. Dulles's main confidantes included Herman Phleger, his legal adviser; diplomat Douglas

MacArthur II, a nephew of the general and later ambassador to Japan; Herbert Hoover Jr., who served over two years as undersecretary after Smith; and veteran diplomat Robert Murphy, who, as deputy undersecretary for political affairs, was the senior career man.

2. Sherman Adams, *Firsthand Report* (New York, 1961), pp. 89–91; *The New York Times,* February 2, 1958; Andrew Berding, *Dulles on Diplomacy* (Princeton, 1965) pp. 16–22. The continuing difficult relationship between secretary and UN ambassador is reflected by the fact that the latter, Henry Cabot Lodge, twice defied departmental instructions and also maintained his own direct line to the president. Michael Guhin, *John Foster Dulles: A Statesman and His Time* (New York, 1972), p. 185.
3. Robert Divine, *Eisenhower and the Cold War* (New York, 1981), pp. 20–23, and others argue that Eisenhower used Dulles as his tool, but this seems oversimplified, particularly when compared to the relationships between other secretaries of state and presidents. Obviously, when Eisenhower chose to intervene, he dominated policymaking, but Dulles had a greater degree of autonomy than his successors. In part, such revisionism is a reaction to earlier works in which Eisenhower's abilities and involvement were understated.
4. Emmet John Hughes, *The Ordeal of Power* (New York, 1963), pp. 103–113; Divine, op. cit., pp. 206–207. See also Fred Greenstein, *The Hidden-Hand Presidency* (New York, 1982), and Bennett Rushkoff, "Eisenhower, Dulles and the Quemoy-Matsu Crisis, 1954–1955," *Political Science Quarterly* (Fall 1981).
5. *The New York Times,* June 10, 1956.
6. According to one compilation, Dulles was away 36 percent of the time compared to Hull's 22 percent, Byrnes's 62 percent, Marshall's 47 percent, and Acheson's 25 percent. In his time, Hull's absences were considered large. Henry Wriston, "The Secretary of State Abroad," *Foreign Affairs* (July 1956), p. 523. C. L. Sulzberger called Dulles a "wandering Presidential agent," and said his actions undermined the role of ambassadors and led to a paralysis of decision making in Washington. *The New York Times,* August 29, 1956.
7. Arthur Krock, *Memoirs: Sixty Years on the Firing Line* (New York, 1968), p. 322.
8. Emmet Hughes, op. cit., pp. 81–83.
9. Cited in John Osborne, "Is the State Department Manageable," *Fortune* (March 1957), p. 267, and "What's the U.S. Foreign Service Worth," *Fortune* (May 1957); Guhin, op. cit., pp. 193–196.
10. Robert Donovan, *Eisenhower: The Inside Story* (New York, 1956), pp. 248–250.
11. Bohlen, op. cit., pp. 309–336; Kennan, Vol. 2, op. cit., pp. 173–177; Donovan, op. cit., pp. 88–89; *FSJ* (May 1953), p. 34. Dulles was an early defender of Hiss; see Guhin, op. cit., pp. 188–190. In 1956, Robert Bowie's nomination as an assistant secretary was opposed by China lobby senators because he favored admission of the People's Republic of China into the UN, but opposition was withdrawn when Bowie convinced them that he opposed it under existing circumstances. *The New York Times,* January 16 and February 8, 1956.
12. Osborne, op. cit., p. 269; Emmet Hughes, op. cit., pp. 119–121.
13. *DOSB,* March 29, 1954, p. 470.
14. The dismissal figures are somewhat misleading, reflecting both administrative changes and the irrationality of the McCarthy-era debate, since, for example, 281 employees were removed through personnel actions in 1945–1946 alone; McCarthy merely claimed that 56 others were cleared. Ralph S. Brown Jr., *Loyalty and Security* (New Haven, 1958), pp. 28–29, 54–60; *The New York Times,* January 6, 1955;

Charlotte Knight, "What Price Security," *Colliers,* July 9, 1954; Letter to Editor, *The New York Times,* January 17, 1954; Hans Morgenthau, "The Impact of the Loyalty-Security Measures on the State Department," *Bulletin of Atomic Scientists,* Vol. 11, No. 4 (April 1955), pp. 134–140; Caute, *The Great Fear* (New York, 1978), pp. 258–259. For a profile of the Foreign Service, see James McCamy and Alexandro Corradini, "The People of the State Department and Foreign Service," *American Political Science Review* (December 1954), pp. 1067–1082. In the late 1940s, about 1200 candidates took the Foreign Service exam each year. The number fell to 807 in 1950, and to 758 the following year. Villard, op. cit., pp. 168–171.

15. Donovan, op. cit., pp. 289–290. McLeod was even unable to distinguish between Socialists and Communists. His selection, like so many department appointments, was accidental. The Chicago neighbor of the inexperienced Undersecretary for Administration Donald Lowrie, a Quaker Oats executive, suggested McLeod. Lowrie found him companiable at lunch and, to his later regret, hired him. Robert Ferrell, *The Eisenhower Diaries* (New York, 1981) pp. 256–257; Emmet Hughes, op. cit., pp. 84–85. Laxness on security is illustrated by the case of John Montgomery, State's desk officer on Finland, found a suicide at home in January 1953. The department was unaware that he was rejected by the army on psychological grounds. In 1940, there were only about four security officers. The number rose to forty-seven in 1945, seventy-seven in 1947, and ninety-four in 1953. House of Representatives Committee on Government Operations, 83rd Congress, "Security and Personnel Practices" (83rd Congress, 1953).
16. Emmet Hughes, op. cit., p. 91; *The New York Times,* January 14, 1954.
17. Abell, op. cit., pp. 253, 289, 475–476; Acheson to Battle, August 6, 1953, in McLellan and Acheson, op. cit., p. 89.
18. Caute, op. cit., pp. 303–321; *DOSB,* November 15, 1954, pp. 752–754; Emmerson, op. cit., pp. 307–333; *The New York Times,* March 2, 1957.
19. David Linebaugh, letter to *FSJ* (January 1955), p. 6; *FSJ* (January 1953), p. 17; Clubb, op. cit., pp. 284–287.
20. *The New York Times,* September 16, 1956.
21. George Kennan, "The Future of our Professional Diplomacy," *Foreign Affairs* (July 1955), p. 568. Dorothy Fosdick, "For the Foreign Service—Help Wanted," *The New York Times Magazine,* November 20, 1955, p. 13. Kennan spoke for many in calling diplomacy a "thankless, disillusioning and physically exhausting profession." George Kennan, "History and Diplomacy as Viewed by a Diplomatist," *The Review of Politics,* Vol. 18, No. 2 (April 1956), p. 170.
22. Dulles explained, "It is on the social occasions that you make far more progress, as a rule, than merely handling formal notes." *The New York Times,* March 24, May 1, and May 17, 1956, and April 8, 30, and May 1, 1957.
23. Cited in Charles Thayer, *Diplomat* (New York, 1959), p. 274; Glenn Wolfe, "Administration is Substantive Work," *FSJ* (April 1959), p. 23.
24. Charles Saltzman, "The Reorganization of the American Foreign Service," *DOSB,* September 27, 1954. The report was published as *Toward a Stronger Foreign Service* (Washington, 1954); see also the *DOSB,* June 28, 1954, pp. 1002–1004, and "The Department of State 1930–1955: Expanding Functions and Responsibilities," *DOSB,* March 21 and 28, 1955.
25. *FSJ* (May 1954), p. 36; Henry Wriston, "Young Men and the Foreign Service," *Foreign Affairs* (October 1954), p. 32. Ironically, when asked to pass on Wriston's qualification for State Department employment in August 1943, Harley Notter, chief

of its Division of Political Studies, found him "just a thoughtful American interested in world affairs, with knowledge gained from general reading and university activity." Cited in Mark Stoler, "Turning the Tables: State Department Opinion of Diplomatic Historians During World War II," *The Society for Historians of American Foreign Relations Newsletter* (September 1982), p. 6.

26. Frank Hopkins, "Policy, Action, and Personnel," *FSJ* (April 1960), p. 42; McCamy and Corradini, op. cit., p. 218. In 1960, only 60 percent of officers had proficiency in at least one of the most spoken world languages. H. Field Haviland, Jr. et al., *The Formulation and Administration of U.S. Foreign Policy* (Washington, 1960), p. 137.
27. Allison, op. cit., pp. 225–226; Haviland, op. cit., p. 113. The main departments with overseas employees were the U.S. Information Agency, Treasury, Commerce, Agriculture, and Health, Education and Welfare.
28. *The New York Times,* April 27, 1957. Embassies varied widely. In Moscow, 100 Americans were isolated from Soviet life, and depended for information on newspapers, other foreign diplomats, and observing the ranking Soviet leaders at major parades. Since the embassy was bugged, sensitive information had to be written out and passed around rather than discussed. In Paris, there was a constant stream of visitors, visa applicants, and congressional delegations. Ten people worked just to keep track of the numerous dispatches. While Bohlen knew most of the Moscow staff, there were 800 Americans in the Paris embassy plus 500 in aid programs—not everyone even recognized the ambassador. Martin Herz, "Life at the Paris Embassy," *FSJ* (February 1952).
29. Villard, op. cit., pp. 178–180. For a defense of nonprofessionals, see Clare Boothe Luce, "The Ambassadorial Issue: Professionals or Amateurs?" *Foreign Affairs* (October 1957); *The New York Times,* March 3, 1957.
30. In 1924, one-third of heads of mission were career people. By 1940 they were a little more than half, and in April 1959, 51 of 74 (compared to 31 of 58 in 1940) were careerists. But of the 15 best West European posts, only 6 of the smaller ones were held by FSOs. *The New York Times,* May 3, 1959. The first black ambassador, Dr. John Murrow, was appointed to Guinea in 1959. See also *The New York Times,* September 10, 1959, and November 19, 1959; October 5, 1956; March 24, 1957; and September 11, 1957.
31. *The New York Times,* May 22, 1958.
32. Robert Cutler, "The Development of the National Security Council," *Foreign Affairs* (April 1956).
33. Raymond Thurston, "The Ambassador and the CIA," *FSJ* (January 1979).
34. Stephen Ambrose with Richard Immerman, *Ike's Spies* (Garden City, N.J. 1981) pp. 221–223; Stephen Schlesinger and Stephen Kinzer, *Bitter Fruit* (Garden City, N.Y., 1982) pp. 106–107; *DOSB,* November 8, 1954, pp. 691, 695.
35. John Moors Cabot, *First Line of Defense* (Washington, 1979), pp. 18, 90–91.
36. Schlesinger and Kinzer, op. cit., pp. 145–146.
37. Ambrose, op. cit., pp. 230–232. On the 1953 Iran coup, see Barry Rubin, *Paved with Good Intentions: The American Experience and Iran* (New York, 1980). On Indonesia, see Allison, op. cit., pp. 307–308.
38. Robert F. Corrigan, "An Appreciation of a Diplomat," *FSJ* (November 1957). The discussion of U.S.-Egypt relations is taken from Barry Rubin, "America and the Egyptian Revolution," *Political Science Quarterly* (Spring 1982).

39. State Department records, 874.2614/6-1656, Byroade to Dulles and Dulles to Byroade, obtained through the Freedom of Information Act.
40. Bohlen, op. cit., p. 458.
41. Emmet Hughes, op. cit., pp. 253–254.
42. *The New York Times,* April 24, 1959.
43. Divine, op. cit., pp. 144–145.
44. Eisenhower, *Waging Peace* (Garden City, N.Y., 1965), p. 324.
45. Ibid., p. 521. Earl Smith, *The Fourth Floor* (New York, 1962), pp. 164–168; Hugh Thomas, "The U.S. and Castro, 1959–1962," *American Heritage* (October–November 1978).
46. Philip Bonsal, *Cuba, Castro and the United States* (Pittsburgh, 1971), pp. 149–150. Harvey Wellman, "The Last Days of an Embassy," *FSJ* (May 1961).
47. *The New York Times,* January 26, October 16 and 17, 1962. With interesting foresight, on June 12, 1961, a senator asked Robert Hill, just returned from being U.S. ambassador to Mexico, what would happen if a shipload of Soviet arms to Cuba came up against an American embargo. Hill replied, "They would back down. . . . They are not prepared for a showdown at this time. You understand I have read intelligence reports that state that Russia has been amazed at the ease with which they have been allowed to take over in Cuba." U.S. Senate, Judiciary Committee, "The Communist Threat to the United States Through the Caribbean" (Washington, 1961), p. 825.
48. Eisenhower, op. cit., pp. 549–552; Bohlen, op. cit., p. 465.
49. Henry Jackson, "Organization for National Security," *FSJ* (January 1962), pp. 21–22.
50. Henry Jackson, *The Secretary of State and the Ambassador* (New York, 1964), p. iii. Paul Nitze. "The 'Impossible' Job of Secretary of State," *The New York Times Magazine,* February 24, 1957, p. 60.
51. *The New York Times,* February 28, 1960.
52. George Kennan, "World Problems and America's Administrative Response," *FSJ* (September 1963), p. 22.
53. Jackson, op. cit., pp. 43, 114.
54. Ibid.
55. Ibid., p. 45.
56. Kennan, op. cit., pp. 24–25. See also John Kenneth Galbraith, *Ambassador's Journal* (Boston, 1969), p. 212.
57. Jackson, op. cit., p. 13.
58. Ibid., pp. 102–103. Kennan wrote that foreign affairs needed "intimate day-by-day, hour-by-hour direction, sensitive to the smallest significant change in the world situation. . . . And this is precisely what the National Security Council cannot do for it." Kennan, op. cit., p. 23.
59. Kennan, ibid., pp. 25–26.
60. William Bragg Ewald, Jr., *Eisenhower the President* (Englewood Cliffs, N.J., 1981) pp. 163–164, 228, 208, and 214. Ewald well describes Dulles as a lieutenant strong on his assigned turf who at times prevailed over the president's preferences, p. 201.

5. *On the Team*

1. Dean Rusk, "At the Pleasure of the President," *The Washington Post,* July 18, 1982. Earlier in his career, Rusk also gave advice to incoming Secretary Dulles.

Rusk compared the secretary of state's position to a four-engine airplane, drawing power from the president, Congress, the career staff, and the public. All four had to be won over and utilized. William Snyder, "Dean Rusk to John Foster Dulles, May–June 1953," *Diplomatic History,* Vol. 7, No. 1 (Winter 1983), p. 80.

2. *FSJ* (July 1962), p. 28.
3. Ralph Dungan, "A Year of Substantial Progress," *FSJ* (April 1963).
4. Berle, op. cit., p. 750; Dean Acheson, "The Eclipse . . ." op. cit.; Arthur Schlesinger, *A Thousand Days* (Boston, 1965), p. 433.
5. Bohlen, op. cit., pp. 489–490.
6. Edward Weintal and Charles Bartlett, *Facing the Brink* (New York, 1967), pp. 148–149.
7. Ibid., pp. 150–160. One of Rusk's aides said he lacked the passion of creative and visionary men. "Dulles, a totally unlovely person, made a place for himself in history, but Rusk, an incredibly decent person, will have none." Douglas Kinnard, *The Secretary of Defense* (Lexington, Ky., 1981), p. 83. A good and detailed profile of the department in the 1960s is Leacacos, op. cit.
8. Benjamin Read, oral history, Lyndon Johnson library, pp. 5–6.
9. Frederick Thayer, "Presidential Policy Processes and 'New Administration,'" *Public Administration Review* (September–October 1971); I. M. Destler, "National Security Advice to U.S. Presidents," *World Politics,* Vol. 29, No. 2 (January 1977); I. M. Destler, "National Security Management: What Presidents Have Wrought," *Political Science Quarterly,* Vol. 95, No. 4 (Winter 1980–81).
10. John Paton Davies, *Foreign and Other Affairs* (New York, 1966), p. 198. In Rusk's words, "The world is a very complicated place. I am skeptical of dealing with complicated situations in easy and dramatic phrases." E. W. Kenworthy, "Evolution of Our No. 1 Diplomat," *The New York Times Magazine,* March 18, 1962.
11. Warren Cohen, op. cit.; J. Robert Moskin, "Dean Rusk: Cool Man in a Hot World," *Look,* September 6, 1966; Michael O'Neill, "The Quiet Diplomat," in Lester Tanzer, *The Kennedy Circle* (Washington, 1961).
12. Dean Rusk, *The Winds of Freedom* (Boston, 1963), p. 69.
13. Ibid., pp. xi–xii. On Rusk's daily schedule, see *DOSB,* April 25, 1966, pp. 651–654; *The New York Times,* January 30, 1961.
14. Ibid., pp. 63–64; Dean Rusk, "Foreign Policy and the Political Officer," *FSJ* (April 1961), pp. 32–33. Bohlen, op. cit., pp. 474–476, 489–490. Schlesinger, op. cit., pp. 430–431. He became ambassador to France and, under Johnson, deputy undersecretary for political affairs.
15. On ambassadors, see *The New York Times,* February 8, April 16, May 10, June 10, July 2, and December 3, 1961.
16. Schlesinger, op. cit., p. 442.
17. Chester Bowles, *Promises to Keep* (New York, 1971), pp. 304, 316.
18. Chester Bowles, "Embassies and Ambassadors," *FSJ* (June 1971).
19. Joseph Kraft, "The Comeback of the State Department," *Harpers* (November 1961), pp. 43–50. *Time,* July 28, 1961; *The New York Times,* July 17, 18, and 20, 1961; Schlesinger, op. cit., pp. 437–447. *Nation,* July 29, 1961, p. 41; *The New Republic,* August 7, 1961, p. 5; Bowles, *Promises . . . ,* op. cit., pp. 308–367.
20. Weintal and Bartlett, op. cit., pp. 170–175. Ball was supported by Ambassador Llewellyn Thompson, a Soviet specialist, China expert Allen Whiting, and ex-Assistant Secretary for Public Affairs James Greenfield, who had moved to the White House.

21. Ibid., p. 152; Schlesinger, op. cit., p. 412.
22. Kraft, op. cit.
23. William Attwood, "The Labyrinth in Foggy Bottom," *Atlantic Monthly* (February 1967), pp. 45–46. William Attwood, *The Reds and the Blacks* (New York, 1967). They included a deputy chief of mission; economic, labor, consular, administrative, information, and two political officers; four for finance and housekeeping; three communications clerks; and five secretaries, as well as a dozen African employees. No room in the embassy was big enough to hold all the Americans. "A conscientious ambassador at a small post," Attwood wrote, "has to be part drill sergeant, part chaplain, and part cruise director."
24. Ibid., "Labyrinth . . . ," pp. 46–47. FSOs complained of low salaries, particularly when it came to making ends meet in Washington, and knew that colleagues of equivalent age were earning more money and having more responsibility in private enterprise.
25. Galbraith, op. cit., pp. 204–205.
26. Schlesinger, op. cit., pp. 414–416. Professor Richard Neustadt, who sometimes advised Kennedy, testified to the Jackson committee in 1963, "So far as I can judge, the State Department has not yet found means to take the proferred role and play it vigorously across the board of national security affairs." Quoted in John Harr, *The Professional Diplomat* (Princeton, 1969), p. 104.
27. U. Alexis Johnson, "Internal Defense and the Foreign Service," *FSJ* (July 1962); George McGhee, "The Changing Role of the American Ambassador," *DOSB*, June 25, 1962. In Johnson's words, these required "the use of all available resources for assisting these new nations in building the kind of society and government that can maintain itself, develop in step with the modern world and, above all, remain free from domination or control by Communist forces hostile to us." While Communism exacerbates these problems, "we face a world," even without their influence, that "would still be wracked by stresses and strains, and would still be fertile ground for political turmoil and even violence," pp. 20–21. This represents a break with the largely bipolar view of the Cold War in the Dulles era.
28. Kraft, op. cit. The result of this philosophy was that FSOs were encouraged to battle for accuracy, but told to avoid excessive or public conflict with those in authority. As one of them wrote, the dilemma was to separate "constructive comment and obstructionism," while protecting one's job and the civil servant approach. "We have generally resolved the dilemma by dissenting only on trivia or in such cautious terms that our dissent is all but inaudible. This accomplishes nothing." William Knight, "On Dissent," *FSJ* (December 1964).
29. Attwood, op. cit., p. 47.
30. Schlesinger, op. cit., p. 416.
31. *Time,* March 15, 1963; *The New York Times,* August 17, 1963. Rusk's first assistant secretary for public affairs, Roger Tubby, had been Truman's press secretary. Kennedy chose him at the suggestion of several journalists, including Walter Lippmann, and Dean Acheson. But Rusk did not give full confidence to a man selected for him. "The assignment sometimes seemed nightmarish," says Tubby. "Often I did not know what cards Rusk held, how he wanted to play them." Tubby had good rapport with the press and initiated regional briefings for media lacking Washington correspondents and background briefings for the foreign press. Tubby to author, October 16, 1982; Roger Tubby, oral history, John F. Kennedy Library.
32. *DOSB,* August 14, 1961.

33. Harr, op. cit., pp. 113ff. Crockett and others also worked to improve contacts between FSOs and Congress, academia, outside foreign policy experts, and the general public through exchange programs and speaking trips. The story is told in detail in Frederick Mosher and John Harr, *Programming Systems and Foreign Affairs Leadership,* (New York, 1970).
34. *The New York Times,* October 1, 1961.
35. Ibid., August 19, 1961; March 24, 1968.
36. Richard Johnson, *The Administration of U.S. Foreign Policy* (Austin, 1971), pp. 146–149. Civil rights leader A. Phillip Randolph answered departmental claims that African countries did not want black diplomats or ambassadors by attributing this to their perception that blacks had little influence on decisions. *The New York Times,* November 26, 1962. In 1968 Barbara Watson, a career employee, became the first black to obtain the rank of assistant secretary of state, as administrator of the Bureau of Security and Consular Affairs.
37. Frederick Dutton, "'Cold War' Between the Hill and Foggy Bottom," *The New York Times Magazine,* September 15, 1963. He cites increasing congressional correspondence with the department on policy questions and the growing number of committees holding international affairs hearings and requesting departmental officials to testify.
38. On Schwartz, see *The New York Times,* March 7 and 23, 1966; on Legion report, Simpson, op. cit., p. 223. The department announced the removal of 18 employees in May 1961—16 homosexuals, 1 for psychological disorders, and 1 for having some contacts with Communists. A June 1963 listing shows 41 regular employees resigning to avoid facing charges. Another 171 applicants or trial employees were rejected before clearance. The largest single cause for firing and forced resignation seems to have been homosexuality. *The New York Times,* May 2, 1961, and June 8, 1963.
39. *The New Republic,* November 23, 1963. See also *U.S. News and World Report,* November 25, 1963; *The New York Times,* November 18, 1965; Charles Stevenson, "The Ordeal of Otto Otepka," *Reader's Digest* (August 1965).
40. *The New York Times,* June 10, 11, 29, 1961.
41. Berle, op. cit. pp. 752, 756–775. Kennedy was dissatisfied with Woodward, who was sent off as ambassador to Spain and replaced by Edwin Martin, a good administrator relatively attuned to White House thinking, who nonetheless restricted Goodwin's activities.
42. Schlesinger, op. cit., pp. 250–251, 266–267.
43. State's participants in the committee's work included Rusk, Ball, U. Alexis Johnson, Martin, Bohlen, and Llewellyn Thompson, the last two as advisers on Soviet policy. Other regulars were McNamara, Bundy, Director of Central Intelligence John McCone, Treasury Secretary Dillon, Presidential Counselor Sorenson, General Maxwell Taylor, chairman of the Joint Chiefs of Staff, Deputy Secretary of Defense Roswell Gilpatric, and Assistant Secretary of Defense Paul Nitze. Vice-President Johnson and UN Ambassador Adlai Stevenson sometimes attended meetings.
44. ABC News Nightline interview of Dean Rusk, conducted by Jay LaMonica on Oct. 7, 1982.
45. Robert Kennedy, *Thirteen Days* (New York, 1968), pp. 78–80.
46. Sorensen interview same as above, note 44. To combat this problem, President Kennedy tried to stay in direct touch with Ball, Martin, Thompson, and other subcabinet officials at State.

47. Walter LaFeber, "Latin American Policy," in Robert Divine, ed., *Exploring the Johnson Years* (Austin, 1981), pp. 65, 82.
48. Ibid., pp. 76–77; Thomas Mann, oral history, Lyndon Johnson Library, p. 18; William Turpin, "Foreign Relations, Yes; Foreign Policy, No," *Foreign Policy* (Fall 1972), p. 57.
49. Robert Nisbet, "Project Camelot: An Autopsy," *The Public Interest* (Fall 1966), pp. 56–57. Harr. op. cit., pp. 105–110.
50. John Tuthill, "Operation Topsy," *Foreign Policy* (Fall 1972). One consul told him, "If the American government had a consulate situated in the middle of the Sahara desert, the consul could be busy 8 hrs a day, 6 days a week, just responding to administrative inquiries from the U.S. government."
51. *The New York Times,* March 3, 1961. See also James Penfield, "The Role of the U.S. in Africa," *DOSB,* June 8, 1959.
52. Michael Samuels and Stephen Haykin, "The Anderson Plan: An American Attempt to Seduce Portugal out of Africa," paper to International Conference Group on Modern Portugal (June 1979).
53. Richard Maloney, *JFK and Africa* (New York, 1984); Madeleine Kalb, *The Congo Cables* (New York, 1982); Anthony Lake, *The 'Tar Baby' Option* (New York, 1976). The Africa, International Organization, and INR bureaus, UN mission, and Legal Adviser's office generally lined up on the liberal side.
54. Both Godley and Carlucci were unusually outspoken FSOs. Godley later became ambassador to Laos, during the peak of the U.S. secret war there, and ambassador to Lebanon; Carlucci, whose role in cutting back the bloated embassy in Brazil was mentioned above, even dared oppose Henry Kissinger on U.S. covert intervention in Portugal as ambassador to Lisbon in 1974.
55. This is a more important variation of "clientitis," which implies that those long stationed in a particular country will tend to take on the prevalent views of that nation.
56. McLellan and Acheson, op. cit., p. 279.
57. Seyom Brown, op. cit., pp. 307–321; Walt Rostow, "The Planning of Foreign Policy," in E. Johnson, ed., *The Dimensions of Diplomacy* (Baltimore, 1965); and Walt Rostow, "The Third Round," *Foreign Affairs* (October 1963).
58. Benjamin Read, oral history, John F. Kennedy Library, p. 48.
59. McLellan and Acheson, op. cit., pp. 279, 281–282. When Ball resigned in 1966, he was replaced by Attorney General Nicholas Katzenbach. Eugene Rostow (Walt's brother) became the other undersecretary. According to Rostow, Katzenbach was supposed to work on administration and congressional relations while Rostow produced ideas and focused on European and economic problems. Eugene Rostow, oral history, Lyndon Johnson Library. On Katzenbach, see *The New York Times,* November 12, 1967, and Victor Navasky, "No. 2 Man at State is a Cooler-Downer," *The New York Times Magazine,* December 24, 1967.
60. Weintal and Bartlett, op. cit., pp. 6–9.
61. Nathaniel McKitterick, "Diplomatic Logjam," *The New Republic,* March 27, 1965. Averell Harriman, oral history, John Kennedy Library, quoted with permission, pp. 102, 124. Weintal and Bartlett, op. cit., pp. 154–155, 165. Harriman oral history, op. cit., p. 121.
62. Smith Simpson, "Who Runs the State Department," *The Nation,* March 6, 1967, pp. 298–299.
63. James Thomson cited in Leslie Gelb with Richard Betts, *The Irony of Vietnam*

(Washington, 1979), p. 16. In this regard, State's problem with the military in the 1960s was similar to its disadvantage competing with the CIA in the 1950s.

64. William Lenderking, "Dissent, Disloyalty, and Foreign Service Finkism," *FSJ* (June 1974), pp. 13–14.
65. Gelb and Betts, op. cit., p. 234; see also David Halberstam, *The Best and the Brightest* (New York, 1972).
66. Ibid., p. 302. See also George Ball, *The Past has a Different Pattern* (New York, 1981) pp. 371–372.
67. Roger Hilsman, oral history, John Kennedy Library, tape 1, p. 20. William Bundy, a Hilsman contemporary at Defense and his replacement as assistant secretary for East Asia, attributes Hilsman's firing to "a lack of confidence, a feeling that he was indiscreet. The one thing I've never heard said, and by implication everything I've heard is to the contrary, is that his departure had anything to do with his views on policy and the conduct of the war. Oral History Lyndon Johnson Library, tape 1, p. 14.
68. Gelb and Betts, op. cit., pp. 234–235. Oral history, Lyndon Johnson Library, tape 1, p. 11.
69. Ball, op. cit., pp. 383–384; George Herring, "The War in Vietnam," in Divine, ed., op. cit. Rusk aide Benjamin Read wrote years later, "The Ball briefs in my opinion . . . stand up infinitely better on reading today in terms of foresight and wisdom than the military briefs which had a desperately simplistic approach—that if you put in such-and-such efforts, you'll get out these results; never taking the obvious next step of what the North Vietnamese would be able to do and had the will to do." Oral history, Lyndon Johnson Library, p. 9.
70. Gelb and Betts, op. cit., pp. 315–316. William Bundy was McGeorge Bundy's brother. Their father was an assistant secretary of state in the Hoover administration.
71. Quoted by Winthrop Brown, oral history, John F. Kennedy Library, p. 17.
72. William Bundy, oral history, Lyndon Johnson Library, tape 1, pp. 32–33.
73. Ibid., tape 2, p. 33.
74. Ibid., tape 3, p. 32; Gelb and Betts, op. cit., pp. 169, 1975.
75. Joseph Kraft, "Dean Rusk Show," *The New York Times Magazine,* March 24, 1968.
76. Ball, op. cit., p. 3.
77. John Campbell, *The Foreign Affairs Fudge Factory* (New York, 1971), pp. 95–96.

6. *The Contemporary State Department*

1. U.S. Department of State, *The Department of State Today* (Washington, 1981). The autonomous AID has an additional 6200 workers, 2500 of whom are in Washington. The USIA has 7750 workers, 850 of whom are Foreign Service Information officers, 3400 are domestic employees, and 3500 are foreign nationals.
2. Ambassador Charles Bohlen: "One of the advantages . . . of having been around government for as long as I have is you tend to know people, and I know Paul Nitze very, very well indeed. And I know Clark Clifford . . . Dick Helms I've known for many years." Oral history, Lyndon Johnson Library, pp. 2–3.
3. George Ball, *The Past has a Different Pattern* (New York, 1981) p. 376.
4. Lyndon Johnson complimented Ball by telling him, "George, you're like the school teacher looking for a job with a small school district in Texas. When asked by the school board whether he believed that the world was flat or round, he replied: 'Oh, I

can teach it either way.' . . . You can argue like hell with me against a position, but I know outside this room you're going to support me." Ibid., pp. 377–378.

5. Chester Bowles, oral history interview, John F. Kennedy Library, p. 59.
6. Charles Maechling Jr., "Foreign Policy-Makers: The Weakest Link?" *FSJ* (June 1976), p. 18.
7. Ibid., p. 18.
8. Ibid., p. 33.
9. Roger Hilsman, oral history interview, John F. Kennedy Library, p. 4.
10. Congressional Research Service, *The Ambassador in U.S. Foreign Policy* (Washington, 1981), p. 3.
11. Ellis Briggs, *Farewell to Foggy Bottom* (New York, 1964), p. 5. In recent decades, about 70 percent of ambassadors have come from the career ranks, though the political appointees tend to receive the most important European and many of the most pleasant assignments. The ambassador is aided by a deputy chief of mission, a minister, and a counselor (in smaller embassies these two functions may be combined). See Jack Perry, "On Being a Deputy Chief of Mission," *FSJ* (August 1978).
12. Dean Mann, *The Assistant Secretaries* (Washington, 1965), p. 255.
13. U.S. Department of State, op. cit.
14. Congressional Research Service, op. cit., p. 4.
15. Thomas Etzold, "Does Macy's Tell Gimbel's?" *FSJ* (August 1977), pp. 7–8.
16. David Willis, "Fudge Watching: A Reporter Views the State Department," *FSJ* (November 1967), pp. 20–21. Willis and other reporters consider Robert McCloskey, PA assistant secretary in the Johnson administration, as "perhaps the best public information officer in the government. . . . He helps us if he can: yet he leaves no doubt that there are lines beyond which he will not go."
17. Leacacos, op. cit., p. 23; see also pp. 19–40.
18. Arnold Beichman, *The 'Other' State Department: The United States Mission to the United Nations* (New York, 1967).
19. See her May 4, 1982, testimony before the House Appropriations Committee's subcommittee on Foreign Operations. "At the present time, none of the Deputy Assistant Secretaries for International Organizations has ever served at the US Mission to the United Nations, nor has the Acting Director of the Office of United Nations Political Affairs, nor have any of that office's Deputy Directors." Only two of the twelve officers within the Office of United Nations Political Affairs in State ever served there.
20. On the job's problems, see Anthony Solomon, oral history, Lyndon Johnson Library. Congress created the special trade representative in 1962 because many members felt State was not energetic enough in defending U.S. economic interests. Solomon's position was strengthened both by the illness of that official and doubts as to the office's permanence. Today, the U.S. Trade Representwtive is attached to the Executive Office of the president. The assistant secretary works with the undersecretary for economic affairs but reports to the secretary. About 40 FSOs were shifted to the Commerce Department beginning in 1980 as the core of its new Foreign Commercial Service. See *The Washington Post,* February 25, 1983.
21. Other sections include the Foreign Service Institute to train department employees in introductory and advanced language classes, area studies, and seminars, and the Language Services Division to translate notes, treaties, documents, and letters. Three other agencies, not dealt with in detail in this book, are associated with State.

The Agency for International Development (AID) runs nonmilitary foreign aid programs and disburses billions of dollars in assistance. The U.S. Information Agency (USIA) makes news, information, and cultural materials about the United States available in other countries, runs the Voice of America radio, and handles exchange programs. The Arms Control and Disarmament Agency (ACDA) deals with strategic weapons issues, including arms limitation talks. All three have their own personnel systems but receive guidance from State. USIA public affairs officers and AID advisers are assigned to embassies abroad.

22. According to some sources, Assistant Secretary William Tyler was sent to an ambassadorial post by Undersecretary George Ball during the Johnson administration because he was considered too soft on Charles de Gaulle, but EUR assistant secretaries have had a far easier time dealing with department superiors and the White House than have, for example, counterparts working on Latin America.
23. Malcolm Toon, "In Defense of the Foreign Service," *The New York Times Magazine,* Dec. 12, 1982.
24. David Newsom, "Miracle or Mirage: Reflections on US Diplomacy and the Arabs," *Middle East Journal* (Summer 1981), pp. 302–330. For the debate over State Department Arabists, see, for example, Curtis Jones, "The Education of an Arabist," *FSJ* (December 1982), and the reply by Harvey Feldman, *The Washington Post,* Feb. 27, 1983; Joseph Kraft, "Those Arabists in the State Department," *The New York Times Magazine,* November 7, 1971; press conference of retiring Ambassador Talcott Seelye, *The New York Times,* September 1, 1981.
25. Leacacos, op. cit., p. 85.
26. Ronald Morse and Edward Olsen, "Japan's Bureaucratic Edge," *Foreign Policy* (Fall 1983), pp. 167–180.

7. *State of Decline, 1969–1976*

1. Preceding and following quotes are from Henry Kissinger, "Bureaucracy and Policymaking," in Henry Kissinger and Bernard Brodie, *Bureaucracy, Politics and Strategy,* Security Studies Paper #17, (Los Angeles, 1968) pp. 1–2.
2. Ibid., pp. 12–13. Kissinger also discussed such traditional themes as overstaffing and excessive paperwork. "On the whole, if we could get rid of the bottom half of the Foreign Service we might be better off. . . . I think the first eight to ten years of a State Department man's career tend to be such as to drive the more imaginative and more purpose-oriented people out of it."
3. Henry Kissinger, "America and the World: Principle and Pragmatism," *Time,* December 27, 1976, p. 42. In his first talk to State Department employees as secretary, Kissinger said, "Now we are in a situation where we have to conduct foreign policy the way many other nations have had to conduct it throughout their history. We no longer have overwhelming margins of safety" or resources. *DOSB,* October 22, 1973, pp. 506–507.
4. Henry Kissinger, *White House Years* (hereafter, Vol. 1) (Boston, 1979), p. 607.
5. Robert Ferrell, op. cit., pp. 360–361. Since this was during and immediately after the McCarthy era, any reluctance by FSOs to return home is quite understandable.
6. *The New York Times,* October 14, 1968; William Safire, *Before the Fall* (Garden City, N.J. 1975), p. 131.
7. Ralph Blumenfeld et al., *Henry Kissinger: The Private and Public Story* (New York, 1974), p. 174.

8. Thomas E. Cronin, *The State of the Presidency* (Boston, 1975), p. 120. The average size of the NSC's professional staff grew from 12 under McGeorge Bundy (1961–1966) to 18 under Walt Rostow (1966–1969), rising to 50 under Kissinger (1969–1975) and falling to the forties for his successor, Brent Scowcroft (1975–1977), and to the thirties in the era of Zbigniew Brzezinski (1977–1981). I. M. Destler, "NSC II: The Rise of the Assistant (1961–1981)," in Hugh Heclo and Lester Salamon, op. cit., p. 264.
9. *The Washington Post,* March 3, 1971; for Rogers's defense of his role, see *The Washington Post,* February 27, 1972.
10. Quotations from his interview with David Frost, *The New York Times,* May 13, 1977.
11. Benjamin Read, oral history, John F. Kennedy Library, pp. 30–31.
12. On the system's background and structure, see Henry Kissinger, Vol. 1, op. cit., pp. 11ff.; I. M. Destler, "National Security Advice . . . ," op. cit.; Commission on the Organization of the Government for the Conduct of Foreign Policy (hereafter, the Murphy Commission *Report),* Vol. 2 (Washington, 1975), pp. 113–118; U.S. Senate Committee on Government Operations, "The National Security Council: New Role and Structure" (Washington, February 7, 1969), and "The National Security Council" (Washington, March 3, 1970); Destler, "National Security Management . . . ," op. cit., p. 576.
13. This and much of the following material is based on interviews with participants.
14. Bert Rockman, "America's Department of State: Irregular and Regular Syndromes of Policy Making," *American Political Science Review,* Vol. 75, No. 4 (Winter 1981), p. 911, called the NSC staff "Less and less composed of graying and grayish anonymous career foreign service officers, and more and more composed of foreign policy intellectuals and prospective high-fliers, many of whom are drawn from America's leading universities"; but many of them were still FSOs who, taken out of the State Department "culture," were expected to perform better. On the staff members, see also Marvin and Bernard Kalb, *Kissinger* (Boston, 1974), pp. 80–95; Roger Morris, *Uncertain Greatness* (New York, 1977), pp. 134–144.
15. Colonel Alexander M. Haig, former deputy superintendent at the West Point Military Academy, had worked for Secretary of the Army Cyrus Vance in the Johnson administration. Recommended by Vance and another colleague, Joseph Califano, he gradually gathered power in the NSC and became Kissinger's deputy in June 1970. When he became army vice-chief of staff in January 1973, he was replaced by Lt. Gen. Brent Scowcroft, an organization man who held the position until succeeding to the top post when Secretary of State Kissinger relinquished it in November 1975. Haig became White House chief of staff in May 1973.
16. He even discouraged President Gerald Ford, Vice-President Nelson Rockefeller, and other cabinet-level officials from going abroad. Robert Hartmann, *Palace Politics* (New York, 1980), p. 84n.
17. In 1969, NSC staffers rated State-chaired studies only "50 to 70 percent acceptable." John Leacacos, "Kissinger's Apparat," *Foreign Policy* (Winter 1971–1972), analyzes and lists these NSSMs.
18. Tad Szulc, *The Illusion of Peace* (New York, 1978), pp. 18, 64.
19. David Landau, *Kissinger: The Uses of Power* (New York, 1978), p. 138. For other critiques, see Morris, op. cit.; Tad Szulc, op. cit.; Seymour Hersh, *The Price of Power* (New York, 1983).
20. I. M. Destler, "The Nixon System: A further look," *FSJ* (February 1974), p. 11.

The Murphy Commission *Report,* Vol. 2, pp. 113–115, points to the tendency to squeeze out the bureaucracy rather than forcing it to think harder and work better.

21. Morris, op. cit., p. 2.
22. William Turpin, op. cit., p. 51.
23. Kissinger, Vol. 1, op. cit., p. 30. He accuses Rogers of sometimes trying to circumvent the NSC, by opening hijacking negotiations with Cuba with only 36 hours' notice to the NSC, for example, or by trying to improve relations with India when the White House had ordered a cooler policy. Henry Kissinger, *Years of Upheaval* (hereafter, Vol. 2) (New York, 1982), pp. 419–420.
24. Kissinger, Vol. 1, op. cit., p. 411. Hillenbrand later became ambassador to West Germany. Snubbed by Kissinger, he complained in a dispatch as his tenure ended in 1976, after 37 years as an FSO: "The Foreign Service is in a state of intellectual disorientation because many of its members no longer have a clear idea of the role it should and can play." *The Washington Post,* November 10, 1976.
25. Kissinger, Vol. 1, op. cit., pp. 425–428.
26. Ibid., pp. 385–387.
27. Charles Bray, cited in William Macomber, *The Angels' Game* (New York, 1975), p. 199.
28. John Leacacos, *Fires . . .* , op. cit., pp. 460–461.
29. Other activists and supporters included several who rose to high positions, including Theodore Eliot (later an ambassador and State's inspector-general), Morris Draper and Walker (deputy assistant secretaries), L. Dean Brown (later an ambassador and deputy undersecretary for management), and Philip Habib, who became undersecretary for political affairs, the highest career position. For Walker's statement, *The New York Times,* May 13, 1969.
30. William I. Bacchus, "Diplomacy for the 70's: An Afterview and Appraisal," *American Political Science Review* (June 1974), p. 738. U.S. Department of State, *Diplomacy for the 70s* (Washington, 1970).
31. One top-ranking State official of the time, later asked about the reform efforts, could not remember them. "That shows how important they were," he said. For Macomber's defense, see "Change in Foggy Bottom," *DOSB,* February 14, 1972.
32. Murphy Commission *Report,* Vol. 6, op. cit., pp. 22–23.
33. Letter from John Harter, *The New York Times,* December 15, 1972.
34. *The New York Times,* October 10, 1971; *The Washington Post,* February 1, 1975.
35. Ibid.; *The New York Times,* December 1, 1971; *The Washington Post,* December 11, 1972.
36. *The New York Times,* April 24, 1972; *The Washington Post,* August 13, 1973.
37. *The Washington Post,* August 26, 1971, January 26, 1972, and August 21, 1975; *The New York Times,* February 26, 1972.
38. Murphy Commission *Report,* Vol. 6, op. cit., pp. 93–96. As time went on, the NSSMs "shifted from broad policy reviews to narrow operational questions" (tuna boat seizures by Latin American states), and responses to specific events. For a critique of the NSSM system, see Elmo Zumwalt, *On Watch* (New York, 1976) pp. 421–422.
39. Chester Cooper, *The Lost Crusade* (New York, 1972), Chapter 11; Paul Kattenburg, *The Vietnam Trauma in American Foreign Policy* (New Brunswick, N.J., 1982), pp. 3–6, 79–184; Halberstam, op. cit., pp. 449–464, 754–756. On recent analyses of the war, see Fox Butterfield, "The New Vietnam Scholarship," *The New York Times Magazine,* February 13, 1983.

40. George Ball, *The Past . . .*, op. cit., p. 336.
41. Kissinger, Vol. 1, op. cit., pp. 1351–1352.
42. On Godley's nomination, see *DOSB*, August 13, 1973, pp. 257–258; Kissinger, Vol. 1, op. cit., pp. 261–264; Szulc, op. cit., p. 19.
43. Interviews; John Claymore (pseudonym), "Vietnamization of the Foreign Service," *FSJ (December 1971); The New York Times*, May 9, 1970, and December 30, 1971.
44. *The New York Times*, September 20, 1971; July 26 and August 6, 1974. *The Washington Post*, April 3, 1974; *U.S. News and World Report*, April 29, 1974.
45. Szulc, op. cit., pp. 253, 281; William Shawcross, *Sideshow: Kissinger, Nixon and the Destruction of Cambodia* (New York, 1979), pp. 53–54, 139, 302; Kissinger, Vol. 1, op. cit., pp. 245–247, 466. According to Kissinger, State used bureaucratic techniques to slow down establishment of a CIA station in Cambodia to estimate and facilitate needed supplies. Only Nixon's direct intervention forced EA's acquiescence. On Rogers's objections to a later incursion into Laos, see Kissinger, Vol. 1, op. cit., p. 999.
46. Interviews; Peter A. Poole, "Pacific Overtures with Marshall Green," *FSJ* (June 1979), pp. 16–20.
47. Shawcross, op. cit., pp. 161, 265–270.
48. Ibid., p. 322.
49. Kissinger, Vol. 1, op. cit., pp. 189–190, 686–691; Szulc, op. cit., p. 347. See also, "Reminiscences: 'Diplomatic Lessons Learned the Hard Way,'" *State Department Newsletter* (February 1977).
50. Interviews; *The Washington Post*, February 27 and April 24, 1976.
51. Kissinger, Vol. 1, op. cit., p. 141, 153–154. For State's analytical work on Communist affairs, see Leacacos, op. cit., pp. 493–499.
52. John Newhouse, *Cold Dawn* (New York, 1973) pp. 42–50, 148–149. *The Washington Post*, June 25, 1974; Szulc, op. cit., pp. 112–113; Kissinger, Vol. I, op. cit., p. 819.
53. Murphy Commission *Report*, Vol. 2, op. cit., p. 114.
54. Morris, op. cit., pp. 213–214.
55. Ibid., pp. 216–226.
56. Ibid., op. cit., p. 226; Hersh, op. cit., pp. 444–464; Kissinger, Vol. 1, op. cit., pp. 842–918, especially 853–856.
57. Christopher Van Hollen, "The Tilt Policy Revisited: Nixon-Kissinger Geopolitics and South Asia, *Asian Survey*, Vol. 20, No. 4 (April 1980), pp. 340–347. Ambassador Van Hollen, a departmental South Asia hand, writes, "It was both unnecessary and unwise to raise the Bangladesh regional crisis to the level of global geopolitics. Kissinger is wrong in concluding that Nixon's willingness to risk war with the Soviet Union, including the deployment of a U.S. aircraft carrier to South Asia, saved West Pakistan and preserved the structure of world peace."
58. *The Washington Post*, January 22, 1972.
59. Philip Oldenburg, "The Breakup of Pakistan," Murphy Commission *Report*, Vol. 7, op. cit., p. 125.
60. *DOSB*, December 8, 1969, p. 509; see also April 13, 1970, p. 498.
61. Interviews; Szulc, op. cit., pp. 356–360; Morris, op. cit., pp. 232–236; Seymour Hersh, "The Price of Power, Kissinger, Nixon, and Chile," *The Atlantic Monthly* (December 1982), and Hersh, *Price . . .*, op. cit., pp. 258–296.
62. Kissinger, Vol. 1, op. cit., pp. 663–668. He wrote, "The philosophical bias of our bureaucracy, the confusion between economic development and foreign policy ob-

jectives, had produced paralysis. The prevalent view . . . was apparently that it was acceptable for a radical candidate to receive substantial funds from Cuba and other Communist sources, but improper for the United States to assist the democratic candidate with the best chance of success, even if his program was less reformist than some might have wished.''

63. Zumwalt, op. cit., p. 322.
64. The editorial continued, ''If the word of American representatives abroad is to stand for anything, promises of confidentiality offered to foreigners by the executive branch under one set of circumstances . . . cannot be lightly broken in open congressional hearings simply because circumstances have changed. . . .'' *The Washington Post,* June 6, 1976; *The Washington Post,* June 12, 1976; John Marks, ''Lying to Congress,'' *The Washington Post,* June 6, 1976.
65. Taylor Branch and Eugene Propper, *Labyrinth* (New York, 1982), p. 347. The U.S. ambassador at the time of the coup, Nathaniel Davis, and two other former embassy officials filed a lawsuit against the film *Missing,* which charged, in highly colored fashion, that U.S. diplomats had done little to help the relatives of an American citizen killed in the coup. Carla Hall, ''The Scars after 'Missing,''' *The Washington Post,* January 18, 1982.
66. On NEA and the 1967 war, see William Quandt, *Decade of Decisions: American Policy Toward the Arab-Israeli Conflict 1967–1976* (Los Angeles, 1977), pp. 25–69. See also Ethan Nadelmann, ''Setting the Stage: American Policy Toward the Middle East, 1961–1966,'' *International Journal of Middle East Studies* (November 1982).
67. Kissinger called Sisco ''Intense, gregarious, occasionally frenetic . . . not a conventional Foreign Service Officer.'' Sisco was ''enormously inventive . . . sometimes offering more solutions than there were problems . . . seized the bureaucratic initiative and never surrendered it.'' He was also the only ranking official who succeeded in getting along with both Rogers and Kissinger and bridging the two regimes. See Kissinger, Vol. 1, op. cit., pp. 348–49. When Sisco resigned as assistant secretary for NEA in the fall of 1973, Kissinger persuaded him to come back as undersecretary for political affairs, succeeding U. Alexis Johnson.
68. Quandt, op. cit., pp. 79–80.
69. Kissinger, Vol. 1, op. cit., pp. 356–376; ibid., p. 93.
70. Interviews; Kissinger, Vol. 2, op. cit., pp. 211–212.
71. Ibid., pp. 200, 1278, 1283.
72. On the oil issue, see Szulc, op. cit., pp. 438–439; see also Yitzhak Rabin, *The Rabin Memoirs* (Boston, 1979), pp. 192–193.
73. *The New York Times,* August 26, 1973; Kissinger, Vol. 2, op. cit., pp. 7–8.
74. Quandt, op. cit., pp. 166–167. Quandt recounts that the contingency plan job was assigned to a junior State Department official who considered it a waste of time. When war broke out several months later, the task was still unfinished. As in so many cases of prediction, ''The flow of information was staggering, but it was also inconsistent.''
75. Kissinger, Vol. 2, op. cit., p. 803. On Kissinger's policy, see Edward Sheehan, *The Arabs, Israelis, and Kissinger* (New York, 1976); Matti Golan, *The Secret Conversations of Henry Kissinger* (New York, 1976).
76. Kissinger, Vol. 2, op. cit., pp. 415–416.
77. In *DOSB,* November 11, 1974, p. 635.
78. Sources for this account include *The Washington Post,* November 14, 1971; Morris,

op. cit.; Laurence Stern, *The Wrong Horse: The Politics of Intervention and the Failure of American Diplomacy* (New York, 1977).

79. Stern, op. cit., pp. 26–27.
80. Ibid; Martin Herz, ed., *Contacts with the Opposition: A Symposium,* Institute for the Study of Diplomacy (Washington, 1979), pp. 68, 71 (Tasca); pp. 50–51 (Doder).
81. *Department of State Newsletter* (October 1980), pp. 12–13; Morris, op. cit., pp. 273–276.
82. Morris, op. cit., p. 276.
83. *The Washington Post,* October 23, 1975; *DOSB,* August 2, 1976, p. 169; Matti Golan, *Shimon Peres, A Biography* (New York, 1982), p. 156.
84. Anthony Lake, *The "Tar Baby" Option: American Policy Toward Southern Rhodesia* (New York, 1976), pp. 130–131. *The New York Times,* August 13, 1974. "While we are anxious to serve in Africa," said a black American stationed in West Africa, "the proof of our success in this field will be our postings . . . to France, Peru, Norway, and China." The five black FSOs who were ambassadors, the seven leading AID missions abroad, and seven of the ten who headed USIA offices overseas were assigned to Africa. At the 1976 National Urban League convention, Kissinger was jeered when he defended State's hiring practices.
85. Lake, op. cit., pp. 125–128; Morris, op. cit., pp. 75, 111.
86. Morris, op. cit., pp. 110–120. *The New York Times,* editorial of March 1, 1975, June 15, 1975, December 14, 1975; *The Washington Post,* September 1 and February 25, 1975; Kissinger, Vol. 2, op. cit., p. 440. See also Nathaniel Davis, "The Angola Decision of 1975: A Personal Memoir," *Foreign Affairs* (Fall 1975).

8. *Divided Counsels: The Carter Administration 1977–1981*

1. *DOSB,* March 21, 1977, pp. 259, 263. He also said, in a spirit of open diplomacy to contrast with Kissinger's realpolitik "deviousness," "I want us to tell the Saudi Arabians and the Syrians and the Egyptians and the Lebanese and the Jordanians and the Israelis the same thing, so that there never is any sense of being misled."
2. Statements backed up with citations from newspapers have usually been confirmed by interviews. *The New York Times,* March 21, 1980, and April 29, 1980; *The Washington Post,* March 16, 1980, and October 18, 1982; Jack Germond and Jules Witcover, *Blue Smoke and Mirrors* (New York, 1981) p. 164n; Destler, "National Security Management . . ." op. cit., p. 573.
3. Kissinger, Vol. 2, op. cit., pp. 265–266. Cyrus Vance, *Hard Choices* (New York, 1983), outlines his views and advice to Carter.
4. Cyrus Vance, oral history interview, Lyndon Johnson Library, p. 11.
5. Zbigniew Brzezinski, *Power and Principle* (New York, 1983) pp. 17, 29–30, and 36. See also Leslie Gelb, "Muskie and Brzezinski: The Struggle Over Foreign Policy," *The New York Times Magazine,* July 20, 1980, p. 27; Destler, "NSC II . . . ," op. cit.
6. Richard Burt in *The New York Times,* April 17, 1978.
7. Leslie Gelb in *The New York Times,* April 29, 1980.
8. *The Washington Post,* July 16, 1980.
9. Burt, op. cit, also claims Shulman wanted to inform Moscow—before European allies were told—of Carter's decision to cancel production on a neutron high-radiation weapon until Brzezinski blocked this action. See also Brzezinski, op. cit., pp. 36–42.

10. Quoted in *The Washington Post,* May 5, 1980. Senator Edmund Muskie, Vance's replacement, was soon making similar statements: "If I were President, I would appoint somebody as Secretary of State and make sure that the NSC role is that of coordinating not anything else." *The New York Times,* October 6, 1980.
11. Vance, op. cit., p. 14; *The Washington Post,* March 16, 1980; *The New York Times,* December 22, 1978; *Department of State Newsletter* (January 1977).
12. *The New York Times,* June 19, 1977.
13. *The New York Times,* January 13, 1977.
14. *The Washington Post,* June 23, 1977; cited in Barry Rubin, "Carter, Human Rights, and U.S. Allies," in Barry Rubin and Elizabeth Spiro, *Human Rights and U.S. Foreign Policy* (Boulder, Colo., 1979, p. 116.
15. *The New York Times,* December 12, 1977, August 15, 16, and 17, 1979. "That was not a lie," Ambassador Young said in explaining what he told the department, "it was just not the whole truth." Part of Young's problem may have come from his replacement of 14 people on his staff. "No one here on the mission has a good grasp of what's going on," said one staff member. *The New York Times,* February 19, 1977.

    In early 1980, Carter decided to support a UN resolution critical of Israel if certain language was deleted. McHenry obtained some changes in the wording, but some statements remained that conflicted with U.S. policy. Vance thought adequate revisions had been made, and Carter accepted his suggestion for an affirmative U.S. vote. Later, Vice-President Mondale and others suggested the vote be retracted lest it hurt Carter's reelection effort; Vance was left to take the blame. The incident made the administration appear doubly inept: on the vote itself and on the embarrassing reversal.
16. Robert Hunter, *Presidential Control of Foreign Policy* (New York, 1982), pp. 8–27. See also Brzezinski, op. cit., pp. 59–78, and his speech in *Department of State Newsletter* (January 1978).
17. Vance, op. cit., pp. 36–39; interviews.
18. Hunter, op. cit., pp. 42–44; Vance, op. cit., pp. 38–39. The weekly reports are presented in Brzezinski, op. cit., pp. 556–570.
19. Dick Kirschten, "Beyond the Vance-Brzezinski Clash Lurks an NSC under Fire," *National Journal,* May 17, 1980, pp. 814–816.
20. Gelb, "Muskie and Brzezinski," op. cit., p. 39.
21. Warren Christopher, "Ceasefire Between the Branches: A Compact in Foreign Affairs," *Foreign Affairs* (Summer 1982), p. 1000; Robert Pastor, "Coping with Congress's Foreign Policy," *FSJ* (December 1975), pp. 15–16. The difference in styles between the two institutions is well described by Congressman Edward Derwinski, who became State's counselor in the Reagan administration: "Congressmen tend to approach issues that fall into black or white categories. The approach of the State Department is to look at all sides, study all options, try for compromise." But it is a different sort of compromise as well: "On the Hill, you call a member and have a five-minute conversation and say, 'O.K. Jack, it's a deal.'" At State, however, the system is slower, more impersonal, ridden by paperwork and complex internal negotiations. Cited in *The New York Times,* July 1, 1983. Examples of the expanding congressional role in foreign policy include the 1973 War Powers Act, the 1974 Turkish arms embargo, the 1974 Nelson-Bingham amendment allowing Congress to veto arms sales, the 1974 Hughes-Ryan amendment requiring congressional notification on covert operations, the 1975 Clark amendment forbidding U.S. involvement

in military operations in Angola without congressional permission, and the 1978 Nuclear Non-Proliferation Act.

22. On the issue's history, see Walter Laqueur and Barry Rubin, *The Human Rights Reader* (New York, 1979); Steven Cohen, "Conditioning U.S. Security Assistance on Human Rights Practices," *The American Journal of International Law* (April 1982); Rubin and Spiro, op. cit.
23. Interviews. On the Human Rights bureau's organization, see *Department of State Newsletter* (March 1978), pp. 24–25. During the Nixon administration a young diplomat asked Roger Morris, "Did you ever know any official whose career has been advanced because he spoke out for human rights?" Roger Morris, "Clientism in the Foreign Service," *FSJ* (February 1974), p. 24; Martin Anderson, "The Cost of Quiet Diplomacy," *The New Republic,* March 19, 1984.
24. Provisions of the law are described in *Department of State Newsletter* (October 1980), p. 3.
25. *The Washington Post,* February 26, 1980.
26. State Department notice, June 21, 1977; Department of State telegram, October 1978; Toon, op. cit.
27. *Department of State Newsletter* (June 1977), p. 5; *The Washington Post,* June 22 and July 7, 1979.
28. *The Washington Post,* December 28, 1979.
29. Vance, op. cit., pp. 13–14.
30. On African issues, see Vance, op. cit., pp. 71–72, 84–85, 256–313; Brzezinski, op. cit., pp. 178–187.
31. Brzezinski, op. cit., pp. 349, 203. On U.S.-Soviet relations, see Brzezinski, op. cit., pp. 146–190; Vance, op. cit., pp. 99–119, 133–139, and 349–367; Jimmy Carter, *Keeping Faith* (New York, 1982), pp. 212–265. On China, see Brzezinski, op. cit., pp. 196–233; Vance, op. cit., pp. 75–83, 109–119; Carter, op. cit., 186–211.
32. Firsthand accounts of Camp David include Brzezinski, op. cit., pp. 83–122, 234–288; Vance, op. cit., pp. 159–255; Carter, op. cit., pp. 267–429. Ezer Weizman, *The Battle for Peace* (New York, 1981); Moshe Dayan, *Breakthrough: A Personal Account of the Egypt-Israel Peace Negotiations* (New York, 1981).
33. For a history of U.S.-Iran relations and the revolution and hostage crises, see Rubin, *Paved . . . ,* op. cit. Firsthand accounts of the revolution are in Carter, op. cit., pp. 431ff.; Hamilton Jordan, *Crisis: The Last Year of the Carter Presidency* (New York, 1982); Brzezinski, op. cit., pp. 354–398, 470–509; Vance, op. cit., pp. 314–333, 368–383; John Stempel, *Inside the Iranian Revolution* (Bloomington, Ind., 1981); and William Sullivan, *Mission to Iran* (New York, 1981). The experiences of hostages are recounted in Richard Queen and Patricia Haas, *Inside and Out* (New York, 1981); Barbara and Barry Rosen with George Feifer, *The Destined Hour* (Garden City, N.Y., 1982).
34. Herz adds, "Only when one has experienced a change of government when the previous opposition is accused of dishonesty, incompetence and worse, does it dawn on the younger diplomat that for people who are out of power it is much easier to be possessed of the purest motives, to be selfless in their devotion to the public weal, incorruptible, and devoted to policies to promise instant solution to the country's problems." Herz says the U.S. position should be that it will talk to the opposition without commitment and that if those groups come to power, Washington will try to do business with them as well. Herz, *Contacts . . . ,* op. cit., pp. v–vi.

35. Nikki Keddie and Eric Hoogland, *The Iranian Revolution and the Islamic Republic* (Washington, 1982), pp. 159–161. Richard Helms, former CIA director and Sullivan's predecessor as ambassador to Iran, writes, "The Embassy had a good general idea of the dissatisfaction and disaffection of various elements of Iranian society. . . . But no man (or woman), Persian or foreigner . . . indicated that these elements were strong enough to destroy the government and end the monarchy." In Herz, *Contact* . . . , p. 23.
36. Students Following the Imam's Line, *Revelations From the Nest of Spies* (Tehran, no date), memorandum from Peter Tarnoff to Zbigniew Brzezinski, January 2, 1979, in Vol. 13b, pp. 32–34. Kissinger, "Bureaucracy and Policymaking," op. cit., pp. 12–13.
37. U.S. House of Representatives, Permanent Select Committee on Intelligence, "Iran: Evaluation of U.S. Intelligence Performance Prior to November 1978" (Washington 1979), pp. 2ff; U.S. policymaking during the crisis is discussed in detail in Rubin, *Paved* . . . , op. cit., pp. 190–336.
38. U.S. embassy dispatch of February 1, 1978, "The Iranian Opposition," in Vol. 12a, pp. 31–38. Compare the attitudes of junior officers in "Consulate Principal Officers Conference," June 5, 1978, pp. 125–128, with the official embassy position in "Iran in 1977–78: The Internal Scene," June 1, 1978, pp. 114–124. Volumes 13 through 16 also deal with U.S. political reporting in 1978 and 1979. These documents, captured and published by the Iranian militants who seized the U.S. embassy, provide many details on U.S. policy toward Iran during this period.
39. Interviews. Precht became an FSO in 1962 and served as a political officer and then as politico-military affairs counselor in Tehran from 1972 to 1976. He was deputy director of the Office of Security Assistance and Sales before taking charge of Iranian affairs. *Department of State Newsletter* (August–September 1980), p. 13. In 1981, Precht was denied the post of ambassador to Mauritania because of conservative Senator Jesse Helms's opposition. Precht became deputy chief of mission in Cairo, a better, though lower-ranking job, not requiring Senate confirmation.
40. *Revelations* . . . , op. cit., Sullivan to Vance, October 19, 1978, Vol. 12b, pp. 140, 143.
41. Ibid., Precht to Sullivan, December 19, 1978, Vol. 13, pp. 16–18. Ball himself records that Brzezinski "admonished me, immediately on my arrival, that I should not talk with [Precht] because he 'leaked'—an instruction I, of course, immediately disregarded." Ball, *The Past* . . . , op. cit., p. 458. See also pp. 461–462.
42. Ibid., Precht to Saunders, December 19, 1978, Vol. 13b, op. cit., pp. 19–24.
43. Bernard Gwertzman, "The Hostage Crisis—Thirty Years Ago," *The New York Times Magazine,* May 4, 1980; *The New York Times,* February 15 and April 7, 1965. During the January 1968 Tet offensive in the Vietnam War, enemy guerrillas briefly occupied the U.S. embassy compound before they were wiped out. Tne embattled defenders had great difficulty obtaining military support from units a few miles away because of communications problems while, ironically, they could send instant situation reports to Washington. *The Wall Street Journal,* November 4, 1981.
44. On the Colombia affair, see Diego and Nancy Ascencio with Ron Tobias, *Our Man is Inside: Outmaneuvering the Terrorists* (New York, 1983).
45. Herz, *Diplomats and Terrorists,* op. cit., p. 7.
46. This description is taken from firsthand accounts in the *Department of State Newsletter* (March 1979) pp. 16–20. On the background of Dubs's murder, see Henry Bradsher, *Afghanistan and the Soviet Union* (Durham, N.C., 1983), pp. 98–100.

47. Read was later rewarded with a Reagan administration appointment as ambassador to Morocco.
48. *The Wall Street Journal,* January 3, 1980.
49. Jordan, op. cit., pp. 44–45.
50. For descriptions of these operations, see the above-cited accounts and *The New York Times,* December 30, 1979, November 14, 1980; *The Washington Post,* November 20, 1979.
51. James Taylor, "Afghanistan: Eyewitness Story of the Soviet Invasion," *Department of State Newsletter* (March 1980), p. 4.
52. *The Washington Post,* April 9, 1980. Most of the 53 hostages held up reasonably well under the isolation and strain. Richard Queen, released in the summer of 1980 because of a serious illness, noted how departmental training had helped prepare him. Shortly before taking his first assignment in Tehran, as a third secretary, he attended a one-day seminar on terrorism. Among the rules suggested for those captured: Attempt to establish personal contact with your captors so that they come to view you as an individual human being, not just as a nameless and faceless "enemy" agent. Develop and maintain a daily regimen, especially an exercise program. Avoid talking about politics and other subjects that can only needlessly antagonize your captors. To prevent problems in the first place, American personnel were advised to avoid routines, remain alert (and report suspicious individuals to the security officer), know the country and its language, maintain a low profile, keep a list of emergency phone numbers available, and avoid crowds or civil disturbances. See Queen's account in *Department of State Newsletter* (March 1981), pp. 15–16.
53. Cyrus Vance, oral history, op. cit., p. 18.
54. The following draws mostly on interviews; see also Vance, oral history, op. cit., pp. 409–410.
55. *The Washington Post,* April 29, 1980.
56. It is a remarkable statement on the extent of competition within the administration that Muskie added, "I took this job not to be second in foreign policy, but to be first." *The Washington Post,* May 4, 1980.
57. Interviews. In an interesting parallel to the controversy over admitting the Shah to the United States, when Nicaragua's dictator, Anastasio Somoza, suffered a heart attack in 1977, Todman wanted to send an Air Force plane to bring him to the United States. Todman asked how it was possible not to give every bit of help to a man who had been a loyal American ally. After Somoza lost power, he took refuge in Florida but was later assassinated in Paraguay.
58. Most of this section is based on interviews with participants.
59. The Carter, Jordan, Vance, and Brzezinski memoirs hardly mention the Nicaragua issue.
60. Interviews; *The Washington Post,* July 19, 1979.
61. *The Washington Post,* March 6, 1980, and May 13, 1980.
62. Brzezinski, op. cit., p. 523.
63. Ibid., p. 44.
64. *The Washington Post,* May 4, 1980.

9. *On-the-Job Training, 1981*

1. Kissinger was once quoted as saying that Allen had almost a third-rate mind. According to a Kissinger staff anecdote, Allen asked him, "I hear you think I have a

third-rate mind?" Kissinger supposedly replied, "Actually, Richard, I don't think about you all that much." Incidentally, William Casey, Reagan's CIA director, was ousted from his position as undersecretary of state for economic affairs by Kissinger. Ironically, Nixon reportedly argued that Haig was a man who could take orders, while another contender for the job, George Shultz, was too independent. *The New York Times,* November 6, 1981. Haig, 56 years old in 1981, served as NATO commander from 1974 to his retirement from the army as a four-star general in 1979. He then became president and chief operating officer of United Technologies Corporation.

2. Richard Allen, "In Foreign Policy, The Change has been Fundamental," *The Washington Post,* January 23, 1983.
3. *The Washington Post,* March 25, 1981; Morton Kondracke, "The Sinister Force Returns," *The New Republic,* November 25, 1981, p. 10.
4. Sources close to Haig said, "Not only had it not been discussed with him, he thought [the announcement] was not going to be issued." *The Washington Post,* March 28, 1981. Alexander Haig, *Caveat* (New York, 1984), pp. 57–61, 74ff, 142–148.
5. *Department of State Newsletter* (February 1982), p. 15.
6. See Crocker's articles, "Making Africa Safe for the Cubans," *Foreign Policy* (Summer 1978), pp. 31–33; (with William Lewis) "Missing Opportunities in Africa," *Foreign Policy* (Summer 1979), pp. 142–161; and "South Africa: Strategy for Change," *Foreign Affairs* (Winter 1980/81), pp. 323–351.
7. *The New York Times,* October 30, 1981.
8. See, for example, *The Washington Post,* November 23, 1980.
9. Ibid., June 16, 1981. Abrams was originally made assistant secretary for international organizations. For his views, see *The New York Times,* October 19, 1982.
10. On the leaks, see *The Washington Post,* June 25, 1981; on South Africa, see *The Washington Post,* March 27, 1981.
11. In the Carter administration, 73 percent of U.S. chiefs of mission were FSOs and only 27 percent were political appointees. John Maclean, "I'd like to Present Our Ambassador, Mr. Klunk," *The Washingtonian* (July 1983); *The New York Times,* May 30, 1981; *The Washington Post,* October 14, 1982. There were also "purges" of personnel involved in the Law of the Sea Conference at a critical moment, see *The Washington Post,* March 8, 1981, and in the Latin America bureau of State, see below. In the first such assignment in department history, Reagan nominated spouses to ambassadorial posts: Carleton Coon to Nepal and Jane Abell Coon to Bangladesh.
12. *Department of State Newsletter* (February 1981), p. 16.
13. *The Wall Street Journal,* October 28, 1980; Dick Kirschten, "His NSC Days May be Numbered But Allen Is Known for Bouncing Back," *National Journal,* November 28, 1981.
14. Haig, op. cit., pp. 130–131. Haig's presidential ambitions—an attitude that helped limit James Byrnes to an equally short tenure as secretary of state—were another problem.
15. *The New York Times,* May 1, 1981. *The Washington Post,* September 13, 1981.
16. See, for example, *Business Week,* October 26, 1981; *The Washington Post,* November 6, 1981; Morton Kondracke, op. cit., p. 12; *The New York Times,* November 4, 1981. Allen denied he was attacking Haig. "Bring on the polygraphs," he told reporters. Haig considered Baker the main source, although ironically the White House chief of staff was probably closer to Haig on policy than were the other aides.

17. *The Los Angeles Times,* November 4, 1981; *The New York Times,* November 6, 1981. See also Richard Allen, "Foreign Policy and National Security," in Richard Holwill, ed., *Agenda '83* (Washington, 1983). *The New York Times,* January 7, 1982.
18. Dick Kirschten, "A Tough Manager," *National Journal,* July 17, 1982, pp. 20–22. Several top officials in interviews specifically referred to Leslie Gelb's article, *The New York Times,* October 19, 1981, as the most accurate contemporary picture of problems with the policy process.
19. *The New York Times,* January 19 and October 18, 1982; *The Washington Post,* March 21, 1982; *Department of State Newsletter* (March 1981), p. 4. In an interview, White House aide Baker said that Clark must "be perceived as having the necessary clout from the President to referee disputes within the national security community." *The New York Times,* January 17, 1982. Meetings of the full NSC, as in previous administrations, remained relatively unimportant, and under Clark the number of meetings declined. As McFarlane described them: "[Clark] explains why the meeting has been called and frames the issues on the agenda. He then calls upon the CIA director for an intelligence presentation of the problem. Then the Cabinet officers comment. . . . After this roundtable discussion, the issues for decision will be framed once again and the options summarized. At this point, the President will either reach decisions or express personal judgments about the information that has been presented. Often, he asks additional questions . . . [The President] then adjourns the meeting." Kirschten, "A Tough Manager," op. cit.
20. Interviews; *The New York Times,* March 16, 1982. According to State Department sources, the career staff's real choice was Rozanne L. Ridgway, U.S. ambassador to Finland, who had served as departmental counselor during the Carter administration.
21. *The Wall Street Journal,* June 29, 1982; Flora Lewis, "Policy Tug-of-War," *The New York Times,* February 25, 1982. Haig, op. cit., pp. 306–313.
22. Morton Kondracke, "Nowhere Man," *The New Republic,* May 16, 1983, p. 16. Murray Weidenbaum, *The Wall Street Journal,* August 29, 1983.
23. James Reston, "Policy and Politics," *The New York Times,* March 16, 1983; *The New York Times,* May 11, 1983; *Department of State Newsletter* (August–September 1982). Even on the economic side, however, Shultz's main interest, he had to accept Reaganaut Richard McCormack as assistant secretary for economic and business affairs.
24. Cited in Kondracke, "Nowhere Man," op. cit., p. 22.
25. Interviews.
26. Richard Kirkland, "Shultz's Full Plate of Foreign Economic Issues," *Fortune,* August 9, 1982, pp. 34–35; *The Washington Post,* December 12, 1981.
27. The story is told in detail in Strobe Talbott, *Deadly Gambits* (New York, 1984). *The New York Times,* December 4, 1981; *The Washington Post,* December 9, 1982, and January 13, 1983. Haig, op. cit., pp. 22–23.
28. For the author's more detailed assessment of U.S. Middle East policy, see Barry Rubin, "The United States and the Middle East," in Colin Legum et al., *Middle East Contemporary Survey,* Vol. V, 1980–81 (London, 1982), and Vol. 6, 1981–82 (London, 1983). See also Les Janka, "The National Security Council and the Making of American Middle East Policy," *Armed Forces Journal* (March 1984), p. 84; Barry Rubin, *The Arab States* . . . , op. cit.; Roy Gutman, Battle Over Lebanon, *FSJ* (June 1984), p. 28.

29. Interviews; *The Washington Post,* September 28, 1982; *The New York Times,* September 9, 1982.
30. Interviews; Karen Elliott House's series in *The Wall Street Journal,* April 14 and 15, 1983, is very useful, but its dependence on King Hussein overstates U.S. diplomatic errors in handling the issue.
31. *The Washington Post,* September 28, 1979.
32. Before Reagan took office, several mid-level officials in different agencies produced a position paper proposing the United States push for negotiations rather than become involved on the side of the Salvadoran government. This was never considered by the administration. The text is published as "Dissent Paper on El Salvador and Central America" (Washington, November 16, 1980) by the U.S. Committee in Solidarity with the People. In contrast to the political appointees' confidence, career FSOs were much more doubtful about Washington's ability to dominate the situation. One aide to Carter-era Deputy Secretary of State Warren Christopher commented, "We're behaving as though we push buttons in Washington and the Salvadorans jump. It's a fantasy, a self-delusion." Quoted in Richard Feinberg, *The Intemperate Zone* (New York, 1983), p. 14. The most important leak by ARA liberals was of the forthcoming Big Pine II military exercise, see below.
33. The Council for Inter-American Security, "A New Inter-American Policy for the Eighties" (Washington, 1980). This is often referred to as the Santa Fe Report. It is interesting to compare it to the report of a study group headed by Ambassador Sol Linowitz, produced just before the Carter administration took office. Fontaine, formerly with the Center for Strategic and International Studies and with the American Enterprise Institute, also served as Reagan's Latin America adviser during the presidential campaign but had relatively little influence in office. Other articles in this vein include Pedro Sanjuan, "Why Don't We Have a Latin America Policy," and Roger Fontaine, Cleto Di Giovanni, Jr., and Alexander Kruger, "Castro's Specter," *The Washington Quarterly* (Autumn 1980).
34. 1980 Republican National Convention Platform, reprinted from *Congressional Record,* July 31, 1980.
35. Jeane Kirkpatrick, "U.S. Security and Latin America," *Commentary* (January 1981), p. 29. See also her article, "Dictatorships & Double Standards," *Commentary* (November 1979), and Nestor Sanchez, "The Communist Threat," *Foreign Policy* (Fall 1983). On the UN staff, see *The New York Times,* March 3, 1981.
36. *The Boston Globe,* April 9, 1981. A Latin American trip by former CIA Deputy Director Vernon Walters on behalf of the Reagan transition team also signaled the change in U.S. policy to local leaders.
37. *The Washington Post,* January 26, 1981; *The Washington Post,* February 2, 1981; *The Washington Post,* February 27, 1981; *The Washington Post,* March 8, 1981; *The Washington Post,* March 12, 1981; Carla Anne Robbins, "A State Department Purge," *The New York Times,* November 3, 1981; *The Boston Globe,* November 23, 1981; *The New York Times,* February 2, 1981. White's quick firing was partly due to his outspoken dissent on Salvadoran policy. He refused a Pentagon-inspired request from the chief of the U.S. military aid group to sign a cable requesting 75 U.S. military advisers and publicly criticized the Salvadoran authorities' half-hearted investigation of the killing of three American nuns by government security forces. White also warned that the violence of the far right was driving moderates into alliance with the left. "The improvement of the socio-economic conditions in this country is the only long-term solution to the grave problems it faced," White said

shortly before his removal. He believed that the left could not win the war but the government could lose it by failing to make political reforms. "What Latin America needs desperately," explained White, "is a non-Communist model for revolution. . . ."

38. The mistrust between the career Latin Americanists and the political choices is demonstrated by two exaggerated, bitter anecdotes. An FSO claimed, "It isn't embarrassing that the secretary of state doesn't know anything about Central America, and it is only modestly embarrassing that the assistant secretary doesn't know very much, but it is very bad when the deputy assistant secretaries and even office directors know so little." The story told among Reagan appointees is that a career officer commented at one meeting, "Let's face it, we have our bastards in El Salvador and the Cubans have theirs and there isn't much difference between the two." Interviews. On the U.S. ambassadors in Central America, see Christopher Dickey, "The Proconsuls," *Rolling Stone,* August 18, 1983.
39. U.S. Congress Permanent Select Committee on Intelligence; "U.S. Intelligence Performance on Central America: Achievements and Selected Instances of Concern," September 22, 1982.
40. For an overview of U.S. policy toward El Salvador and Nicaragua, see Raymond Bonner, *Weakness and Deceit* (New York, 1984), and Robert Leiken, ed., *Central America: Anatomy of Conflict* (New York, 1984).
41. Stephanie Harrington, "Salvadoran Runaround," *The New Republic,* December 12, 1981; *The Washington Post,* April 2, 1981; *The Los Angeles Times,* March 8, 1981. U.S. Department of State, "Communist Interference in El Salvador: Documents Demonstrating Communist Support of the Salvadoran Insurgency" (Washington, February 23, 1981). Haig, op. cit., pp. 117–140.
42. Interviews; *The New York Times,* March 13, 1981; *The New York Times,* March 20, 1981; *The Washington Post,* March 1, 1981; *The Washington Post,* March 14, 1981.
43. Interviews; *The New York Times,* April 30, 1981; *The New York Times,* July 17, 1981; *The New York Times,* November 18, 1981; *The New York Times,* December 6, 1981; *The Washington Post,* November 15, 1981; *The Boston Globe,* September 26, 1981.
44. Interviews; *The New York Times,* November 5, 1981; *The New York Times,* December 3, 1981; *The New York Times,* December 5, 1981; *The New York Times,* March 4, 1982.
45. Interviews; *The New York Times,* November 5, 1981; *The New York Times,* November 23, 1981; *The Washington Post,* November 22, 1981; *The Washington Post,* November 22, 1981; *The Washington Post,* December 10, 1981; *The Boston Globe,* November 22, 1981. Enders's testimony to Senate Foreign Relations Committee, December 14, 1981, U.S. Department of State, Current Policy #352. An April 1982 NSC working paper, "U.S. Policy in Central America and Cuba Through Fiscal Year 1984, Summary Paper," was apparently purposely leaked to the press to stress the administration's "moderate" goals: creating "stable, democratic states" while "not allowing the proliferation of Cuba-model states which would provide platforms for subversion, compromise vital sea lanes and pose a direct military threat at or near our borders." The paper also portrayed U.S. efforts as relatively successful in the region, but stymied by U.S. public and congressional opinion that "jeopardizes our ability to stay the course." For the published text, see *The New York Times,* April 7, 1983.
46. Shultz's first State Department briefing on Central America, June 26, 1982, maintained that "the trend of events in Central America is now running in our favor." It was optimistic about the Salvadoran military, the effect of elections there, the suc-

cess of arms interdiction, and the degree of cooperation from other Latin American states. This assessment ran contrary to the more worried view that would emerge elsewhere in the administration.

47. Interviews; *The New York Times,* December 4, 1982. *The New York Times,* April 5, 1982; *The New York Times,* November 9, 1982.
48. Interviews; *The Washington Post,* March 6, 1983; see also her speech, U.S. Mission to the UN #28 (83), May 9, 1983.
49. *The Washington Post,* Feburary 10, 1983; *The New York Times,* February 11, 1983; *The New York Times,* February 11, 1983. U.S. embassy personnel in Central America had already been complaining that their reports were being neglected. These events demonstrated an even further downgrading of such analyses. The embassy in Nicaragua, for example, was far more doubtful of the importance of any military buildup there. Interviews; *The Washington Post,* March 8, 1983; *The Boston Globe,* April 20, 1983. Interviews; Morton Kondracke, ''Interoffice Interference,'' *The New Republic,* May 23, 1983, and ''Enders' End,'' *The New Republic,* June 27, 1983; *The New York Times,* March 23, 1983.
50. *The New York Times,* March 10, 1983; *The Washington Post,* June 28, 1983.
51. Stephen Rosenfeld in *The Washington Post,* June 3, 1983; see also *The Washington Post,* June 12, 1983; for Shultz's statement, see State Department, Public Affairs #199, June 2, 1983.
52. Motley's background is in *The Washington Post,* June 11, 1983. In Spain, Enders succeeded Ambassador Terence Todman, Jimmy Carter's first assistant secretary for Latin America, who had been sent there after criticizing that administration's human rights policy.
53. *The New York Times,* April 22 and May 23, 1983; the text of the speech is in U.S. State Department Current Policy #482, April 27, 1983. On Guatemala, see *The New York Times,* April 7, 1983.
54. Interviews.
55. Jeane Kirkpatrick in *The Washington Post,* June 20, 1983.
56. *The Washington Post,* August 7, 1983.
57. *The New York Times,* September 12, 1983.
58. For the best analysis of U.S. policymaking on the Falklands War and the Grenada invasion, see *The Economist,* March 3 and March 10, 1984.
59. The most interesting articles on Clark's shift and McFarlane's appointment include: Charles Mohr in *The New York Times,* October 18, 1983; Lou Cannon, *The Washington Post,* October 22, 1983; and Don Oberdorfer, *The Washington Post,* October 23, 1983.
60. Zbigniew Brzezinski, ''Who Makes Foreign Policy,'' *The New York Times Magazine,* September 18, 1983.
61. Kennedy-Johnson: Dean Rusk, McGeorge Bundy, and Walt Rostow. Nixon-Ford: William Rogers, Henry Kissinger in both posts, Brent Scowcroft. Carter: Zbigniew Brzezinski, Cyrus Vance, Edmund Muskie. Reagan: Richard Allen, Alexander Haig, George Shultz, William Clark, Robert McFarlane.
62. Cited in *The Washington Post,* June 24, 1984.
63. Brzezinski, ''Who Makes Foreign Policy,'' op. cit.

10. *State Department People*

1. Charles Bohlen, oral history interview, Lyndon Johnson Library, pp. 7–8.
2. Post and Hallie Ermine Wheeler, *Dome of Many Coloured Glass* (New York, 1955), p. 529.

3. John Emmerson, op. cit., p. 99.
4. John Campbell, "An Interview with George F. Kennan," *FSJ* (August 1970), p. 19.
5. Ibid., pp. 17–18. Cited in Robert Nisbet, "Project Camelot: An Autopsy," *The Public Interest* (Fall 1966), p. 45.
6. U.S. Department of State, "Rules Regarding Freedom of Expression for Foreign Service Officers," April 6, 1983.
7. Henry Wriston, "The Secretary and the Management of the Department," in Price, op. cit., p. 76. Andrew Scott, "The Department of State: Formal Organization and Informal Culture," *FSJ* (August 1969), pp. 14–15; Scott, "The Department of State: Formal Organization and Informal Culture," *International Studies Quarterly* (March 1969.)
8. Henry Villard, op. cit., pp. 2–3; John Paton Davies, op. cit., p. 168.
9. John Campbell, "An Interview . . . ," op. cit. p. 19.
10. State Department, *A Short History,* op. cit., p. 37. Arthur Schlesinger, op. cit., pp. 411–412. Roger Morris, *Uncertain Greatness,* op. cit., p. 25.
11. Campbell, *Fudge Factory,* pp. 139–140.
12. Christopher Argyris, "Do You Recognize Yourself?" *FSJ* (January 1967), pp. 22–25. Christopher Argyris, "Some Causes of Organizational Ineffectiveness Within the Department of State" (Washington, 1967).
13. Sandy Vogelgesang, "Feminism in Foggy Bottom: 'Man's World, Woman's Place?' " *FSJ* (August 1972); Women's Action Organization, *Action News* (Fall 1980), pp. 4–5; Susanna Wingfield, "Old Tabus Against Women Linger On," *FSJ* (February 1969). In discussing Latin American policy, one official recently characterized Hispanic-American representation as "grossly inadequate." In 1977, there were 36 Hispanic officers in the Foreign Service and another 34 holding reserve appointments. *DOSB,* October 31, 1977; for an account of recruitment procedures and the Foreign Service training program, see B. Drummond Ayres Jr., "A New Breed of Diplomat," *The New York Times Magazine,* September 11, 1983.
14. U.S. Commission on Civil Rights, "Equal Opportunity in Presidential Appintments," June 1983; statistics on female FSOs are cited in Women's Action Organization statement to House of Representatives Foreign Affairs Committee, June 16, 1982.
15. Gelb and Betts, op. cit., p. 238.
16. Henry Owen, oral history interview, Lyndon Johnson Library, p. 9.
17. Charles Taft, former director of wartime economic affairs, *The New York Times,* February 10, 1946.
18. For a good discussion of these contemporary problems, see William Bacchus, "Staffing State: Three Dilemmas," *FSJ* (December 1981).
19. *Department of State Newsletter* (May 1979), p. 2.
20. John Campbell, "An Interview . . . ," op. cit., p. 22. "Excess people in the Government—unnecessary personnel—cause a kind of damage that multiplies throughout the system," said Kennan. "You might even put it mathematically: the damage caused by unnecessary people is equal to the square of their number."
21. Commission on the Organization of the Government for the Conduct of Foreign Policy, *Report,* Volume 2 (Washington, 1975), pp. 141–169. John Krizay, "Reporting Glut: Clogging the Department's Arteries," *FSJ* (February 1977).
22. Ibid.
23. For an early discussion of this problem, see Thomas Donovan, "Political Reporting Trends," *FSJ* (November 1963), pp. 42–44.

24. William Cochran, "A Diplomat's Moments of Truth" *FSJ* (September 1953), pp. 23, 62. See also William Landerking, "Dissent, Disloyalty, and Foreign Service Finkism," *FSJ* (May 1974).
25. Cecil Griffith, "The Secret Sin of the Foreign Service," *FSJ* (September 1967).
26. Interview.
27. Morris, "Clientism . . . ," op. cit., p. 24.
28. Ibid. For a different view of the Biafra situation, see Martin Peretz, "Unreliable Sources," *The New Republic,* September 12, 1983.
29. George Kennan, "History and Diplomacy . . . ," op. cit., pp. 170–171.

11. *The Policymakers*

1. Charles Bohlen, oral history interview, Lyndon Johnson Library, p. 7.
2. Walter Laqueur, "What We Know About the Soviet Union," *Commentary,* (February 1983), p. 14; Sir William Hayter cited in Robert McClintock, "Thoughts on an American Diplomatic Style," *FSJ* (February 1962). But British diplomats face criticisms like their U.S. counterparts. Former Foreign Minister Roy Hattersley said, "The Foreign Office finds it difficult to believe that it *can* be wrong." "What the Foreign Office does best is inertia," says another writer. A third concludes, "The Ministry of Agriculture looks after farmers—the Foreign Office looks after foreigners." Edward Pearce, "The Department of Lethargy," and Patrick Cosgrave, "Our Man in Grenada," in *Encounter* (July–August 1984), pp. 40–45.
3. Bess Demaree, "Why Americans Hate the State Department," *Saturday Evening Post,* August 19, 1950, p. 23.
4. George Ball, *Diplomacy for a Crowded World* (New York, 1976), p. 194.
5. Rockman, op. cit., p. 919. See also I. M. Destler, "A Job that Doesn't Work," *Foreign Policy* (Spring 1980).
6. Charles Maechling, Jr., "Foreign Policy-Makers: The Weakest Link?" *FSJ* (June 1976), p. 18. See also David Raynolds, "Who Makes Foreign Policy, and How?" *FSJ* (August 1967).
7. Maechling, op. cit.
8. Raymond Hare, oral history interview, John F. Kennedy Library, p. 3; George Ball, op. cit., p. 194.
9. John Campbell, *The Foreign Affairs Fudge Factory* (New York, 1971) pp. 6, 53–54.
10. Kissinger, Volume 1, op. cit., pp. 28–30.
11. Rockman, op. cit., p. 915.
12. I. M. Destler, "State and Presidential Leadership," *FSJ* (September 1971), p. 31. Cited in Campbell, . . . *Fudge Factory,* p. 8.
13. Henry Jackson, "Organizing for National Security," *FSJ* (January 1962), pp. 21–22. Other recommendations included: clearer understanding of national interests, making it easier for private citizens to serve in government, the proper use of the NSC, a stronger Bureau of the Budget, and better congressional organization to reduce fragmentation in committee jurisdictions and hearings. The record on most of these points is quite discouraging.
14. *The Washington Post,* August 23, 1970; Dean Acheson, "The President and the Secretary of State," in Don Price, ed., *The Secretary of State* (Englewood Cliffs, N.J. 1981), p. 43. See also Smith Simpson, "A Lesson in Diplomacy for Henry Kissinger," *The Washington Post,* January 2, 1982.

15. Charles Bohlen, oral history interview, John F. Kennedy Library, p. 32. Another veteran FSO, Raymond Hare, gave a succinct interpretation of this viewpoint, "You serve a president, an administration, whether it be Republican or Democrat. I think you condition your mind that way. And . . . you don't focus much on the American political scene, perhaps not as much as you should." Hare added, "It takes a little time to be appreciated but part way through an administration, you usually find that the career people get into a harmonious adjustment with the people who come in from the outside." Raymond Hare, oral history interview, John F. Kennedy Library, p. 3.
16. Interview.
17. *The Washington Post,* June 27, 1976.
18. Letter to author from Ambassador William Sullivan, August 9, 1983.

Bibliography

Books

Abell, Tyler. *Drew Pearson's Diaries 1949–1959*. (New York, 1974).

Acheson, Dean. *Present at the Creation. My Years in the State Department*. (New York, 1969).

———. *Sketches from Life of Men I Have Known*. (Westport, Conn., 1961).

Adams, E. *Great Britain and the Civil War*. (London, 1925).

Adams, Sherman. *Firsthand Report*. (New York, 1961).

Allison, John. *Ambassador from the Prairie*. (Boston, 1973).

Ambrose, Stephen, with Richard Immerman. *Ike's Spies: Eisenhower and the Espionage Establishment*. (Garden City, N.Y., 1981).

Ascencio, Diego and Nancy Ascencio with Ron Tobias. *Our Man is Inside: Outmaneuvering the Terrorists*. (New York, 1983).

Attwood, William. *The Reds and the Blacks*. (New York, 1967).

Ball, George. *Diplomacy for a Crowded World*. (New York, 1976).

———. *The Past has a Different Pattern*. (New York, 1981).

Bamford, James. *The Puzzle Palace*. (New York, 1982).

Baram, Philip. *The State Department in the Middle East, 1918–1945*. (Philadelphia, 1978).

Beale, Howard. *Theodore Roosevelt and the Rise of America to World Power*. (Baltimore, 1956).

Beichman, Arnold. *The 'Other' State Department: The United States Mission to the United Nations*. (New York, 1967).

Bemis, Samuel. *The Diplomacy of the American Revolution*. (Bloomington, Ind., 1957).

———, ed. *The American Secretaries of State and their Diplomacy*. (New York, 1933).

Bendiner, Robert. *The Riddle of the State Department*. (New York, 1942).

Berding, Andrew. *Dulles on Diplomacy*. (Princeton, 1965).

Berle, Beatrice, and Travis Jacobs, eds. *Navigating the Rapids: The Diaries of Adolf Berle*. (New York, 1973).
Bishop, James. *FDR's Last Year*. (New York, 1974).
Blum, John. *The Price of Vision: The Diary of Henry Wallace, 1942–1946*. (Boston, 1972).
Blumenfeld, Ralph, et al. *Henry Kissinger: The Private and Public Story*. (New York, 1974).
Bohlen, Charles. *The Transformation of American Foreign Policy*. (New York, 1969).
———. *Witness to History 1929–1969*. (New York, 1972).
Bonner, Raymond. *Weakness and Deceit. U.S. Policy and El Salvador*. (New York, 1984).
Bonsal, Philip. *Cuba, Castro and the United States*. (Pittsburgh, 1971).
Bowles, Chester. *Promises to Keep*. (New York, 1971).
Boyd, Julian P., et al. *Papers of Thomas Jefferson*. (Princeton, 1961).
Braden, Spruille. *Diplomats and Demagogues*. (New Rochelle, N.Y., 1971).
Bradsher, Henry. *Afghanistan and the Soviet Union*. (Durham, N.C., 1983).
Branch, Taylor, and Eugene Propper. *Labyrinth*. (New York, 1982).
Briggs, Ellis. *Farewell to Foggy Bottom*. (New York, 1964).
Brown, Ralph S., Jr. *Loyalty and Security*. (New Haven, 1958).
Brown, Seyom. *The Faces of Power*. (New York, 1968).
Brzezinski, Zbigniew. *Power and Principle*. (New York, 1983).
Burns, James MacGregor. *Roosevelt: Soldier of Freedom*. (New York, 1970).
Byrnes, James. *All in One Lifetime*. (New York, 1958).
———. *Speaking Frankly*. (New York, 1947).
Cabot, John Moors. *First Line of Defense*. (Washington, D.C., 1979).
Campbell, John. *The Foreign Affairs Fudge Factory*. (New York, 1971).
Campbell, Thomas, and George Herring. *The Diaries of Edward Stettinius, Jr. 1943–46*. (New York, 1975).
Carter, Jimmy. *Keeping Faith*. (New York, 1982).
Caute, David, *The Great Fear: The Anti-Communist Purge under Truman and Eisenhower*. (New York, 1978).
Childs, J. Rives. *American Foreign Service*. (New York, 1948).
Clubb, O. Edmund. *The Witness and I*. (New York, 1974).
Cohen, Warren. *Dean Rusk*. (Totowa, N.J., 1980).
Commission on the Organization of the Government for the Conduct of Foreign Policy. (June, 1975). 7 volumes.
Compton, James. *The Swastika and the Eagle*. (Boston, 1967).
Congressional Research Service, *The Ambassador in U.S. Foreign Policy*. (Washington, D.C., 1981).
Cooper, Chester. *The Lost Crusade*. (New York, 1972).
Craig, Gordon, and Felix Gilbert. *The Diplomats 1919–1939*. (Princeton, 1933).
Cronin, Thomas E. *The State of the Presidency*. (Boston, 1975).
Dallek, Robert. *Franklin D. Roosevelt and American Foreign Policy, 1932–45*. (New York, 1979).
———. *Democrat and Diplomat. The Life of William E. Dodd*. (New York, 1968).
Daniels, Jonathan. *The Man of Independence*. (New York, 1950).
Davies, John Paton. *Foreign and Other Affairs*. (New York, 1964).
Dayan, Moshe. *Breakthrough: A Personal Account of the Egypt-Israel Peace Negotiations*. (New York, 1981).

de Conde, Alexander. *The American Secretary of State: An Interpretation*. (New York, 1962).
de Novo, John. *American Interests and Policies in the Middle East 1900–1939*. (Minneapolis, 1978).
de Santis, Hugh. *The Diplomacy of Silence: The American Foreign Service, the Soviet Union and the Cold War, 1933–1947*. (Chicago, 1980).
Divine, Robert. *Eisenhower and the Cold War*. (New York, 1981).
———, ed. *Exploring the Johnson Years*. (Austin, Texas, 1981).
Dodd, William Jr., and Martha Dodd. *Ambassador Dodd's Diary, 1933–1938*. (New York, 1941).
Donovan, Robert. *Conflict and Crisis*. (New York, 1977).
———. *Eisenhower: The Inside Story*. (New York, 1956).
Duberman, Martin. *Charles Francis Adams*. (Boston, 1961).
Dulles, Eleanor. *A Memoir*. (Englewood Cliffs, N.J., 1980).
Einstein, Lewis. *A Diplomat Looks Back*. (New Haven, 1968).
Eisenhower, Dwight D. *Mandate for Change*. (New York, 1963).
———. *Waging Peace*. (Garden City, N.Y., 1965).
Emmerson, John. *The Japanese Thread: A Life in the US Foreign Service*. (New York, 1978).
Esherick, Joseph. *Lost Chance in China: The World War Two Dispatches of John S. Service*. (New York, 1974).
Etzold, Thomas H., and John L. Gaddis. *Containment: Documents on American Policy and Strategy, 1945–1950*. (New York, 1978).
Ewald, William Bragg, Jr. *Eisenhower the President*. (Englewood Cliffs, N.J., 1981).
Feinberg, Richard, ed. *Central America: International Dimensions of the Crisis*. (New York, 1982).
———. *The Intemperate Zone*. (New York, 1983).
Ferrell, Robert. *The Eisenhower Diaries*. (New York, 1981).
Fitzpatrick, John, ed. *The Writings of George Washington*. (Washington, D.C., 1944).
Gaddis, John L. *Russia, the Soviet Union and the United States*. (New York, 1980).
———. *The United States and the Origins of the Cold War: 1941–1947*. (New York, 1972).
Galbraith, John Kenneth. *Ambassador's Journal*. (Boston, 1969).
Gelb, Leslie, and Richard Betts. *The Irony of Vietnam*. (Washington, D.C., 1979).
Germond, Jack, and Jules Witcover. *Blue Smoke and Mirrors*. (New York, 1981).
Gilbert, Felix. *To the Farewell Address: Ideas of Early American Foreign Policy*. (Princeton, 1961).
Glad, Betty. *Charles Evans Hughes and the Illusion of Innocence*. (Urbana, Ill., 1966).
Golan, Matti. *Shimon Peres, A Biography*. (New York, 1982).
———. *The Secret Conversations of Henry Kissinger*. (New York, 1976).
Graebner, Norman, ed. *An Uncertain Tradition: American Secretaries of State in the Twentieth Century*. (New York, 1961).
Greenstein, Fred. *The Hidden-Hand Presidency*. (New York, 1982).
Griffith, Robert. *The Politics of Fear*. (Lexington, Ky., 1970).
Guhin, Michael. *John Foster Dulles: A Statesman and His Time*. (New York, 1972).
Haig, Alexander. *Caveat*. (New York, 1984).
Halberstam, David. *The Best and the Brightest*. (New York, 1972).
Haldeman, Robert, with Joseph Dimona. *The Ends of Power*. (New York, 1978).

Hamby, Alonzo. *Beyond the New Deal: Harry S. Truman and American Liberalism.* (New York, 1973).

Harr, John. *The Professional Diplomat.* (Princeton, 1969).

Harriman, W. Averell and Elie Abel. *Special Envoy to Churchill and Stalin, 1941–1946.* (New York, 1975).

Hartmann, Robert. *Palace Politics.* (New York, 1980).

Haviland, H. Field, Jr., et al. *The Formulation and Administration of U.S. Foreign Policy.* (Washington, D.C., 1960).

Heclo, Hugh, and Lester Salamon, eds. *The Illusion of Presidential Power.* (Boulder, Colo., 1981).

Heinrichs, Waldo, Jr. *American Ambassador: Joseph C. Grew and the Development of the United States Diplomatic Tradition.* (Boston, 1966).

Heller, Francis. *The Truman White House: The Administration of the Presidency 1945–1953.* (Lawrence, Kans., 1960).

Hersh, Seymour. *The Price of Power.* (New York, 1983).

Herz, Martin, ed. *Contacts with the Opposition: A Symposium.* Institute for the Study of Diplomacy. (Washington, D.C., 1979).

———, ed. *Diplomats and Terrorists: What Works, What Doesn't.* (Washington, D.C., 1983).

Hilsman, Roger. *The Politics of Policy Making in Defense and Foreign Affairs.* (New York, 1971).

Holwill, Richard. *Agenda '83.* (Washington, D.C., 1983).

Hughes, Emmet John. *The Ordeal of Power.* (New York, 1963).

Hull, Cordell. *Memoirs.* 2 volumes. (New York, 1948).

Hunt, Gaillard. *The Department of State of the United States.* (New Haven, 1914).

———, ed. *Writings of Madison.* (New York, 1903).

Hunter, Robert. *Presidential Control of Foreign Policy: Management or Mishap?* (New York, 1982).

Hurewitz, J. C. *The Struggle for Palestine.* (New York, 1968).

Ilchman, Warren. *Professional Diplomacy in the U.S., 1779–1939.* (Chicago, 1961).

Isserman, Maurice. *Which Side Were You On?* (Middletown, Conn., 1982).

Jackson, Henry. *The Secretary of State and the Ambassador.* (New York, 1964).

Jessup, Philip. *The Birth of Nations.* (New York, 1974).

Johnson, E., ed. *The Dimensions of Diplomacy.* (Baltimore, 1965).

Johnson, Richard. *The Administration of U.S. Foreign Policy.* (Austin, Texas, 1971).

Johnston, Henry P., ed. *Correspondence and Papers of John Jay.* (New York, 1890).

Jones, Joseph. *The Fifteen Weeks.* (New York, 1955).

Jordan, Hamilton. *Crisis: The Last Year of the Carter Presidency.* (New York, 1982).

Kahn, David. *Hitler's Spies.* (New York, 1978).

Kalb, Madeleine. *The Congo Cables.* (New York, 1982).

Kalb, Bernard, and Marvin Kalb. *Kissinger.* (Boston, 1974).

Kattenburg, Paul. *The Vietnam Trauma in American Foreign Policy.* (New Brunswick, N.J., 1982).

Kaufmann, William. *The McNamara Strategy.* (New York, 1964).

Keddie, Nikki, and Eric Hoogland. *The Iranian Revolution and the Islamic Republic.* (Washington, D.C., 1982).

Kennan, George F. *Memoirs, 1925–1950,* Vol. 1. (Boston, 1967).

———. *Memoirs, 1950–1963,* Vol. 2. (Boston, 1972).

———. *Realities of American Foreign Policy.* (Princeton, 1954).

———. *Russia Leaves the War*. (Princeton, 1956).
Kennedy, Robert. *Thirteen Days*. (New York, 1968).
Kinnard, Douglas. *The Secretary of Defense*. (Lexington, Ky., 1981).
Kissinger, Henry. *White House Years*, Vol. 1. (Boston, 1979).
———. *Years of Upheaval, Vol. 2. (New York, 1982).*
———, *and Bernard Brodie. Bureaucracy, Politics and Strategy*. Security Studies Paper No. 17. (Los Angeles, 1968).
Kitchen, Helen. *U.S. Interests in Africa*. (New York, 1983).
Krock, Arthur. *Memoirs: Sixty Years on the Firing Line*. (New York, 1968).
Lake, Anthony. *The "Tar Baby" Option: American Policy Toward Southern Rhodesia*. (New York, 1976).
Landau, David. *Kissinger: The Uses of Power*. (New York, 1978).
Langer, William, and S. Everett Gleason. *Challenge to Isolation*. (New York, 1952).
Laqueur, Walter, and Barry Rubin, eds. *The Human Rights Reader*. (New York, 1979).
Latham, Earl. *The Communist Controversy in Washington*. (Cambridge, Mass., 1966).
Leacacos, John. *Fires in the In-Basket*. (Cleveland, Ohio, 1968).
Leahy, Fleet Admiral William D. *I Was There*. (New York, 1950).
Legum, Colin, et al. *Middle East Contemporary Survey*, Vol. 5, 1980–81. (London, 1982).
———. *Middle East Contemporary Survey*, Vol. 6, 1981–82. (London, 1983).
Leiken, Robert. *Central America: Anatomy of Conflict*. (New York, 1984).
Lilienthal, David E. *The Journal of David E. Lilienthal, Vol. 2: The Atomic Energy Years, 1945–1950*. (New York, 1964).
Lockhart, R. Bruce. *British Agent*. (New York, 1933).
Loewenheim, Francis, et al. *Roosevelt and Churchill: Their Secret Wartime Correspondence*. (New York, 1975).
Louis, William Roger. *Imperialism at Bay*. (New York, 1978).
Macomber, William. *The Angels' Game*. (New York, 1975).
Maloney, Richard. *JFK and Africa*. (New York, 1984).
Mann, Dean. *The Assistant Secretaries*. (Washington, D.C., 1965).
May, Ernest. *The Ultimate Decision: The President as Commander-in-Chief*. (New York, 1960).
May, Gary. *China Scapegoat: The Diplomatic Ordeal of John Carter Vincent*. (Washington, D.C., 1979).
McLellan, David. *Dean Acheson: The State Department Years*. (New York, 1976).
———, and David Acheson. *Among Friends: Personal Letters of Dean Acheson*. (New York, 1980).
Merlie, Frank, and Theodore Wilson, eds. *Makers of American Diplomacy*. (New York, 1974).
Messmer, Robert. *The End of an Alliance: James F. Byrnes, Roosevelt, Truman and the Origins of the Cold War*. (Chapel Hill, N.C., 1982).
Miller, Merle. *Plain Speaking: An Oral Biography of Harry S. Truman*. (New York, 1974).
Millis, Walter. *The Forrestal Diaries*. (New York, 1957).
Morison, Elting E. *Turmoil and Tradition: A Study of the Life and Times of Henry L. Stimson*. (New York, 1964).
Morris, Richard. *Great Presidential Decisions*. (Greenwich, Conn., 1966).
Morris, Roger. *Uncertain Greatness*. (New York, 1977).

Mosher, Frederick, and John Harr. *Programming Systems and Foreign Affairs Leadership.* (New York, 1970).
Murphy, Robert. *Diplomat Amongst Warriors.* (Garden City, N.Y., 1964).
Newhouse, John. *Cold Dawn.* (New York, 1973).
Offner, Arnold. *American Appeasement: U.S. Foreign Policy and Germany, 1933–1938.* (Cambridge, Mass., 1968).
Owsley, F. *King Cotton Diplomacy.* (Chicago, 1931).
Oye, Ken, et al. *The Eagle Defiant.* (Boston, 1983).
Peterson, M., ed. *The Portable Thomas Jefferson.* (New York, 1975).
Phillips, Cabell. *The Truman Presidency.* (New York, 1966).
Price, Don, ed. *The Secretary of State.* (Englewood Cliffs, N.J., 1960).
Price, Raymond. *With Nixon.* (New York, 1977).
Pruessen, Ronald. *John Foster Dulles: The Road to Power.* (New York, 1982).
Quandt, William. *Decade of Decisions: American Policy Toward the Arab-Israeli Conflict 1967–1976.* (Los Angeles, 1977).
Queen, Richard, and Patricia Haas. *Inside and Out.* (New York, 1981).
Rabin, Yitzhak. *The Rabin Memoirs.* (Boston, 1979).
Rosen, Barbara, and Barry Rosen, with George Feifer. *The Destined Hour.* (Garden City, N.Y., 1982).
Rostow, Walt. *View from the Seventh Floor.* (New York, 1964).
Rubin, Barry. *The Arab States and the Palestine Conflict.* (Syracuse, N.Y., 1981).
———. *Paved with Good Intentions: The American Experience and Iran* (New York, 1980).
———, and Elizabeth Spiro. *Human Rights and U.S. Foreign Policy.* (Boulder, Colo., 1979).
Rusk, Dean. *The Wings of Freedom.* (Boston, 1963).
Safire, William. *Before the Fall.* (Garden City, N.Y., 1975).
Schaller, Michael. *The U.S. Crusade in China, 1938–1945.* (New York, 1979).
Schlesinger, Arthur. *The Vital Center.* (Boston, 1949).
———. *A Thousand Days.* (Boston, 1965).
Schlesinger, Stephen, and Stephen Kinzer. *Bitter Fruit.* (Garden City, N.Y., 1982).
Schulzinger, Robert. *The Making of the Diplomatic Mind.* (Middletown, Conn., 1975).
Shawcross, William. *Sideshow: Kissinger, Nixon and the Destruction of Cambodia.* (New York, 1979).
Sheeehan, Edward. *The Arabs, Israelis, and Kissinger: A Secret History of American Diplomacy in the Middle East.* (New York, 1976).
Sherwood, Robert. *Roosevelt and Hopkins.* (New York, 1948).
Sisson, Edgar. *One Hundred Red Days: A Personal Chronicle of the Bolshevik Revolution.* (New Haven, 1931).
Smith, Earl. *The Fourth Floor.* (New York, 1962).
Spaulding, E. Wilder. *Ambassadors Ordinary and Extraordinary.* (Washington, D.C., 1961).
Stempel, John. *Inside the Iranian Revolution.* (Bloomington, Ind., 1981).
Stern, Laurence. *The Wrong Horse: The Politics of Intervention and the Failure of American Diplomacy.* (New York, 1977).
Stevenson, William. *A Man Called Intrepid.* (New York, 1976).
Stuart, Graham. *The Department of State.* (New York, 1949).
Students Following the Imam's Line. *Revelations from the Nest of Spies.* (Tehran, no date).

Stupak, Ronald. *The Shaping of Foreign Policy*. (Miami, Ohio, 1969).
Sullivan, William. *Mission to Iran*. (New York, 1981).
Sykes, Christopher. *Crossroads to Israel*. (Bloomington, Ind., 1973).
Szulc, Tad. *The Illusion of Peace*. (New York, 1978).
Tanzer, Lester, ed. *The Kennedy Circle*. (Washington, D.C., 1961).
Tatum, Edward. *The United States and Europe: 1815–1823*. (Berkeley, Calif., 1936).
Thayer, Charles. *Diplomat*. (New York, 1959).
Theoharis, Athan. *The Yalta Myths*. (Columbia, Mo., 1970).
Thomas, Charles. *Allies of a Kind: the United States, Britain and the War Against Japan: 1941–1945*. (New York, 1978).
Truman, Harry. *Year of Decision: 1945*. (New York, 1955).
U.S. Department of State. *The Department of State Today* (Washington, D.C., 1981).
———. *Foreign Relations of the U.S.*, (FRUS), annual volumes, (1933, 1934, 1936, 1938). (Washington, D.C., 1949).
———. *A Short History of the U.S. Department of State 1781–1981*. (Washington, D.C., 1981).
———. *Peace and War: United States Foreign Policy 1931–1941*. (Washington, D.C., 1943).
Vance, Cyrus. *Hard Choices: Critical Years in America's Foreign Policy*. (New York, 1983).
Villard, Henry. *Affairs at State*. (New York, 1965).
Welles, Sumner. *Seven Decisions That Shaped History*. (New York, 1951).
Weil, Martin. *A Pretty Good Club*. (New York, 1978).
Weinstein, Allen. *Perjury*. (New York, 1978).
Weintal, Edward, and Charles Bartlett. *Facing the Brink*. (New York, 1967).
Weizman, Ezer. *The Battle for Peace*. (New York, 1981).
West, Rachel. *The Department of State on the Eve of the First World War*. (Athens, Ga., 1978).
Weyl, Nathaniel. *The Battle Against Disloyalty*. (New York, 1951).
Whalen, Richard. *The Founding Father: The Story of Joseph P. Kennedy*. (New York, 1964).
Wheeler, Post, and Hallie Ermine Wheeler. *Dome of Many Coloured Glass*. (New York, 1975).
White, D. *The Federalists: A Study in Administrative History*. (New York, 1948).
Wilson, Hugh. *A Career Diplomat*. (Westport, Conn., 1973).
Woodward, Robert, and Carl Bernstein. *The Final Days*. (New York, 1976).
Wriston, Henry. *Toward a Stronger Foreign Service* (Washington, D.C., 1956).
Zumwalt, Elmo. *On Watch*. (New York, 1976).

Articles

Acheson, Dean. "The Eclipse of the State Department." *Foreign Affairs*. (July 1971).
———. "The President and the Secretary of State," in Don Price, ed. *The Secretary of State*. (Englewood Cliffs, N.J., 1981).
Allen, Richard. "Foreign Policy and National Security: The White House Perspective," in Richard Holwill, ed. *Agenda '83*. (Washington, D.C., 1983).
Anderson, David. "The Ambassador as Administrator." *FSJ*. (September 1961).
Argyris, Christopher. "Do You Recognize Yourself?" *FSJ*. (January 1967).

Arneson, R. Gordon. "Anchor Man of the Department: Alvey Augustus Adee." *FSJ*. (August 1971).

Attwood, William. "The Labyrinth in Foggy Bottom." *Atlantic Monthly*. (February 1967).

Ayres, B. Drummond. "A New Breed of Diplomat." *New York Times Magazine*, September 11, 1983.

Bacchus, William I. "Diplomacy for the 70's: An Afterview and Appraisal." *American Political Science Review* 68, No. 2. (June 1974).

———. Staffing State: Three Dilemmas." *FSJ*. (December 1982).

Bowie, Robert. "The Secretary and the Development and Coordination of Policy," in Don Price, ed. *The Secretary of State*. (Englewood Cliffs, N.J., 1960).

Bowles, Chester. "Embassies and Ambassadors." *FSJ*. (June 1971).

Brenner, Michael, "The Problem of Innovation and the Nixon-Kissinger Foreign Policy." *International Studies Quarterly* 17, No. 3. (September 1973).

Brzezinski, Zbigniew. "Deciding Who Makes Foreign Policy." *New York Times Magazine*, September 18, 1983.

Bundy, McGeorge. "The H-Bomb: The Missed Chance." *New York Review of Books*, May 13, 1982.

Burns, Richard. "James Byrnes," in Norman Graebner, ed. *An Uncertain Tradition: American Secretaries of State in the Twentieth Century*. (New York, 1961).

Butterfield, Fox. "The New Vietnam Scholarship," *The New York Times Magazine*, February 13, 1983.

Cabot, Thomas D. "I Worked for State." *FSJ*. (October 1952).

Campbell, Colin. "In Search of Executive Harmony: Cabinet Government and the US Presidency—the Experience of Carter and Reagan." (Unpublished paper).

Campbell, John. "An Interview with George F. Kennan." *FSJ*. (August 1970).

Catledge, Turner. "Secretary Byrnes: Portrait of a Realist." *The New York Times Magazine*, July 8, 1945.

Christopher, Warren. "Ceasefire Between the Branches: A Compact in Foreign Affairs." *Foreign Affairs*. (Summer 1982).

Clark, G. Edward. "Key to Continuity—The Local." *FSJ*. (January 1955).

Claymore, John. "Vietnamization of the Foreign Service." *FSJ*. (December 1971).

Cochran, William P., Jr. "A Diplomat's Moments of Truth." *FSJ*. (September 1953).

———. "Our Third-Rate Diplomacy: Is it good enough?" *FSJ*. (March 1963).

Cohen, Steven. "Conditioning U.S. Security Assistance on Human Rights Practices." *The American Journal of International Law*. (April 1982).

Corrigan, Robert. "An Appreciation of a Diplomat." *FSJ*. (November 1967).

Council for Inter-American Security. "A New Inter-American Policy for the Eighties." (Washington, D.C., 1980).

Crocker, Chester. "Making Africa Safe for the Cubans." *Foreign Policy*. (Summer 1978).

———. "South Africa: Strategy for Change." *Foreign Affairs*. (Winter 1980/81).

———, and William Lewis. "Missing Opportunities in Africa." *Foreign Policy*. (Summer 1979).

Cutler, Robert. "The Development of the National Security Council." *Foreign Affairs*. (April 1956).

Davis, Nathaniel. "The Angola Decision of 1975: A Personal Memoir." *Foreign Affairs*. (Fall 1975).

de Conde, Alexander. "George Marshall," in Norman Graebner, ed. *An Uncertain Tra-*

dition. American Secretaries of State in the Twentieth Century. (New York, 1961).

Destler, I. M. "A Job that Doesn't Work." *Foreign Policy.* (Spring 1980).

———. "National Security Advice to U.S. Presidents." *Politics* 29, No. 2. (January 1977).

———. "National Security Management: What Presidents Have Wrought." *Political Science Quarterly* 95, No. 4. (Winter 1980–81).

———. "The Nixon System: A further look." *FSJ.* (February, 1974).

———. "NSC II: The Rise of the Assistant 1961–1981," in Heclo and Salamon, eds. *The Illusion of Presidential Power.* (Boulder, Colo., 1981).

———. "State and Presidential Leadership." *FSJ.* (September 1971).

Demaree, Bess. "Why Americans Hate the State Department." *Saturday Evening Post* 223, No. 8, August 19, 1950.

Dickey, Christopher. "The Proconsuls." *Rolling Stone,* August 18, 1983.

Donovan, Thomas. "Political Reporting Trends." *FSJ.* (November 1963).

Dungan, Ralph. "A Year of Substantial Progress." *FSJ.* (April 1963).

Dutton, Frederick. "'Cold War' Between the Hill and Foggy Bottom." *New Times Magazine,* September 15, 1963.

Elson, Robert. "The New Strategy in Foreign Policy." *Fortune.* (December 1947).

Etzold, Thomas. "Does Macy's Tell Gimbel's?" *FSJ.* (August 1977).

Fontaine, Roger, et al. "Castro's Specter." *The Washington Quarterly* 3, No. 4. (Autumn 1980).

Ford, Franklin. "Three Observers in Berlin: Rumboldt, Dodd, and Francois-Poncet," in Gordon Craig and Felix Gilbert, eds. *The Diplomats 1919–1939.* (Princeton, 1933).

Fosdick, Dorothy. "For the Foreign Service—Help Wanted." *The New York Times Magazine,* November 20, 1955.

Frost, Arthur. "Nathaniel Hawthorne: Consul at Liverpool." *FSJ.* (August 1958).

Gelb, Leslie. "Muskie and Brzezinski: The Struggle Over Foreign Policy." *The New York Times Magazine,* July 20, 1980.

Graebner, Norman. "John Quincy Adams," in Frank Merlie and Theodore Wilson, eds. *Makers of American Diplomacy.* (New York, 1974).

Graves, Harold. "State Department, New Model." *The Reporter,* November 8, 1949.

———. "The President and the Secretary of State," in Don Price, ed. *The Secretary of State.* (Englewood Cliffs, N.J., 1960).

Griffith, Cecil. "The Secret Sin of the Foreign Service." *FSJ.* (September 1967).

Gutman, Roy. "Battle Over Lebanon." *FSJ.* (June 1984).

Gwertzman, Bernard. "The Hostage Crisis—Thirty Years Ago." *The New York Times Magazine,* May 4, 1983.

Harrington, Stephanie. "Salvadoran Runaround." *The New Republic,* December 12, 1983.

Herring, George. "The Truman Administration and the Restoration of French Sovereignty in Indochina." *Diplomatic History* 1, No. 2 (Spring 1977).

———. "The War in Vietnam," in Robert Divine, ed. *Exploring the Johnson Years.* (Austin, Texas, 1981).

Hersh, Seymour. "The Price of Power: Kissinger, Nixon, and Chile." *The Atlantic Monthly.* (December 1982).

Herter, Christian A. "Testimony before the 'Jackson Subcommittee.'" *FSJ.* (August 1960).

Herz, Martin T. "Life at the Paris Embassy." *FSJ*. (February, 1952).

Hess, Gary. "The First American Commitment in Indochina." *Diplomatic History* 2, No. 4. (Fall 1978).

Hughes, H. Stuart. "The Second Year of the Cold War." *Commentary*. (August 1969).

Hopkins, Frank. "Policy, Action, and Personnel." *FSJ*. (April 1960).

Immerman, Richard. "Guatemala as Cold War History." *Political Science Quarterly* 95, No. 4. (Winter 80–81).

Immerman, Robert M. "The Formulation and Administration of U.S. Foreign Policy." *FSJ*. (April 1960).

Jackson, Sir Geoffrey. "Premonitions and Forewarnings," in Martin Herz, ed. *Diplomats and Terrorists: What Works, What Doesn't*. (Washington, D.C., 1983).

Jackson, Henry. "Organizing for National Security." *FSJ*. (January 1962).

———. "Organizing for Survival." *Foreign Affairs*. (April 1960).

Janka, Les. "The National Security Council and the Making of American Middle East Policy." *Armed Services Journal*. (March 1984).

Johnson, U. Alexis. "Internal Defense and the Foreign Service." *FSJ*. (July 1962).

Johnson, Robert. "The National Security Council: The Relevance of its Past to its Future." *Orbis*. (Fall 1969).

Jones, Curtis. "The Education of an Arabist." *FSJ*. (December 1982).

Kaufmann, William. "Two Ambassadors," in Gordon Craig and Felix Gilbert, eds. *The Diplomats 1919–1939*. (Princeton, 1933).

Keisling, Phil. "The Tallest Gun in Foggy Bottom." *Washington Monthly*. (November 1982).

Kennan, George F., "The Future of our Professional Diplomacy." *Foreign Affairs*. (July 1955).

———. "History and Diplomacy as Viewed by a Diplomatist." *The Review of Politics* 18, No. 2. (April 1956).

———. "America's Administrative Response." *FSJ*. (October 1963).

———."World Problems and America's Administrative Response." *FSJ*. (September 1963).

———, et al. "Planning in the Department." *FSJ*. (March 1961).

Kenworthy, E. W. "Evolution of Our No. 1 Diplomat." *The New York Times Magazine*, March 18, 1962.

Kessel, John H. "The Structures of the Carter White House." (Unpublished paper).

——— "The Structures of the Reagan White House." (Unpublished paper presented to The American Political Science Association, September 1983).

Kimball, Warren, and Bruce Bartlett. "Roosevelt and Prewar Commitments to Churchill: The Tyler Kent Affair." *Diplomatic History* 5, no. 4. (Fall 1981).

Kirkland, Richard I., Jr. "Shultz's Full Plate of Foreign Economic Issues." *Fortune*, August 9, 1982.

Kirkpatrick, Jeane. "Dictatorships and Double Standards." *Commentary*. (November 1979).

———. "U.S. Security and Latin America." *Commentary*. (January 1981).

Kirschten, Richard. "A Tough Manager." *National Journal*, July 17, 1982.

———. "Beyond the Vance-Brzezinski Clash Lurks an NSC Under Fire." *National Journal*, May 17, 1980.

———. "His NSC Days May be Numbered But Allen is Known for Bouncing Back." *National Journal*, November 28, 1981.

Kissinger, Henry. "America and the World: Principle and Pragmatism." *Time*, December 27, 1976.

———. "Bureaucracy and Policymaking," in Kissinger and Bernard Brodie, *Bureaucracy, Politics and Strategy*, Security Studies Paper #17. (Los Angeles, 1968).

Knight, Charlotte. "What Price Security." *Colliers*, July 9, 1954.

Knight, William. "On Dissent." *FSJ*. (December 1964).

Kondracke, Morton. "Enders' End." *The New Republic*, June 27, 1983.

———. "Interoffice Interference." *The New Republic*, May 23, 1983.

———. "Nowhere Man." *The New Republic*, March 16, 1983.

———. "The Sinister Force Returns." *The New Republic*, November 25, 1981.

Kraft, Joseph. "The Comeback of the State Department." *Harpers*. (November 1961).

———. "Dean Rusk Show." *The New York Times Magazine*, March 24, 1968.

———. "Those Arabists in the State Department." *The New York Times Magazine*, November 7, 1971.

Krizay, John. "Reporting Glut: Clogging the Department's Arteries." *FSJ*. (February 1977).

Krock, Arthur. "Washington Hasn't Enough Time to Think." *The New York Times Magazine*, December 9, 1945.

LaFeber, Walter. "Latin American Policy," in Robert Divine, ed. *Exploring the Johnson Years*. (Austin, Texas, 1981).

Landerking, William. "Dissent, Disloyalty, and Foreign Service Finkism." *FSJ*. (May 1974).

Laqueur, Walter. "What We Know About the Soviet Union." *Commentary*. (February 1983).

Leacacos, John. "Kissinger's Apparat." *Foreign Affairs*. (Winter 1971–72).

Lees, Lorraine. "The American Decision to Assist Tito, 1948–1949." *Diplomatic History* 2, No. 4. (Fall 1978).

Lenderking, William. "Dissent, Disloyalty, and Foreign Service Finkism." *FSJ*. (June 1974).

Luce, Clare Boothe. "The Ambassadorial Issue: Professionals or Amateurs." *Foreign Affairs*. (October 1957).

Maclean, John. "I'd Like to Present Our Ambassador, Mr. Klunk." *The Washingtonian*. (July 1983).

Maddox, William. "The Foreign Service in Transition." *Foreign Affairs*. (January 1947).

Madison, Christopher. "U.S. Testing Ground." *National Journal*, November 28, 1981.

Maechling, Charles, Jr. "Foreign Policy-Makers: The Weakest Link?" *FSJ*. (June 1976).

Martin, James V., Jr. "The Quiet Revolution in the Foreign Service." *FSJ*. (February 1960).

May, Ernest R. "The Development of Political-Military Consultation in the U.S." *Political Science Quarterly* 70, No. 2. (June 1955).

McCamy, James L. "Rebuilding the Foreign Service." *Harper's Magazine*. (November 1959).

———. and Alexandro Corradini. "The People of the State Department and Foreign Service." *American Political Science Review* 48, No. 4. (December 1954).

McClintock, Robert. "Thoughts on an American Diplomatic Style." *FSJ*. (February 1962).

McKitterick, Nathaniel. "Diplomatic Logjam." *The New Republic*, March 27, 1965.

Morgenthau, Hans. "The Impact of the Loyalty-Security Measures on the State Department." *Bulletin of Atomic Scientists* 11, No. 4. (April 1955).

Morris, Roger. "Clientism in the Foreign Service." *FSJ*. (February 1974).

Moskin, J. Robert. "Dean Rusk: Cool Man in a Hot World." *Look*, September 6, 1966.

Nadelmann, Ethan. "Setting the Stage: American Policy Toward the Middle East, 1961–1966." *International Journal of Middle East Studies*. (November 1982).

Navasky, Victor. "No. 2 Man at State is a Cooler-Downer." *The New York Times Magazine*. December 24, 1967.

Nelson, Anne Kasten. "National Security I: Inventing a Process," in Hugh Heclo and Lester Salamon, eds. *The Illusion of Presidential Government*. (New York, 1981).

Newsom, David. "Miracle or Mirage: Reflections on U.S. Diplomacy and the Arabs." *Middle East Journal*. (Summer 1981).

Nisbet, Robert. "Project Camelot: An Autopsy." *The Public Interest*. (Fall 1966).

Nitze, Paul. "The 'Impossible' Job of Secretary of State." *The New York Times Magazine*, February 24, 1957.

Oldenberg, Philip. "The Breakup of Pakistan." Murphy Commission *Report*, Vol. 7. (June 1975).

O'Neill, Michael. "The Quiet Diplomat," in Lester Tanzer, ed. *The Kennedy Circle*. (Washington, D.C., 1961).

Osborne, John. "The Importance of Ambassadors." *Fortune*. (April 1957).

———. "Is the State Department Manageable?" *Fortune*. (March 1957).

———. "What's the U.S. Foreign Service Worth?" *Fortune*. (May 1957).

Pastor, Robert. "Coping with Congress' Foreign Policy." *FSJ*. (December 1975).

Peretz, Martin. "Unreliable Sources." *The New Republic*, September 12, 1983.

Perry, Jack. "On Being a Deputy Chief of Mission." *FSJ*. (August 1978).

Pollack, Herman. "Office of the Secretary of State." *FSJ*. (February 1961).

Poole, Peter A. "Pacific Overtures with Marshall Green." *FSJ*. (June 1979).

Raynolds, David. "Who Makes Foreign Policy, and How?" *FSJ*. (August 1967).

Rockman, Bert. "America's Department of State: Irregular and Regular Syndromes of Policy Making." *The American Political Science Review* 75, No. 4. (Winter 1981).

Rostow, Walt. "The Planning of Foreign Policy," in E. Johnson, ed. *The Dimensions of Diplomacy*. (Baltimore, 1965).

———. "The Third Round." *Foreign Affairs*. (October 1963).

Rubin, Barry. "America and the Egyptian Revolution." *Political Science Quarterly* 97, No. 1. (Spring 1982).

———. "Carter, Human Rights, and U.S. Allies," in Barry Rubin and Elizabeth Spiro, eds. *Human Rights and U.S. Foreign Policy*. (Boulder, Colo., 1979).

———. "The Reagan Administration and the Middle East," in Ken Oye et al. *The Eagle Defiant*. (Boston, 1983).

———. "The United States and the Middle East," in Colin Legum et al. *Middle East Contemporary Survey*, Vol. 5, 1980–1981 (London, 1982); and Vol. 6, 1981–1982 (London, 1983).

Rusk, Dean. "Foreign Policy and the Political Officer." *FSJ*. (April 1961).

Rushkoff, Bennett. "Eisenhower, Dulles and the Quemoy-Matsu Crisis, 1954–1955." *Political Science Quarterly* 96, No. 3. (Fall 1981).

Samuels, Michael, and Stephen Haykin. "The Anderson Plan: An American Attempt to

Seduce Portugal out of Africa." Paper delivered at the International Conference Group on Modern Portugal. (June 1979).

Sanchez, Nestor. "The Communist Threat." *Foreign Policy*. (Fall 1983).

Sanjuan, Pedro. "Why Don't We Have a Latin America Policy." *The Washington Quarterly* 3, No. 4. (Autumn 1980).

Schlesinger, Arthur, Jr. "Origins of the Cold War." *Foreign Affairs*. (October 1967).

———. "America Experiment or Destiny?" *American Historical Review* 82, No. 3. (June 1977).

Scott, Andrew. "The Department of State: Formal Organization and Informal Culture." *FSJ*. (August 1969).

Service, John. "Pertinent Excerpts." *FSJ*. (October 1969).

Simpson, Smith. "Who Runs the State Department." *The Nation*, March 6, 1967.

———. "Perspectives of Reform." Part Two, *FSJ*. (September 1971).

Snyder, William. "Dean Rusk to John Foster Dulles, May–June 1953." *Diplomatic History* 7, No. 1. (Winter 1983).

Spain, James. "The Country Director: A Subjective Appraisal." *FSJ*. (March 1969).

Stephens, Richard. "The Impact of Administration on the Foreign Service." *FSJ*. (February 1953).

Stevenson, Charles. "The Ordeal of Otto Otepka." *Reader's Digest*. (August 1965).

Stoler, Mark. "Turning the Tables: State Department Opinion of Diplomatic Historians During World War II." *The Society of Historians of American Foreign Relations Newsletter*. (September 1982).

Thayer, Charles W. "Our Ambassadors: An Intimate Appraisal of the Men and the System." *Harper's Magazine*. (September 1959).

———. "The Lesson of 'The Ugly American.'" *FSJ*. (December 1958).

Thayer, Frederick. "Presidential Policy Processes and 'New Administration.'" *Public Administration Review*. (September–October 1971).

Thomas, Hugh. "The US and Castro 1959–1962." *American Heritage*. (October–November 1978).

Thurston, Raymond L. "The Ambassador and the CIA." *FSJ*. (January 1979).

Toon, Malcolm. "In Defense of the Foreign Service." *The New York Times Magazine*, December 12, 1982.

Trask, Roger. "The Impact of the Cold War on United States-Latin America Relations 1945–1949." *Diplomatic History* 1, No. 3. (Summer 1977).

Turpin, William. "Foreign Relations, Yes; Foreign Policy, No." *Foreign Policy*. (Fall 1972).

Tuthill, John. "Operation Topsy." *Foreign Policy*. (Fall 1972).

U.S. Department of State. "Communist Interference in El Salvador: Documents Demonstrating Communist Support of the Salvadorean Insurgency." (Washington, D.C., February 23, 1981).

———. "Rules Regarding Freedom of Expression for Foreign Service Officers." (Washington, D.C., April 6, 1981).

———. "The Department of State Today." (Washington, D.C., 1981).

U.S. Committee in Solidarity with the People, "Dissent Paper on El Salvador and Central America." (Washington, D.C. 1980).

U.S. House Committee on Government Operations. "Security and Personnel Practices." 83rd Congress, 1953.

U.S. House of Representatives, Permanent Select Committee on Intelligence, "Iran:

Evaluation of U.S. Intelligence Performance Prior to November 1978." (Washington, D.C., 1979).

———. "U.S. Intelligence Performance on Central America: Achievements and Selected Instances of Concern." September 22, 1982.

U.S. Senate Committee on Government Operations. "The National Security Council." 91st Congress, 1970.

———. "The National Security Council: New Role and Structure." 91st Congress, 1970.

U.S. Senate Foreign Relations Committee. "State Department Employee Loyalty Investigation." 81st Congress, 1950.

———. "Recruitment and Training for the Foreign Service of the United States." 85th Congress, 1958.

U.S. Senate Judiciary Committee "The Communist Threat to the United States Through the Caribbean." 86th Congress, 1960.

U.S. Subcommittee on National Policy Machinery. "Organizing for National Security: The Secretary of State and the National Security Policy Process." 87th Congress, 1961.

U.S. Subcommittee on National Security and International Operations. "Specialists and Generalists: A Selection of Readings." 90th Congress, 1968.

———. "The Secretary of State and the Problem of Coordination: New Duties and the Procedures of March 4, 1966." 89th Congress, 1966.

U.S. Subcommittee on National Security Staffing and Operations. "Administration of National Security: The American Ambassador." 88th Congress, 1964.

———. "Administration of National Security: The Secretary of State." 88th Congress, 1964.

———. "The Ambassador and the Problem of Coordination." 88th Congress, 1963.

Van Hollen, Christopher. "The Tilt Policy Revisited: Nixon-Kissinger Geopolitics and South Asia." *Asian Survey* 20, No. 4. (April 1980).

Vogelgsang, Sandy. "Feminism in Foggy Bottom: 'Man's World, Woman's Place?'" *FSJ*. (August 1972).

Weisman, Steven R. "The Influence of William Clark: Setting a hard line in Foreign Policy." *The New York Times Magazine,* August 14, 1983.

Wellman, Harvey R. "The Last Days of An Embassy." *FSJ*. (May 1961).

Wiegele, Thomas. "Decision-Making in an International Crisis." *International Studies Quarterly* 17, No. 3. (September 1973).

Willis, David. "Fudge Watching: A Reporter Views the State Department." *FSJ*. (November 1967).

Wingfield, Susanna. "Old Tabus Against Women Linger On." *FSJ*. (February 1969).

Wittner, Lawrence. "The Truman Doctrine and the Defense of Freedom." *Diplomatic History* 4, No. 2. (Spring 1980).

Wolfe, Glenn. "Administration is Substantive Work." *FSJ*. (April 1959).

Wriston, Henry. "The Secretary and the Management of the Department." in Don Price, ed. *The Secretary of State*. (Englewood Cliffs, N.J., 1960).

———. "The Secretary of State Abroad." *Foreign Affairs*. (July 1956).

———. "Young Men and the Foreign Service." *Foreign Affairs*. (October 1954).

Libraries

Dwight Eisenhower Library
Gerald Ford Library
Lyndon Johnson Library
John F. Kennedy Library
Franklin Roosevelt Library
Stettinius Papers, University of Virginia Library
Laurence Steinhardt Papers, Library of Congress
Harry Truman Library

Journals

Department of State Bulletin (DOSB)
Department of State Newsletter
Diplomatic History
Foreign Service Journal (FSJ)
Fortune
The Nation
The New Republic
Newsweek
The New York Times
Time
U.S. News and World Report
The Washington Post

Index